BAD IDEAS?

BAD IDEAS?

An arresting history of our inventions

ROBERT WINSTON

BANTAM PRESS

LONDON • TORONTO • SYDNEY • AUCKLAND • JOHANNESBURG

TRANSWORLD PUBLISHERS
61–63 Uxbridge Road, London W5 5SA
A Random House Group Company
www.rbooks.co.uk

First published in Great Britain
in 2010 by Bantam Press
an imprint of Transworld Publishers

A CIP catalogue record for this book
is available from the British Library.

ISBNs 9780593060278 (cased)
9780593060285 (tpb)

Addresses for Random House Group Ltd companies outside the UK
can be found at: www.randomhouse.co.uk
The Random House Group Ltd Reg. No. 954009

The Random House Group Limited supports The Forest Stewardship
Council (FSC), the leading international forest-certification organization. All our
titles that are printed on Greenpeace-approved FSC-certified paper carry the FSC logo.
Our paper procurement policy can be found at
www.rbooks.co.uk/environment

Typeset in 11/14pt Sabon by
Falcon Oast Graphic Art Ltd
Printed and bound in Great Britain by
Clays Ltd, Bungay, Suffolk

2 4 6 8 10 9 7 5 3 1

Mixed Sources
Product group from well-managed
forests and other controlled sources
www.fsc.org Cert no. TT-COC-2139
© 1996 Forest Stewardship Council
FSC

To Kathy Sykes,
Professor of Sciences and Society at the University of Bristol,
who has been an inspiration to so many people

'It is not thy duty to complete the work,
But neither art thou free to desist from it.'
R. TARFON. *Aboth:* 2,15 (*c*.70 CE)

Contents

Preface and Acknowledgements

'Science becomes dangerous only when it
imagines that it has reached its goal.'
GEORGE BERNARD SHAW, 'Preface on Doctors'
(1909) to *The Doctor's Dilemma*

During the late 1990s I had the privilege of chairing the House of
Lords Science and Technology Committee. During the three years
of my chairmanship, we set up a number of inquiries into scientific issues, from the disposal of nuclear waste to the medicinal uses
of cannabis. But the one closest to my heart was the inquiry into
science and society, which lasted for most of 1999 and part of
2000. The inquiry was launched because it seemed that the scientific community had an uneasy relationship with many members of
the public, with policy-makers, with journalists and to some extent
with government.

Among various concerns about public mistrust of science, there
had been a major debacle over the use of genetically modified crops,
and there had been the unedifying photograph of a cabinet minister
feeding his little daughter a beefburger in an effort to persuade people
that British beef was free of bovine spongiform encephalopathy. It
seemed that human embryo research was almost the only scientific
issue on which the public appeared to trust the scientists and the
information those scientists were giving them, and to recognize its
value for healthcare.

So, having taken copious evidence, we argued in our report that there needed to be much better public engagement with science; that merely attempting to increase 'public understanding of science' was really inappropriate. And since then a number of colleagues have become much more involved with thinking about and responding to the difficult issues raised by implementing advancing human technology. But until quite recently, the idea of listening to the public and responding to public concerns seemed almost anathema to many scientists. So various far-thinking people, and notably my friend and colleague Professor Kathy Sykes of Bristol University, raised the idea of dialogue with the public on these complex issues.

The thesis behind this book – namely, that the key to successful living in a society dominated by advancing technology lies in better public engagement with science and technology – will seem strange to many people. Yet it seems increasingly obvious that if we are to avoid harm from the increasingly powerful tools we have, we need to have much better methods of control. This control cannot be exercised solely by governments: history shows, as this book recounts, that governments do not always use scientific knowledge wisely. Nor can control simply be left to my community – the scientists – even though it is clear that most scientists are strongly altruistic and genuinely committed to improving the health and welfare of the society in which they live. So in my view, people from all sections of the community have a responsibility to learn and understand more about science in order that, in democratic societies at least, they will have a more powerful say in how science is used.

This book has been difficult to write, and in the first place I am deeply in debt to my publishers, Transworld, and particularly Sally Gaminara, who had sufficient faith in me to delay publication by a whole year while I struggled to make a very imperfect text into something slightly less imperfect.

My old friend Matt Baylis has as usual been a tower of strength and energy in seeking out so many matters of interest in the research for this book. We had been discussing the ideas behind the book for some years and had even pitched it as the theme for a television series. Matt's wide knowledge, his remarkable enthusiasm for all things scientific and historical, and his wonderful sense of humour have meant it has been a continued joy to work with him on this difficult project.

It is my very good fortune to have had Gillian Somerscales as an editor for another of my books. Her intelligence and clarity of thought have made a massive difference to my imperfect text, and as usual her input has been made with tact and with great insight.

Lira, my wife, has been, as ever, remarkable. She has been forbearing and tolerant when I am preoccupied by writing and thinking, and always deeply encouraging. Her wisdom in suggesting improvements for the text has helped make this a better book. Rachel Ward, my assistant and secretary, has been amazingly supportive. Her invaluable interest in and enthusiasm for this project was hugely helpful in persuading me to continue with it in the usual moments of depression.

Other colleagues and friends have read bits of the draft manuscript and made most helpful suggestions. Many others have helped in detailed conversations about its focus and its argument and about crystallization of my ideas in general. I am indebted to Carol Readhead, James Wilsdon, Sheba Jarvis, Brian Wynne, Anne Cooke, Jacqui Roche, the late Lord Porter of Luddenham, and various members of the Societal Issues Panel, and particularly Peter Ferris, of the Engineering and Physical Science Research Council.

Finally, I am so grateful to my old friend and agent, Maggie Pearlstine, who has always taken such a strong interest in my various projects and without whom this book would never have been published.

Introduction

Too Clever by Half?

HUMANS ARE CLEVER. Our cleverness and our ability to design and use tools have led to our increasing domination over the planet on which we live – though ultimately, as most of us realize, we cannot control or regulate it. As humans have developed more and more complex ideas, more and more sophisticated mechanisms, we have improved our lives in a multitude of ways and have increased our influence over much of our environment. But nearly all technologies are increasingly threatening. The achievements we rightly celebrate could also be bringing us closer to our own destruction – and not only our destruction, but also the demise of many species of animals and plants that share the planet with us. One obvious concern is the frightening threat of significant climate change and the alarming consequences of global warming. The great majority of scientists now agree on the basis of the evidence that this impending crisis is of human making. It may well have been started by our implementation of the technology of farming, but the data now strongly support the view that it is being brought about progressively by technologies which increase our recent dependence on fossil fuels.

I have set out in this book to present, chapter by chapter, a thematic history of human inventiveness and an account of its positive and negative aspects. By way of introduction to some of my themes, I start with stone.

Stones – our use of them – are what separate us from all other species. The earliest stone implements fashioned by hominids are perhaps more than 2 million years old. A flint fractured by chance in the right place, and others more deliberately worked into sharpened stones with a hard cutting edge, gave humans their unique power and led ultimately to their domination over the planet. Human technology, which enabled humankind to control its environment – even to the point of modifying the evolution of its own species – stemmed essentially from the development of the stone hand-axe. With the worked edge of this implement our hominid ancestors could deflesh the bones of animals they scavenged and, when they were lucky and hunting in groups, even occasionally kill their prey. Early man, hunting on his own, would have had few chances to kill a large animal for food. But hunting in groups, with its need for rapid communication between individuals, is thought to have been a powerful driver for the development of the human brain. Moreover, nearly 80 per cent of that brain consists of fatty substances called lipids, and an enriched diet containing plenty of animal fats and proteins would have supplied the vital fuel allowing the large brain of *Homo sapiens* to evolve. What I find remarkable is that the unique intellectual prowess of human beings was gained as a result of the earliest technology we invented. Humans are surely the only species that have changed their own evolution in this way.

Of course, the sharp hand-tool was not only a means of gaining sustenance, but could also be used as a weapon. With it, aggressive members of our species could gain power over other members of the same species. And those with a better weapon, improved technology, and the skills and cunning to use it, would tend to be the individuals who survived to pass on their genes so that their descendants could achieve greater mastery of the environment around them.

One surprising thought is how long each initial advance in our earliest technology took. Crafted, sharpened hand-tools have been around for over 2 million years. The earliest spears that have been dug up, excavated from a site in Schöningen, Germany, are 400,000 years old and show a beautifully worked wooden point.[1] Yet it seems that for a very long time nobody thought how vastly the power of either implement could be increased with a most simple device. Attach a sharp stone to the end of a stout piece of wood, and with this lever you can hugely increase the speed, force, effectiveness and safety with which the axe can be wielded. Fix a stone, sharpened with a deadly edge, at the end of a carefully weighted stick, and you have a weapon – a dart, arrow or a spear – that could be lethal at a considerable distance. Yet it seems that our ancestors did not get around to this critical refinement for around a million years after the first stone hand-tools were crafted. And when this major technological advance did finally occur, it may well have been developed in different parts of the globe more or less simultaneously, during the same period of our existence.

ATLATL AND AXE: WHAT SCIENCE TEACHES US ABOUT ITS BEGINNINGS

At first glance, an atlatl hardly looks like a major advance in any history of technology. Yet this rather unprepossessing – indeed, now largely forgotten – object gave humans even greater power over their environment. In itself the delicate-looking atlatl is not really a weapon at all. In its simplest form it is merely a notched stick, or a piece of bone or carved antler, into which the actual weapon – a stone-tipped spear or a dart – fits. When the atlatl is swung hard in an overhand arc, the dart is given huge momentum on its rapid journey through the air to hit its target.

Atlatls, in various forms, have probably been around for longer than bows and arrows. Atlatls made of finely carved reindeer antlers sculpted into animal shapes have been dug up in France and dated by archaeologists to be as much as 17,000 years old. In America they are associated with the Clovis culture, from about 14,000 BP. Note that I use the abbreviation 'BP', which stands for 'before present'. The 'present' in 'before present' does not refer to the year 2009, when I am writing, but to the year 1950. There's a reason for this.

All living things on this planet constantly take up carbon, and this process stops only at death. So any substance that has once been living tissue, such as a piece of bone or in this case an antler, contains carbon which has been acquired from the environment. It so happens that carbon has several forms or 'isotopes'. The two most common isotopes are carbon-12 and carbon-14. Carbon-14 is radioactive and that radioactivity is lost over time at a predictable rate – its half-life (the time it takes for half the atoms in any quantity of the substance to decay into another element) is about 5,700 years. Carbon-12 is not radioactive: it remains carbon. So chemists can measure the ratio of carbon-14 to carbon-12 in a biological artefact and by doing so get a very accurate assessment of its age.[2]

Why 1950, then? That was the year when humans started to contaminate the global environment significantly with their numerous attempts to manufacture the most powerful weapon ever devised. Repeated detonations of atom bombs and hydrogen bombs tested in isolated parts of the world altered the ratio of the radioactive isotopes globally, including isotopes of carbon – so we can no longer confidently say in what proportion carbon-12 and carbon-14 exist. For many decades now we have lived with the threat of nuclear weaponry and the risk that an irresponsible government (or indeed other forces) might use such devices recklessly. Remarkably, recent news from North Korea confirms that human beings are still exploding these terrible weapons. Because the effects of all these explosions have persisted, and will continue to do so for a very long time, for biological dating purposes we define 'the present' as 1950.

Well before humans mastered farming, early hunter-gatherers made a series of incremental improvements to the atlatl technology. One significant technical advance was, to my mind, quite counterintuitive. It turns out that if the stick from which the dart is made is not rigid but whippy, it is more lethal. As the darts tend to be 2 metres or more in length, they store considerable kinetic energy when they bend as they leave the notch in the atlatl. At first thought it seems that an arrow which bends significantly as it flies through the air would be inaccurate and likely to miss its target. But it turns out that as the stored energy is released from such supple sticks, their momentum is increased as they hit their victim. And the accuracy can be astonishing, with the power to kill at well over 40 metres from the thrower's arm. Even a child using an atlatl can throw a dart a considerable distance with surprising precision. So, with time, early humans also learned that a flexible atlatl, like a flexible dart, could give added impetus to the launch.

An intuitive appreciation of physics led to another technological advance: the development of a weight, attached to the middle of the atlatl. Such a weight, provided the launcher is carefully balanced, adds considerably to the thrust that the human thrower gives. With a properly balanced and weighted atlatl of the right length, it is possible for an averagely built man to propel a substantial dart more than 120 metres. The position and size of the weight are important, and no doubt our ancestors experimented – science is not a new skill and they must have used considerable trial and error to optimize this physical system. It is interesting that the weight may have had other advantages. Swinging an atlatl through the air causes a substantial noise from its vibration, and a weight roughly halfway along the haft dampens this effect, allowing the hunter the great advantage of stealth. And the obvious material for the weight, of course? Stones of varying sizes have always been freely available.

How do we know that early humans used a machine like a bow, or a lever like an atlatl, to propel a projectile so forcibly? Much Palaeolithic cave art shows pictures of humans repeatedly pierced by thin whippy sticks. Jean Clotte, the French cave specialist about whom I have written in *The Human Mind*,[3] describes some revealing drawings in the Cosquer Cave.[4] Here, as in other caves in the Pyrenees, the human figures are pierced by very thin, crooked sticks much longer than the bodies of their human victims. Such long sticks would be quite unsuitable for use as arrows, and their thinness and curvature would have made them useless as hand-held spears. The most likely way of launching them would have been with an atlatl.

Dr L. Bachechi from the University of Pisa points out that the age of the origin of such weapons is very difficult to guess but nevertheless suggests some clues.[5] Together with his colleagues he made an interesting finding when examining the bones of a young adult woman which had been excavated from a cave in Messina in Italy in 1942, during the Second World War. She was about 1.6 metres in height and experts have dated her skeleton to around 13,760 BP. The most interesting feature of these bones is what is deeply embedded in her hip. Stuck into the pelvic bone is a very thin, sharp stone flake, about 5 centimetres long. This projectile would have entered the woman's buttock from behind – possibly she was running away when she was hit. The hipbone around the stone flint has undergone typical thickening and sinuses have formed, so it is clear that she did not die as a result of this injury and that the flake led to a chronic abscess. This caused bone inflammation – osteomyelitis – a painful injury from which pus would have drained almost certainly for some months, probably longer. There is, of course, no way that a hand-held spear could have entered the solid bone of this part of the hip and then been broken off. A sliver of flint this delicate and sharp must have had considerable momentum to burrow so deep through the fleshy buttock and then the tough pelvic bone. Such momentum could have only been produced at the projectile's launch – either with a bow or, more probably, with a spear-thrower of some kind.

Much information can be gained from osteology – the study of bones – and the skeletons of early humans are revealing. Persistent use of a hand-held spear, for example, led hunter-gatherers to develop prominent muscles in their upper arms, usually on the right side. As all the muscles in our arms and wrist are attached to bones at some point, we can see the extent of this hypertrophy by examining the bones from the shoulder, upper arm and forearm. Living bone is dynamic and responds to pressures on it by thickening and growing stronger. So where strong muscles have repeatedly pulled, an elevated ridge where the muscle was attached develops; and this remains long after death and after the muscle has decomposed.

From such observations, the osteologist can suggest what kind of weapon a particular skeletal arm is likely to have used. Dr Thor Gjerdrum, of the University of California at Santa Barbara, has examined thickening of the ulna (the larger bone) of the right forearm.[6] This is where the supinator muscle, which gives a twisting movement needed for throwing, is attached. He reports that many early hunter-gatherers had considerable hypertrophy of this region, amounting to an asymmetry between the right and left forearms so marked that it is now only seen in professional baseball players. It is interesting that the extreme torsion of the forearm required to use an atlatl most effectively is very comparable to the action involved in pitching a baseball. Later humans who did not throw spears or use an atlatl, but preferred the bow and arrow as a weapon, do not show this marked thickening of the ulna.

From reading bones, osteologists can also make an intelligent guess as to the position in which the spear was habitually held. An overarm position might be most suitable for hunting or killing an animal, but a spear held underarm, pointing upwards, would be more effective when facing a human enemy – and it seems likely that ancient man used this position. It is remarkable to consider that the technology of the spear has persisted until the modern day. British troops, going over the top on the Somme in the First World War, marched steadily towards the German lines with their bayonets held underarm, pointing upwards. The advantages of this position are several. It allows repeated thrusts to ensure the permanent immobilization of an opponent; it also makes the weapon's withdrawal physically easier, partly because the human attacker can use his naturally powerful back muscles and biceps to maximum effect and partly because, by contracting his abdominal muscles and rotating his trunk, he can lean back in preparation for the next thrust. An underarm position is also better defensively. A grip with the spear held above the shoulder exposes the attacker's chest and abdomen, making him very vulnerable if his blow is parried. To this day, infantrymen are still taught to fix their bayonets and trained to thrust them repeatedly upwards as they encounter their foe. Were Palaeolithic men any different? It seems unlikely that they would have exposed their soft parts willingly.

TOWARDS A SCIENTIFIC CITIZENRY

Various themes, then, run through this book. One is a concern about the threat that lies latent in many of our discoveries and increases as our technology becomes more powerful and more widely used. Ever since the hand-axe human progress has, in one sense, been downhill. The hand-axe led to the battle-axe, and the atlatl led to the catapult with which the boy David propelled a stone with sufficient velocity to kill the Philistine giant Goliath. Although iron replaced stone in most human societies long ago, the ancient arrow led to the Battle of Crécy where Edward III and his tiny force of archers destroyed the mighty ironclad horseman and infantry of France. Virtually every major idea that we have had – be it to do with farming, living in cities, writing, communications, the uses of fire, transport, weapons, even medicine – has at one level at least made humankind more vulnerable. As with the stone hand-axe, nearly all the wonderful technological advances that have enabled us to live in difficult and dangerous environments also have their threatening or negative aspects – hardly ever fully recognized at the time of their invention. So, as we shall see, farming resulted in human diseases that had not afflicted humankind previously, as well as loss of genetic diversity and threats to the environment of the planet. City-dwelling greatly increased the vulnerability of humans to infection: when people first lived in cities they had a shorter lifespan and were almost certainly less healthy than when they had maintained them-selves as hunter-gatherers. Writing, one of the greatest gifts ever granted to humans, has spawned some communications that debase our thinking, promote political instability and threaten our privacy. We do not need the menace of advances in weapons technology to remind ourselves how precarious our existence on the planet is, because peaceful technologies like transport are damaging our environment just as effectively, and perhaps irreversibly. Scientific advances in nanotechnology and synthetic biology could risk the un-controlled production of poisonous substances or infectious organisms against which we may have little or no defence. The development of robotics carries the risk of dehumanizing our relationships, and advances in genetics may see a time when manipulations of the human genome require us to redefine the very essence of what it is to be a human being.

Another issue concerns the regulation of science and who controls this work and the knowledge it generates. In chapter 2, I recount in some detail the career and fate of Nikolai Vavilov, the great Russian plant geneticist. His story epitomizes several themes which run through this book: in particular, how scientific knowledge may be abused by scientists themselves who are consumed by ambition and working in a very competitive environment; and how govern-ments grasp science to use it for their own purposes and not necessarily for the good of their citizens. Vavilov, of course, lived in a country ruled by one of the most malign totalitarian regimes of

modern times. But unfortunately, as we shall see, even demo-
cratically elected governments cannot always be trusted to use
science wisely, or even necessarily for the betterment of their
citizens.

A more encouraging theme that recurs throughout the book is
that a very large proportion of discoveries and inventions have all
kinds of beneficial applications that were not remotely envisaged
when they were first made. We sometimes think that stone tools
were abandoned after the Stone Age technology. But the ancient
Egyptians refined the technology to make extremely elegant knives.
Discoveries of beautiful razors made of chert – a very hard crys-
talline silicate – have been found at Giza; these date from the Old
Kingdom of Egypt (2575–2134 BCE). And presumably no early
hominid thought, when he or she fashioned a hand-axe, that the
cutting edge might be used for engraving the artwork that has been
seen on prehistoric bone atlatls.

It is also important to understand – indeed, it is axiomatic – that
many human technological advances are made in unconnected
places around the globe, independently and more or less simultaneously. This certainly seems true of hand-axe technology,
archaeological finds from far-flung places suggesting that as the
hominid brain developed, pre-humans developed the manufacture
and mapping of stone implements independently. The design,
certainly, may vary in different parts of the world, but the technology is essentially the same. And this is undoubtedly the case for
much of modern technology. Scientists are human and like to con-
gratulate themselves (and occasionally each other) that they have
been first with a revolutionary idea which has changed the world.
But most of the time, once human knowledge develops in a particular field or area, scientists everywhere are in a position to make
the next leap forward. This is undoubtedly true in my own field of
in vitro fertilization. Although we British pride ourselves on having
been the first to produce a 'test-tube baby', that first birth could
have easily occurred in Melbourne, or Montreal, or even possibly
Baltimore. Perhaps one key factor in its happening in Manchester
was the serendipity with which Robert Edwards timed the transfer
of the embryo that led to the birth of Louise Brown.

As our world becomes more sophisticated, we are becoming more concerned about the dangers human societies may face from our inventiveness. *Homo sapiens* has existed on the planet for no more than about 100,000 years, and during that time the genes which help to define the capability of our brain have not significantly changed. But in the last 400 years, a tiny fraction of that time, the human mind has expanded extraordinarily.

It is exactly 400 years since William Shakespeare's collection of sonnets was published in 1609. Since then we have invented the telescope, the microscope, the steam engine, vaccination, the telephone, aircraft, television and the computer. Humans have also produced the H-bomb and landed on the moon. Now, using the techniques of synthetic biology, we are trying to create new life forms – a technology which holds great promise but may also be very threatening to life on this planet. No doubt William Shakespeare could have leaned over the parapet of London Bridge 400 years ago, speculating with a fair degree of certainty that the city would be likely to look pretty much the same over the next few decades. Now it is a foolish 'expert' who attempts to predict what the technological capability of our society will be even in five years' time because our knowledge is growing exponentially.

This massive growth in human knowledge is truly awe-inspiring but also, understandably, seems quite frightening to many people. Humans do not find it easy to live with uncertainty and there is a growing perception, particularly in more sophisticated societies, that our pursuit of scientific knowledge and its practical application may be very dangerous. In the last ten years or so we in Britain have seen public suspicion of genetic engineering expressed in vehement protests by ordinary people about genetically modified crops. We see increasing anxiety about using nuclear power, and government is undecided over whether we should develop our nuclear industry further and what we should do with the nuclear waste we have already accumulated. There has been considerable press hysteria over human cloning, many parents have rejected the triple vaccine against measles, mumps and rubella, and mishandling of bovine spongiform encephalopathy and foot-and-mouth disease have made people doubt the integrity of the foods they eat. Attacks on humane animal experimentation have taken little or no account of the value of this important use of biology to the health and welfare of our fellow citizens – as well as the animals we keep for domestic purposes.

I do not quite take the view of Lord Rees, recent President of the Royal Society, that we are necessarily facing Armageddon.[7] We surely have at our disposal sufficient mechanisms to ensure that our technology can be controlled and sufficient ability to harness our resources successfully. As the distinguished Cambridge engineer Alec Broers asserted during his recent Reith Lectures,

> Technology too can provide solutions to these problems but only if people choose to implement them . . . Technology can solve our problems but only if the public engage with it . . . Now we are at risk of permanently endangering our planet. Our aim for this century should be to make comparable progress in protecting our environment. Technology will truly triumph if we succeed.[8]

I am no Luddite, and I most certainly am not pessimistic about the future of humanity. Overall, we live longer, more fulfilled and probably much happier lives now than at any time in humanity's past, and our use of technology has been a key factor in attaining those levels of welfare. Our inventiveness and the knowledge it has brought us are remarkable gifts – but we need to develop them and exploit them with wisdom. We also need to recognize that the technology we develop and control has a huge effect on all our fellow citizens and that we must be responsible for trying to ensure it does not do harm to human society and to the planet on which we live. Scientists have a major role to play in all this, and this book, though written in the hope of reaching the widest possible audience, is at least in part a call to my fellow scientists.

The problems, dilemmas and dangers produced by technology are not ultimately, in my view, always going to be solved only by further technological innovation. In this respect, I both agree and disagree with what Lord Broers hinted at in his Reith Lectures. It is insufficient for scientists to state that 'Technology can solve our problems but only if the public engage with it.' Scientists need to be equally engaged, and in the last chapter of this book I suggest ways in which lay people – members of the public – and scientists may engage more closely with each other. I include a manifesto for scientists, and a number of aphorisms about science for non-scientists. Better understanding of these issues on both sides, I believe, may help ensure that we use our science and our techno-logical prowess more wisely and for long-term good.

Scientists such as myself, I feel, may need to consider that the science we pursue is not *our* science. Our new knowledge is gained in the name of society and mostly is paid for by various members of the public. It is as much their science as ours. Its application may have beneficial effects, but people who are not scientists will have to bear the consequences when the effects of what we have established are harmful. Of course, we have to facilitate the engagement of the public to which Alec Broers refers. But we need to be more effective at communicating with the public so that they understand better what we are doing; and, most importantly, we also need to learn to listen to the public when they voice their fears and reservations about science. Moreover, mere listening is not enough. The majority of scientists are not yet ready to accept that the public should have greater involvement in decisions involving scientific research. But if we are to be good citizens, we need to recognize that by listening and responding better we are likely to make our science more relevant to the societies in which we live. We may even find that public dialogue may actually increase the quality of our work.

I am suggesting that every citizen has a part to play in under-standing scientific achievement and ensuring it is used for good. Indeed, responsible citizenship implies that non-scientists have an obligation to learn and understand more about science and tech-nology. This understanding, I believe, is one of modern humanity's most powerful weapons in ensuring our continued welfare. So this book is not merely about the negative aspects of our technology – it is a celebration of our extraordinary ingenuity. And if we are to ensure the health and welfare of generations to come, this ingenuity must be focused in part on making wise choices about how we promote, protect and use our ideas.

Chapter 1

The Brightest Heaven of Invention

A GROUP OF around 200 people who call themselves The Straight-Boned Ones live deep in the rainforests of north-western Brazil. They are a short, dark-skinned folk, who hunt with bows and arrows. Their huts are made by propping palm fronds on top of sticks, and contain no furniture or decorations, save a raised platform of thin branches as a bed. The average family owns very few possessions, merely a couple of pots and pans, a bow and arrow, a machete and some bags made of palm leaves. For personal adornment, they sport necklaces fashioned from the steel ring-pulls they obtain from trading brazil nuts with outsiders. Unlike other groups in the region, they do not cultivate manioc, nor do they preserve meat or fish by salting or smoking them. Each day, they take only what they require from the surrounding rainforest. Most significantly, the idea of preparing for the future never occurs to them. This remote people, known to the rest of the world as the Pirahã, seems to contradict almost everything we believe about being human.

PEOPLE LIKE US?

Across the planet, ever since we first invented the technology to navigate it, anthropologists, explorers and missionaries have encountered peoples whose way of life seemed basic by comparison with their own – people who lacked more possessions than could be carried, or who could not count beyond two or three. Early European explorers concluded that such people were base and savage, less intelligent, even less human than themselves. But as explorers encountered more indigenous peoples in far-flung parts of the globe, closer inspection revealed that their cultures were every bit as vibrant and complex as our own. Their languages contained a mind-boggling array of inflections, markers, tenses and tones. They possessed a body of myths and legends every bit as wonderful as – and often sharing elements in common with – our European fairy tales, our classical hero stories and our gospels. Not only that, but they could frequently recite them without the aid of written texts, possessing skills of memory that put us to shame. Equally impressive was their knowledge of the surrounding terrain, and the possession by certain individuals of a localized encyclopaedia of useful and harmful herbs, roots and berries. Their social systems were complex, with an array of rights and restrictions governing each individual's conduct and relationships. They were, quite clearly, our equals.

The Pirahã seem to be our equals, too. They walk on two legs, they use tools, they communicate with language and they live in groups. And yet, at the same time, their way of seeing, describing and using the world around them seems radically different from anything we associate with being human. Their language is un-related to any other tongue – as, indeed, are Basque and Japanese – but, unlike any other language, consists of just eight consonants (seven for women) and three vowels. At the same time, it possesses such a rich catalogue of tones, stresses and syllable-lengtheners that it often sounds to outsiders like singing, humming or whistling.

Professor Dan Everett, who at one time was an academic at the University of Manchester, is a linguist who has been living and working with the Pirahã for more than twenty-five years. He has discovered other unusual facets to their culture. They have no collective memory extending beyond the recollections of living members, and nothing approaching a creation myth. When pressed about what might have existed before they and the forest did, they reply, 'It has always been this way.'

The lack of a creation myth is, again, unusual in a human society but not unique, even in the Amazon region. It is striking, though, to find that trait in a group that also has no fixed terms for colour. If shown a red cup, Pirahã will say something like, 'it looks like blood' or 'or it looks like the berry we use to get dye from'. They possess no numbers, either – just terms for 'one', 'two' and 'many'. Their tongue lacks words for 'all', 'each', 'every', 'most' and 'few' – concepts which, according to some schools of linguistic thought, are the common building blocks of human cognition.[1] Most radical of all, Everett claims the Pirahã language has no evidence of recursion.

Recursion is the means by which humans insert one idea into another. For example, 'the man in a top hat' and 'the man is walking down the street' can be combined into the statement 'the man in a top hat is walking down the street'. This, according to the renowned MIT linguist Noam Chomsky, is the facility which enables humans to create an infinite variety of meanings from a finite resource of sounds and rules. It is seen as a universal aspect of human language and yet, according to Dan Everett, the Pirahã do not employ it. If they wanted to communicate a message like, 'I saw the dog that was down by the river getting bitten by a snake', they would break it up into discrete units, such as, 'I saw the dog. The dog was at the river. The dog was bitten by a snake.'

For Everett, the lack of recursion is consistent with the whole ethos of this remarkable people. They break their sentences up into simple, one-point units, because this is also how they perceive their world. For these people, only the immediately observable is real. When someone disappears out of sight because they have walked around a bend in the river, the Pirahã use the term *xibipío* – meaning the person has 'gone out of experience'. They employ the same phrase to talk about a flickering candle flame – the light, in their view, passes in and out of experience.

This is one of the reasons, according to Everett, why the Pirahã have no myths of their own, and why they have proved so resistant to being converted to Christianity (although his former wife, Keren, combines efforts to spread the Gospel with her own linguistic studies among the Pirahã). Told that these stories concerned a saintly man who died 2,000 years ago, the Pirahã became indifferent to their symbolic content. Their immersion in the immediate also accounts for their failure to build up food stocks, and for the simplicity of their drawings. When *New Yorker* journalist John Colapino came by seaplane to visit, he was startled to see a small boy, within hours of his arrival, playing with a surprisingly accurately constructed model of the plane. It seemed to confound Everett's claims that the Pirahã could make only crude, rudimentary sketches. But within hours Colapino saw the seaplane model abandoned, crushed at the side of a path. Every time a seaplane arrived, Everett explained, its arrival would spawn a brief rush of imitative enthusiasm, but as soon as that faded in short-term memory, all interest was lost.

Most people's reactions to Everett's findings fall into one of two types (leaving aside the especially heated reaction from the linguistic community, for reasons we will look at later). One reaction is to assume that he was duped by the Pirahã – as many anthropologists reporting on 'primitive' cultures have been over the years. A famous example was the renowned anthropologist Margaret Mead, whose teenage guides to the ins and outs of the Samoans' extraordinarily exciting sexual habits – upon which Mead based a career-launching book – later admitted they had made it all up. But in this instance, given the consistency of Everett's findings over time, and with those of other visitors, a case of tribal mickey-taking seems unlikely.

The second reaction is to assume that the Pirahã are somehow not quite the same as us. Since they're so isolated, and forbid marriage outside their tiny grouping, it would be plausible to suppose that they suffer from some kind of retardation due to inbreeding. But this should result in physical as well as mental symptoms, and there are no signs of either among the Pirahã. In any case, they regularly keep their gene pool refreshed by permitting women to sleep with outsiders.

There is, of course, a third option, which is to assume that the Pirahã represent some mid-point or missing link between modern humans and their forebears – but that is not, happily, a position any intelligent twenty-first-century person would entertain. To conclude that the Pirahã are less than human, or less able humans, simply because they do not meet our current criteria of humanity, is arrogance bordering upon idiocy. As Everett points out, a Pirahã male can go into the depths of the jungle without food, clothes or tools and emerge three days later with a self-made basket of fruit and nuts. If a Brazilian logger or an American anthropologist were asked to do the same thing, the end result would be a corpse. The Pirahã are as intelligent and as inventive as all of us. They weave baskets from leaves, they make fires, they set traps for animals. But they are still radically, vividly, strikingly not like us. And that in turn raises the question of what we really mean by 'us' in the first place.

The answer to that question is not among the objectives of this book. This book is a history of, and an inquiry into, human invention, from stone hand-axes to hydrogen bombs. It attempts to trace the common thread linking all of our ideas – ideas that have been of immeasurable benefit to humanity, yet on occasion threaten to destroy us. But not one of those ideas could have sprung into light without its creators having been, for want of a better word, special; that is, endowed with traits and abilities not shared with other species. No history of human invention, therefore, can begin without a look at what makes our species so perfectly equipped to invent – and why that specialist equipment came about. But the Pirahã, to whom we shall return, will caution us against making any sweeping assumptions.

Two legs good

I had the idea for this book while reluctantly dead-heading some rhododendrons in my garden. Bending awkwardly to my task – even though I am quite a keen gardener – I found myself, rather churlishly, thinking how much easier it might be if I walked on all fours. That observation led, in turn, to my realizing that the task I was doing, and my discomfort while doing it, were intimately linked. It was because I am a bidepal animal that the task caused me pain – but also that I was doing it in the first place. To be human is to be something of a gardener: to see one's environment as something to be shaped and altered, beautified or put to use. The entire history of human invention rests upon that premise. And that premise may itself rest on the fact of our walking upright on two legs. Standing straight gives us a different perspective on the world around us. It encourages us to think of ourselves as being above, superior to and separate from our environment – a necessary precursor to our being able to shape and change it. It also enables us to see far ahead, spatially and temporally – to see danger, or fruit trees, or desirable mates, long before they are right in front of us; and this ability gives us, in turn, the ability to create strategies, to imagine, to invent, create, design and dream. Standing upright, in a palpable sense, decouples us from the immediate, favours abstraction. But when and why did this change take place?

Complex research into DNA, and the rate at which it mutates, indicates that around 6.6 million years ago some 0.71 per cent of genes in the common ancestor we shared with chimpanzees changed, giving rise to a new branch on the family tree and eventually to the species we now know as human. Somehow, that tiny shift gave rise to space shuttles and skyscrapers, penicillin and the ovens of the Holocaust. More prosaically, it gave rise to increased brain capacity, a lowered larynx and changes in the structure of the pelvis.

To ask, 'Why did these changes take place?' is not only to open, but also to tangle oneself in, an academic can of worms. But here are some observations. It seems that climate changes occurring around 6.6 million years ago forced some chimpanzees out of the treetops and on to the forest floor. A range of factors in this new environment might have encouraged certain adaptations. Once on land, these pioneering chimps were at increased risk of being eaten by large, fast predators. Walking increasingly on two legs enabled them to see danger from further away even when there were fewer trees to act as observation posts; to cover greater areas of territory; to have their hands free to carry weapons; and, arguably, to make themselves appear bigger to scare off those predators.

Walking upright on two legs, bipedalism, may also have made movement more efficient once we moved out on to the hot plains. A creature on four legs exposes more of its body to the sun, and thus requires more moisture to keep going, than one on two legs, which exposes only the top of its head and shoulders. A recent study by David Raichlen and colleagues at the University of Arizona suggests that walking on two legs uses up around 75 per cent less energy – a conclusion they arrived at by rigging humans and chimps up to treadmills and measuring the amount of oxygen they produced.[2]

These changes coincided with a substantial increase in certain areas of the brain, which in turn required extra calories in order to fuel it. The brain is an extremely 'costly' organ – taking just 2 per cent of the body's space but up to 20 per cent of its resources, even when at rest. It is hard to fuel such a demanding engine on a diet of fruit and leaves, so increases in brain capacity are linked with a shift to eating meat, and therefore to hunting – which in turn created the need for tools to kill and prepare our food source, for group living to facilitate the lengthy and dangerous business of hunting, and for the development of language in order to make those groups run more efficiently.

That is one theory, at least. This sort of 'cascade' perspective on human development is very attractive – but not, unfortunately, very likely. Archaeological evidence (though always hotly disputed, at least by archaeologists) suggests bipedalism emerged some 6 million years ago, while the earliest evidence for tool use is from about 2 million years ago. Plenty of other creatures – from ants to wolves – live in large and complex societies, but they have not felt the need to develop language. To these observations we must add a basic fact about evolutionary biology, which is that physical changes that emerge to suit one purpose can be coopted for a different one later. Probably, permanent bipedalism emerged for a variety of reasons and conferred a number of advantages on our species. But perhaps the greatest boon, as far as this history is concerned, was the way walking on two legs lifted up our heads, so that we could glimpse distant horizons, and freed up our hands, and our opposable thumbs, to make and manipulate.

Homo mechanicus

In 2004 participants in a study partially funded by *National Geographic* magazine were stunned by the footage their motion-sensitive cameras had captured. Chimps in the Congo's Nouabale-Ndoki National Park were observed fishing for termites with sticks. This in itself is nothing new. Back in 1960 the legendary primatologist Jane Goodall had observed chimps in Tanzania carefully stripping and preparing twigs for use as termite-fishing tools. But these Congolese chimps were doing something never seen before. The video footage revealed the chimps using a short stick to puncture the crusty surface of the nest, then a longer probe-device actually to fish the juicy termites out. They would modify these fishing probes beforehand, pulling each one through their teeth to fray the ends like a paintbrush and so increase the number of termites captured per 'dip'.

It may sound easy, but termite-fishing, according to this study, was a complex craft that took years of practice. The footage showed infants watching agog as elder chimps withdrew their long sticks, swarming with large, shiny black termites. In this little drama, the researchers felt they might have been watching the very origins of human interaction itself.

They were certainly watching something that challenged our previous notions about tool use among primates. Many of the earth's creatures employ some form of tool. Birds, for example, build nests from twigs, discarded bottle-tops and newspapers. Marmots plug up their winter burrows with dung. Two things, however, are thought to make human tool use unique. First, only humans use one tool to make another tool. Second, only humans use a specific type of tool for a specific purpose. Footage of these chimp 'toolkits' challenges at least one of our assertions. If our hairy cousins use two different types of sticks for two different jobs, then clearly we need to redraw the rules about what defines human tool use.

But there is still a wide gulf between these simple, two-piece kits and the array of instrumentation developed by our ancestors. New finds may upset the picture, but we can roughly trace a line, beginning about 2.5 million years BP and extending to about 40,000 BP, that connects the first crudely hewn pebbles to a standardized toolkit, including animal bones as well as stones, and adapted for purposes as diverse as tree-felling and sewing. The line is marked by stages along the way. Choppers and cutting flakes 1.7 million years ago gave way to symmetrical and surprisingly elegant hand-axes 300,000 years later. These, in turn, ceded ground to the 63-piece kits of the primitive men who were living 250,000 years ago at Levallois-Perret, just 6 kilometres from the centre of what is now Paris, just north of the Périphérique, and who fashioned 'compound tools' by attaching different objects together. By 40,000 years ago, men and women anatomically no different from you and me were sewing clothes, fashioning items like bead necklaces for personal adornment and statuettes for symbolic purposes, burying their dead in standardized fashion and trading items across wide distances. Most of these stages in the human journey to ever greater inventiveness, furthermore, were preceded by leaps in brain size – from the 500 millilitres of the first upright-walking hominids to the 1,450 millilitres of modern man.

During one of these leaps we acquired a fascinating bit of brain function that assisted us not just with the manufacture of tools, but also with a range of further developments and inventions integral to social as well as technological progress. Mirror neurons are a cluster of brain networks in the frontal lobes which become activated when we watch other humans performing certain tasks – with the emphasis on *humans* and *certain*. It seems they fire only when we are watching goal-oriented activity, like someone shelling a nut or peeling a banana – not when we are watching purposeless action or mimicry, and not, interestingly, if a robot is performing the same task. It seems this mechanism evolved specifically to help us learn from other humans.

We can see how mirror neurons would have assisted us in the manufacture and use of tools – we watched, so we did. They would also have been a vital component in the sort of complex society we began to live in after our descent on to the savannah – helping us to pick up instant cues from the behaviour of our fellow band members as to the way we should act. But lest we get too excited about this remarkable 'human' facility, we should remember that monkeys have mirror neurons, too. They respond to one another in much the same way and yet, for reasons we don't fully grasp, monkeys have not developed complex toolkits. Humans are the only true inventors.

It is easy to see how these early inventions would have revolutionized early human life. Able to hunt and prepare protein- and calorie-rich meat instead of subsisting on fruit and leaves, we were able to fuel our increasingly demanding brain. This bigger brain increased our efficiency, which meant less time foraging for food, less exposure to dangerous predators, and therefore enhanced chances of survival. When we look at the earliest human inventions in the glass cases of a museum, they look very uninspiring – dull, flat, crude objects, only dimly appropriate for their stated purpose. It is easy to underestimate the seismic shift in perspective that must have surrounded their conception, creation and use.

Fashioning tools requires considerable brain power. As well as an extraordinary degree of motor control and complex visual apparatus, it needs an ability to imagine, to think outside the moment. Picture yourself, starving in the wilderness and trying to extract some life-saving fragments of buffalo flesh from a carcass. To arrive at a solution, you must lift yourself out of your current desperate situation and form an idea, in abstract, of the sort of device that might best enable you to get at the flesh. You then have to look around you, at the surrounding rocks and trees, to discern which among them might contain the best raw material for the job. You must then formulate and carry out a sequence of planned actions – collecting the right stone, angling it the right way before you strike it with another stone, taking the most suitable of the resultant chips and preparing it further. Underlying all of this must be an idea of being a separate entity from the environment, an individual whose differing actions bring about differing results, an individual with the power to use one's surroundings to one's own ends. We could not have made tools, or done any sort of inventing, without this underlying sense of self – a sense, as I indicated earlier, that may have been encouraged by our ability to stand on two legs.

It is often said that we in the industrialized, overcrowded, demystified West have lost some crucial contact with our planet; that we live in an overcomplicated, man-made cocoon, alienated from the passage of the seasons and from the other species with whom we share so much in common. This sense of separateness, it is claimed, has given rise to our callous disregard for our planet. We pollute the seas, strip the forests, mess up the polar ice-caps, so the thinking goes, because we don't really see ourselves as belonging to, or in an interdependent relationship with, our earth.

The people who trot out this sort of argument also tend to view people of non-industrial societies in the opposite fashion – as the respectful priests tending Nature's shrine. Clad in their bark cloths and animal skins, they view the trees and the rivers as their companions. This school of thought tactfully overlooks evidence from places like New Zealand, Madagascar and Easter Island, whose indigenous inhabitants have in the past cheerfully hunted birds and large mammals to extinction and stripped the forests to the point of ecological meltdown. Such an outlook also, I think, overlooks the fact that to be human is to see oneself as separate from and having power over one's environment. It began, perhaps, when we first lifted our heads and stood on two legs. It must certainly have been confirmed when we first picked up stones and used them to accomplish a task. Contained within those initial acts of creation was the potential, also, for destruction. But perhaps our capacity to invent, to change our world, can be traced to some more basic aspects of our biology.

SACRED GOSSIP

On the island of Tanna, in the South Pacific archipelago of Vanuatu, speech is a sacred activity. To qualify as 'Big Man' – a phrase denoting a person of worth throughout the islands of Melanesia – one must be, among other things, a good speech-maker. To speak is, in many cases, synonymous with doing: a vital road (or water-source) is referred to as 'the way [or tap] which Chief Somebody-or-other spoke'. When a child speaks its first word, ceremonies are held to mark its formal entry into the community. Their term for those unable to speak is *iapou* – meaning weak, sick or exhausted.

As a Jew, I can appreciate the symbolic weight the Tannese lend to language. In former times, when a Jewish child learned Hebrew, his or her teacher might place a drop of honey upon the first letter, to emphasize how sweet an inheritance it was. The Jewish colloquial tongue of eastern Europe, Yiddish, is a conflation of mostly German, Russian and Hebrew – with elements of any of several other languages thrown in, depending on the origin of the speaker. This rather comic but poignant language is often nicknamed the *mama-loshen* – the 'mother-tongue', a term used, in this instance, to denote an intense love, the giving of life itself. And in countless cultures, the act of speaking is invested with performative power. In British marriage ceremonies, the phrase 'I now pronounce you man and wife' carries legal authority, as does, in certain Islamic countries, the act of saying *Talaq* – 'I divorce you' – on three separate occasions.

None of this may sound particularly surprising to readers who are using language, for all sorts of purposes, all day long. They may feel differently when they consider how it first came about. In a lecture inanely entitled 'Why Children Are Stupid', an academic acquaintance of mine used to point out to first-year undergraduates how much of human society depended upon two basic physiological changes.

Compared with other mammals, we have a phenomenally large brain for creatures our size, and a large head to contain it. At the same time, walking on two legs means we have to have a fairly narrow pelvis – and although the female pelvis is slightly wider than the male, this still makes the act of giving birth to a large-headed child painful and dangerous. Indeed, it is often observed that the most perilous journey most of us take is that of just four or five inches down the vagina in the process of being born. The size of our brain, in turn, has an impact on gestation times. Bearing in mind our body size and the stage of development at which other mammals are born, the average human pregnancy should last at least twenty-one months. It doesn't – not just because under such poor terms no woman would let any romantically inclined man near her, but because to give birth to an infant the size of a one-year-old would risk fatal damage to the mother's internal organs.

So, uniquely among mammals, humans give birth to their babies at a stage of development when they are totally unable to fend for themselves. We are far less mature at birth than the offspring of other mammals, and this has implications for the kind of society we live in. The production of babies unable to survive without parental assistance for a long time favours the forming of lasting bonds between men and women. The formation of these enduring bonds is in turn fostered by another unique aspect of human behaviour – the fact that women have a concealed ovarian cycle. In contrast to other mammals, the time when a woman is most fertile is not publicly displayed through scent or the swelling of organs – indeed, most women are rather unsure when they are ovulating. This is one of the reasons why humans are a notoriously infertile species. But hidden or 'occult' ovulation may have a subtle purpose: it assists group cooperation, because it cannot be a source of aggression and competition among males (although women themselves still can, as a quick glance in a pub at closing time will prove). It also encourages males to be interested in females at all times, instead of only at the time of ovulation – which promotes stable pair-bonding, and is better for rearing children.

Given the risks involved, given the number of adaptations necessary to cope with our production of such large-headed, incapable infants, we might ask: 'Why bother?' Our biology, surely, would only have embarked upon such a costly evolutionary strategy if the benefits outweighed those costs. So what could those benefits possibly have been?

Primate researchers Robert Seyfarth and Dorothy Cheney from the University of Pennsylvania uncovered some interesting evidence while recording the various cries and calls of vervet monkeys.[3] Vervets are sociable beasts, spending up to 20 per cent of each day grooming one another. The activity is particularly important among females, who live in tightly knit clans of four or five, along with their offspring and one dominant male. Seyfarth and Cheney discovered just how vital the grooming process was when they played the distress call of a female vervet to a clan member who had just been grooming her. The second monkey became intensely agitated, and looked towards the source of the call, as if determining what to do. When they played the distress call to a vervet that had not been in a grooming partnership with the maker of the call immediately previously, the result was indifference. Subsequent studies failed to elicit the responsive behaviour from other types of monkeys, such as lemurs and galagos, whether they'd been grooming or not. Seyfarth and Cheney realized that this behaviour seemed to be found in the species living in the largest groups.

Their findings tied in with those of Frans de Waal, who has described 'reconciliation behaviour' in gelada baboons. In one instance, a dominant male savaged a female who had been on the verge of an act of infidelity. Her mother didn't intervene, but looked on, concerned and alert, throughout the attack. Afterwards, de Waal noticed the mother giving a series of soft grunts, followed by an intensive grooming session. Grooming and vocalization seemed to go hand in hand with the upkeep of social relationships.

And these relationships, inevitably, become a vital consideration for a species that chooses to live in large groups. It is thought that we first entered this mode of living when we descended from the trees into the open, hostile environment of the savannah. We were able to thrive, and fend off attacks from predators, by living in groups. It meant there were more pairs of eyes and ears on the look-out for danger. Bigger groups were more intimidating to predators, better able to outwit them in a chase. But, like every other neat idea in evolution, group living had its downside. Groups mean more competition for resources, more potential for faction-fighting and feuds. Grooming alliances, as seen in vervets and geladas, are one way of countering the problems of social living.

Grooming promotes hygiene. Grooming is pleasurable – stimulating the production of endorphins, the body's natural opiates. But grooming is also costly – at least in terms of time that could otherwise be spent eating or mating or sleeping. And it seems that the amount of time a species has to spend grooming increases along with two other things: the size of the group in which it lives, and the size of its brain – or rather, not the brain as a whole, but, according to evolutionary biologist Professor Robin Dunbar, a Fellow of Magdalen College, Oxford, just the neocortex.[4] In humans and the higher primates, this outer sheath of neural tissue takes up between 50 and 80 per cent of the total volume of the brain, as opposed to around 30 per cent in other mammals. The size of the neocortex, says Dunbar, has nothing to do with diet, or the size of the territory occupied, or the distance travelled each day to find food. It does, however, bear an observable relationship to the size of the groups lived in. The purpose of all this extra, calorie-guzzling, birth-canal-straining computer power, it seems, is to enable us to form and maintain long-lasting social relationships.

By comparing cortical size with group size across a number of species, Dunbar arrived at the conclusion that we humans should be living in bands of around 150 individuals (slightly fewer than the band of Fellows harmoniously appointed at Magdalen College). Interestingly, 150 is roughly the number of living descendants that two individuals could produce at the rate and lifespan seen in contemporary hunter-gatherer societies. The American Hutterites – an exclusive religious brethren descended from sixteenth-century Anabaptists fleeing persecution in Europe – live in groups of 150; when a group exceeds that number, it breaks up into smaller ones. When the founder of Mormonism, Brigham Young, was planning to lead his followers from Chicago to Utah, he divided them into groups of 150.

These findings may be coincidental, but it is interesting to consider how group size plays into the issue of grooming time. Among our primate cousins, baboons and chimps, the largest grooming groups of fifty to fifty-five individuals invest almost a quarter of their waking hours just in the act of picking over one another's fur. Multiply this to fit with our cortical size and that figure of 150 and you get a required grooming time for humans of around 6.4 hours a day.

Language might have been the solution. It is a way of staying in contact with one another in large groups, without the time-intensive, costly business of grooming. This splendid device, a kind of 'Grooming Version 2.0', is built on the back of our need to live in increasingly large, complex societies, and necessitated a huge spurt in brain size. But, regardless of its original purposes, language would also have fostered and furthered our capacity to invent, and increased our ability to use symbols and abstract ideas. It provided a means of commenting on the world, sharing notions about it. UCLA's Patricia Greenfield believes it was intricately bound up with the use of tools – the brain areas recruited in the manufacture of, say, a hand-axe, or a scraper, using many of the areas also involved in speaking and comprehension.[5] On a conceptual level, to see oneself as a tool-maker perhaps requires the same basic subject–verb–object string of reasoning that we also use to build sentences, as in 'I make the hand-axe.' And the emergence of standardized tools, across large geographical areas, from 700,000 years ago, suggests a sophisticated means of information-sharing.

Virtually no human idea or invention has ever emerged in isolation. The history we all learn in junior school loves absolutes, delights in simple distortions like 'James Watt invented the steam engine' and 'Pasteur discovered bacteria'. As I write, I have just been asked to take part in a series of television programmes on 'Scientists Who Changed the World'. This is, of course, a nonsense – and the recurrence of such nonsense is one of the reasons why I decided to write this book. I am quite reluctant to contribute to the series, but I fear I won't resist the temptation to be on the screen once more. There is a popular concept – promoted by print media and broadcasters who should know better – that every scientific announcement of some 'progress' is a 'breakthrough', and that usually one scientist is the key person responsible. We scientists do little to dispel this misconception as, being human, we find it pleasant to receive praise and public plaudits. But very few scientific discoveries or inventions are the work of one individual or even one team of people. Many of the best inventions emerge from different groups of humans working at a distance from each other geographically – and not collaborating, but competing. There is always a rush to be first in print with new findings because almost invariably someone, somewhere in the world, is working on the same problem and might get the credit that you think is due to you.

The truth about scientific discovery, as we shall see time and time again throughout this book, is that it is always more complex than a solo effort. And of course, nearly all advances are built on previous work by other investigators and researchers. Whether we're looking at crop domestication, anaesthesia or the printing press, there is always some precursor, some dialogue between people across time and space, some traceable evidence of contact. No idea emerges in isolation, because we humans are social animals, and communication is central to that society. Language may be both the reason we developed brains that invent, and the means by which those inventions were made possible. No language – no invention.

So can we pinpoint when language might have come about, using evidence from archaeology? Robin Dunbar is a bit of a hero of mine but, having argued persuasively for the size of the group as a marker for language, he doesn't provide archaeological examples of our ancestors living in groups of 150. He settles instead for a look at the brain size of fossilized humans, and – using that to postulate a range of factors including group size, grooming time and language – concludes that language must have emerged alongside the appearance of *Homo sapiens* as a species, around half a million years ago.

This seems a little weak. But so, of course, are most attempts to pinpoint the birth of language. The human brain is divided into two never quite identical parts: the left hemisphere dominates in right-handed people, and vice versa. In most people, understanding and producing language take place in the left hemisphere, and for that reason some scientists argue that the first asymmetrical human skulls provide us with a possible date for the origins of language. They put the date at 250,000 years ago, which, interestingly, coincides roughly with the result arrived at by a very different route.

Richard Klein, professor of anthropological sciences at Stanford University, looks to genes rather than bones in his own attempts to understand the emergence of language. Klein draws attention to some key work by Cecilia Lai at the University of Oxford which reported genetic studies of the 'KE' family – a large clan from London, three generations of whom have severe difficulties with speech and language comprehension. This excellent study suggests that this family – along with other individuals with similar difficulties around the world – have a mutation on a gene known as FOXP2.[6] Sadly – and this in no way detracts from the researchers' really excellent work – the London *Times*, the *New York Times* and various other newspaper editors called this a 'scientific break-through', when in fact, unsurprisingly, a number of groups around the world are working on this gene, each contributing important pieces to the intellectual jigsaw. It seems that this mutation causes difficulties with the fine motor control necessary for forming sounds; brain imaging in people affected also reveals significant impairment in the areas used for understanding. Klein believes a number of changes in the human brain, including for instance the development of the FOXP2 gene, may have occurred over a short period of time in our evolutionary history, giving us an enhanced ability to communicate.[7]

The team at Leipzig's Max Planck Institute for Evolutionary Anthropology who first identified the gene – dubbed, perhaps over-enthusiastically, the 'language gene' – have argued that it first appeared around 200,000 years ago and spread throughout the human population quickly, within the space of 10,000–20,000 years.[8] This rapid adoption of the 'language gene' suggests that it must have conferred an important survival advantage on those who had it – possibly, as Robin Dunbar argues, the ability to maintain large social groups.

We shall always have a problem dating our development of language because talking doesn't leave traces. In 1989 Israeli scientists discovered an anatomically modern hyoid bone – the bone that connects the larynx or voicebox to the lower jaw – at a Neanderthal site, and this gave rise to some enthusiastic claims about the ability of these proto-humans to speak.[9] The evidence from this site, at Kebara in Israel, seemed to be confirmed when, in 2004, Spanish archaeologists concluded from Neanderthal bones that their hearing system was perfectly attuned to the frequencies of human speech. The Neanderthals, stalking the earth between 130,000 and 30,000 years ago, lived in large groups and may have had language. Disputed evidence suggests that they buried their dead with flowers, looked after their elderly, and engaged in various forms of symbolic behaviour, too. But we are not entitled to infer the use of language from their remains just because some facets of their behaviour seem touchingly close to ours. An ear able to hear human speech isn't an ear that must, of necessity, have heard it. And the most distinct physiological feature enabling humans to produce the wide range of sounds ideally needed for language is our lowered larynx – a muscular structure that doesn't survive in the archaeo-logical record. Studies have shown that, as the larynx descends in males during childhood and puberty – giving them a deeper voice – there's no change in the shape or position of the hyoid bone.

Another theory bearing on the same question involves drawing inferences about the position of the larynx from the angle at the base of the skull. In 1971 Edmund S. Crelin and Philip Lieberman of Yale University Medical School and the University of Connecticut published their measurements of Neanderthal skull angles. Noting that these proto-humans lacked the low larynxes of the *Homo sapiens* who replaced them, they suggested that their language was fairly basic. Combining this supposition with the fact that their toolkits were simple, and barely changed over tens of thousands of years, they argued that the lack of language skills could have been a factor in the Neanderthals' demise. Handy as this theory is, it was challenged by later, more long-range studies indicating that there is no reliable connection between the angles at the base of the skull and the position of the larynx.

Others have concentrated their efforts on the brain, arguing that we can tell from plaster casts of the inside of human skulls when our brains first developed the areas associated with language production and comprehension. Certainly, if you took my brain apart now, you would find two patches in the left hemisphere associated with these tasks – known as Broca's and Wernicke's areas. But the precise shape and location of these patches would be different from the shape and location of the same patches in your brain, as would be the density of the neural connections inside them, between them and to other areas of the brain. You could discern very little about my linguistic abilities from the traces they left upon the grooved underside of my skull.

The still relatively small number of fossils dating back to our earliest beginnings can tell us much about our brain size, our use of tools and our fondness for living in groups. All of these traits, in turn, are what we'd expect to accompany the use of language. But that is as far as it goes. Perhaps that is why, as far back as 1866, the Linguistic Society of Paris declared that it wanted no more of its members to submit speculative articles about the origin of language. A more relevant question for our inquiry is not when but *how* language – the most vital catalyst for human inventiveness – ever came about.

BAMBOOZLING AND BELONGING

Listening to some exchanges at Question Time in the House of Commons just before I wrote this chapter, it struck me how frequently many of us use language for purposes other than those of straightforward communication. Among certain honourable members there is a tendency to use phrases like 'at this point in time' or 'based on my current appreciation of the situation' that convey precious little meaning at all. The purpose, it seems, is to advertise one's membership of a group of important people by using long-winded terms – and, in this context, to stay on one's feet talking for the longest possible amount of time, to prevent one's opponents from getting their own points made. Of course, this never happens in the House of Lords, which is a model of brevity. But the same trend is apparent in many walks of life – as, for example, in some police statements read in court. 'I was proceeding along the street in a westerly direction' conveys the same information as 'I went towards the chippy past Boots', but does so in a certain, formal, complicated way that advertises the speaker's membership of an important institution, and participation in the serious business of law enforcement. Lest I should seem to be dishing out criticism unfairly, I should add that some of the worst offenders are members of my own profession when they write scientific papers, and I have to admit that I am not blameless in this respect myself.

This point underlines a basic fact about language. We may imagine we are creatures at the pinnacle of our evolution, using this great invention of ours to convey an infinity of complicated meanings. But perhaps it emerged for some other purpose – a purpose for which it's still being used by numerous members of the animal and bird kingdoms.

W. Tecumseh Fitch at the University of St Andrews has published some interesting findings on the nature of speech production.[10] In a number of species, he argues, speech is used as a means of identification. Many animals and birds can distinguish the sounds made by relatives and familiar neighbours from those made by strangers. Sounds may also provide clues about the size and shape of individuals we cannot see. The length of the vocal tract is related to body size, with larger animals, as we might expect, being able to produce deeper sounds. Our primate ancestors might have been able to deduce a great deal about the creatures in the forest around them simply by studying the noises they made. This might also explain why the human larynx is positioned lower than that of any other mammal. The deep-set position of this organ in humans allows us to make sounds deeper than one would expect for our body size, and this could have lent us an advantage. In the same way as I might put on a fierce-sounding voice if I suspected the presence of burglars in my house, perhaps our primate ancestors learned to see off predators by pretending to be bigger than they were.

When he died of heart problems in 1980, the actor Peter Sellers left behind him a sizeable fortune and a string of broken marriages. His comic genius seemed to go hand in hand with a certain degree of inner torment – the actor once famously roaring at his infant son, 'You have no father! Your father does not exist!' What endures for me, and for his many fans, however, is an appreciation of his skills as a mimic. From the weedy Bluebottle of *The Goon Show* to the hapless Hrundi V. Bakshi of *The Party*, from the brutal tones of factory foreman Fred Kite in *I'm All Right Jack* to the exaggerated Gallicisms of the Clouseau films, Sellers built his reputation on being able to copy other people. This talent made him – like a lot of successful comics – rich, and earned him the attentions of a considerable number of beautiful women. On a somewhat more modest scale, one of the people I most admired during my schooldays was a classmate who was able to do hauntingly perfect impressions of our entire teaching staff, complete with gestures and twitches. Despite his own spotty complexion, bottle-thick glasses and uselessness at all sports, this individual's talent rendered him highly popular. But why?

It might have something to do with the way mimicry first came about. It is integral to our use of language – part of the process whereby children, from their earliest days, begin to acquire that range of sounds and grammar rules they will need in order to communicate. But it is not restricted to humans. Mimicry is more prevalent in songbirds and dolphins than in our primate cousins, and in both of those cases it serves a distinct purpose. Among male songbirds, it seems those with the largest repertoire of the right kind of sounds attract the greatest number of mates. Male canaries, for example, produce sexy or unsexy 'syllables', and the males that produce the most complex and sexy songs induce the females to perform what the biologists rather primly call 'copulation solicitation' displays.[11] Well, many of us go clubbing, but it is generally too noisy to hear one's partner singing. In the songbird it may be that a hefty personal album collection is an indicator of high intelligence and hence good genes. Or it may be that a bird able to imitate others' songs can engineer the impression that its territory is already occupied by more than one of its kind. To opportunist rivals seeking quick coupling, a wide-ranging repertoire is a way of saying, 'Plenty of us blokes here already – don't bother.'

The novelist Nancy Mitford mentions the discomfort her brother, Tom, felt on first joining the army. Raised in a stately home with a large and exuberant brood of sisters for company, Tom didn't realize he was different until he was in the Officers' Mess and a corporal brought him some sugar. 'How perfectly sweet of you,' Tom thanked him, to a chorus of beady looks and raised eyebrows from his fellow soldiers.

The sharp end of one of language's most vital features was turned by his sister Nancy to considerable profit with the publication of her book *Noblesse Oblige*.[12] In that light-hearted tome, Nancy set out the rules of word usage which defined one's membership of the better classes. 'Lavatory' was to be used instead of 'toilet'; 'looking glass' and not 'mirror'; 'napkin' in preference to 'serviette'. Similarly, film fans will be familiar with the classic scene, usually in the early moments of a Seventies horror flick, when the man enters a country pub, orders a drink in a well-to-do London accent, and within seconds finds a number of fierce-looking, inbred locals staring at him. We use language to state our identity, and to discern who shares that identity with us. We have always done this. The Book of Judges relates a story from the war between the peoples of Ephraim and Gilead. After routing their rivals, the Gileadites set up a blockade by the River Jordan, to question anyone passing as to their identity. In the days before passports, they needed to be more inventive, so they asked any passing strangers to pronounce their word for grain, which was 'shibboleth'. The Ephraimites, lacking a 'sh' sound in their language, instantly gave away their identity: though the numbers are almost certainly greatly exaggerated, in the biblical account 42,000 of them were slaughtered.[13]

But 'we' – in some form or other – may have been using language in this way long before the days of the Old Testament. There is evidence of dialects in many species different from us. Killer whales, who swim the oceans in large groups, enjoy cooperative behaviour and compete with rival groups. Social systems like this place a premium on being able to indicate who belongs and who doesn't. For whales, certain calls, and specific ways of making those calls, provide a handy 'passport' – in just the same way as I find myself modifying some of my vowels when I cross the elegant threshold of London's Garrick Club. Language may have evolved partly for this purpose – although, as a system of identification, it is hardly fool-proof. Endowed as we are with such sophisticated equipment for mimicry, we have always had the ability to fake it.

In my quest to discover what makes us invent, some may feel I have overstated the case for language. It is, after all, not unique to humans. But if other animals can 'speak' and comprehend too, then what is it that is present in our cognitive make-up, but lacking in theirs? The different ways we use language can, I believe, provide the key.

MONKEY SAY, MONKEY DO

In the early part of the twentieth century, German audiences flocked to see 'Clever Hans', a horse with outstanding mathematical abilities. His owner, Herr von Osten, a horse trainer and high-school maths teacher, claimed that Hans could add, subtract, multiply, understand fractions, tell time, keep track of the calendar and understand complex sentences in German. To prove it, von Osten would ask Hans questions in front of the crowds – a typical one being: 'If the eighth day of the month comes on a Tuesday, then what is the date of the following Friday?' To this Hans would respond by tapping his foot eighteen times. Unfortunately – for Herr von Osten at least – the abilities of Hans the Horse were challenged. Renowned psychologist Oskar Pfungst studied the two of them at work and realized that Hans was picking up the correct answers from the number of times his owner moved his head – allegedly involuntarily. This was still clever stuff for a horse, but not quite as smart as it seemed, and in time the audiences drifted away.

The tale of Clever Hans highlights an important point about animal communication. It can be complex, sophisticated and highly efficient, but it is not quite what our imagination would like it to be. Back in the 1960s, when Jane Goodall first recorded the behaviour of chimps in Tanzania, it was thought that the range of sounds they made was limited. It now seems this has more to do with the limitations of our own hearing.

Advanced computer technology has made it possible for us to play back animal calls and analyse them for subtle variations of length, pitch and tone to which our own ears remain deaf. The results have been surprising. Using this method, Drs Cheney and Seyfarth discovered that vervet monkeys have at least four different calls.[14] One is made when approaching a vervet subordinate in the hierarchy, another when approaching a superior; another indicates that the caller has spied a group in the distance, a fourth that he or she is moving on to open grassland. Further studies on the same species have revealed yet more variations in call, depending upon whether the caller has spotted an eagle, a snake or a leopard.

This evidence suggests complexity, but it is still a long way from language. In the vervet calls, one cry is linked to one meaning only. What is thought to make human talk different is an ability to detach sounds from this prison of single meanings, to recombine them into different compounds called 'words' with which we can express an infinity of meanings through basic rules we call 'grammar'. But there may be a precedent for this in the animal kingdom, too.

Kate Arnold and Klaus Zuberbühler, studying putty-nosed monkeys in the forests of Nigeria, were surprised to hear their subjects combining alarm calls to make new meanings.[15] These monkeys usually make a 'pyow' sound to indicate the presence of leopards and a short 'hack' to warn of eagles overhead. More surprisingly, they were observed to make 'pyow-hack' sounds which, when played back to a troupe, encouraged them to leave the area. It is a long way from the wit and wisdom of Winston Churchill, perhaps, but it suggests an ability to combine two sounds according to a rule, and to give that combination a new meaning – a procedure similar to the formation of words.

Of course, we've seen chimps performing language stunts that seem even more impressive. The likes of Washoe, Sherman and Kanzi – chimpanzees reared in captivity and trained in the use of

ASL (American Sign Language) – do demonstrate some extraordinary capabilities. They have been able to understand numbers, as well as addition and subtraction. They can grasp the nature of basic relations, such as bigger than or smaller than, on top of, or underneath. They can also obey simple two- or three-stage instructions.

Washoe, now approaching middle age, was the first non-human to learn symbolic gestures and use them to communicate. The first word she learned – as it is in many children – was 'more'. She now has a repertoire of around 200 individual signs, and has a fondness for combining them in meaningful ways. She has referred to her toilet as 'dirty good', and the refrigerator as 'open food drink', even though her teachers always used the correct terms. Having been reared by humans, Washoe allegedly was terrified at her first sight of other chimps and, when asked to define them, said they were 'black bugs'. She recovered from this shock sufficiently to adopt the orphaned Loulis and teach him ASL herself – the first instance of such a high order of communication between primates.

Even some dogs can master the rudiments of human language.[16] Rico, a border collie with a talent for 'understanding' words, was spotted on a TV show by researchers at the Max Planck Institute. The engaging beast's party trick was to be able to fetch any of some 200 different objects when its name was pronounced. The Planck team, led by Dr Julia Fischer, wanted to see if Rico could learn new words – which they tested by placing an unfamiliar toy among the pile of 200 objects, and using its name. Rico consistently picked out the right one. Wary of the example of Hans the Horse, the researchers wanted to be sure that Rico wasn't responding to some unconscious visual clues from themselves – so they double-checked their study by making sure both dog and owner were out of sight when they asked for the objects to be fetched. The collie did not disappoint. Tested four weeks later, he got half the words right – a retrieval rate on a par with that seen in three-year-old humans.

Rico, of course, may just be an exceptionally bright dog. But his species – and specifically his breed – has evolved in close cooperation with humans for tens of thousands of years, and is probably predisposed to being able to understand our language. Other studies have indicated that dogs are very skilful at following the direction of the human gaze – a trick at which they even out-perform primates.

The ability of certain animals to understand us is impressive. It is in the *production* of language that they seem to fall down. One crucial stumbling block – the faculty our primate relatives lack – may be the recursion mechanism we encountered in discussing the Pirahã tribe. Harvard University's Marc Hauser has shed some interesting light on this subject.[17] It seems that monkeys can grasp basic rules about word patterns, but may be unable to progress to the next level. They can understand simple rules, such as the one which dictates that 'the' or 'a' is always followed by a noun; but they are unable to understand phrase constructions like '*if* A happens, *then* so will B'. Hauser calls this the 'critical bottleneck of cognition'.

His group carried out two types of tests on cotton-top tamarin monkeys, in which sequences of one-syllable words were called out by human voices. In the first test, random words were called out in a rigid pattern of alternating male and female voices. Whenever the rule was broken, the monkeys looked at the loudspeaker – indicating they knew something was wrong. In the second test, the rule dictated that the male voice would call out groups of one, two or three words, as long as the female voice did the same. This is recursion, as it involves a rule within a rule, an idea within an idea. Confronted with this test, the monkeys all failed miserably. It seems this level of abstraction is what gave human brains the advantage in a number of arenas – and, most crucially, led to our ability to develop language.

The same factor may also lie behind our unique inventiveness as a species. This ability to think in the abstract, to consider 'What if?', lies behind all creativity. Human thinking can decouple itself from the immediate, consider past, future and possible. It is, to revisit an idea from earlier in this chapter, akin to standing on two legs instead of four, and may even be intimately bound up with that change.

This argument is appealing, but we need to remember the Pirahã. They are self-evidently human, superbly inventive – but their language shows no evidence of recursion. That is partly why W. Tecumseh Fitch has also been working among this remote Brazilian tribe, and he challenges Dan Everett's findings. Ever since they were first unleashed upon the academic world in 2005,

Everett's views on the Pirahã have stirred controversy, because they go to the very heart of what concerns us about the relationship between thought and language.

The existence of the Pirahã – or what Everett claims to have observed about them – challenges a long-held assumption based on the work of Noam Chomsky, the grandfather of modern linguistic theory. Chomsky believes that a system of grammar, based on recursion, is universal to all humans – something akin to a rule-system within the brain, though not occupying any one particular region of it. Everett considered himself a Chomskyan – until research on the Pirahã for his PhD thesis uncovered too much data that didn't fit.

Puzzled, he found himself drawn to the work of a discredited theorist, Edward Sapir, who died in 1939. Sapir believed that languages differed as much as cultures did – in fact, that they differed precisely *because* cultures did. This chimed with what Everett had observed about the Pirahã and their peculiarly truncated, immediate world-view. Perhaps they had no way of expressing abstractions – ifs, buts, whens and maybes – because their simple, jungle-based hunter-gatherer culture predisposed them to living entirely in the moment. Their language reflected the way they thought – and it was sharply different from the way we thought, sharply different from the way we thought *all* humans thought.

Everett's critics have, perhaps inevitably, argued that there must be recursion in the Pirahã language – he's just missed it. Maybe the debate will never be settled; but for now, The Straight-Boned Ones stand as a vivid and fascinating example of the sheer variety of human existence. Perhaps most important, they also caution us against drawing up the definitions of humanity too sharply. And maybe that very nebulous quality is one of the things that make humans such natural inventors. We are a supremely plastic species – perhaps the most plastic of all, our brain and body forever changing in response to the differing environments we find ourselves in. And perhaps our inventive qualities are part of that adaptive process: the ultimate evolutionary tool, enabling us not just to change swiftly in response to the world, but to change the world as well.

Chapter 2

Appetite for Destruction

ALTHOUGH APPLES appear in Greek mythology, at one time it was thought that it was the ancient Romans who farmed and propagated this fruit, and introduced it to western and northern Europe. But in the late nineteenth century Scandinavian scholars challenged this assumption by presenting archaeological evidence of older fossilized apple cores which they had discovered in the Baltic regions. And they cited ancient Nordic myths that predate Roman civilization, in which the gods dined on apples. But the last word on this whole subject should really go to the great Soviet geneticist Nikolai Vavilov, who spent his life studying edible crops only to fall foul of Stalin's government and, after years of persecution and months of torture, die a horrible death in a Siberian prison.

AN APPRENTICESHIP IN ALTRUISM

Vavilov's life began well, in a successful, well-to-do family where humane values and academic excellence were very much prized.

Having graduated in biological sciences in Moscow, he travelled to Western Europe and worked with William Bateson at the University of Cambridge. By then, his mentor Bateson was a leading figure in the science of heredity. Some years before the two met, Bateson had come across some obscure publications by a largely unknown Austrian monk, Gregor Mendel. Mendel's work, which had been done in the 1850s and then largely forgotten, greatly impressed Bateson.[1] What Mendel had done was carefully to cross-pollinate thousands of pea plants in his monastery garden and meticulously record their inherited attributes. His findings turned out to be among the most momentous discoveries in the whole of biology. Although his work was largely ignored during his lifetime, what Mendel clearly demonstrated was that certain characteristics in his pea plants – their height, the colour of the peas, the location of the flowers on the stem and whether the peas were wrinkled or smooth – were inherited in a precise, mathematically predictable way. Mendel had discovered the unchanging particulate nature of inheritance. (We now know, of course, that the particles whose action he had described are genes, the major functional component of DNA.) When Bateson came across a paper written by Mendel some forty years earlier, he was elated, recognizing immediately the massive importance of this forgotten research. Bateson began to understand that what applied to Mendel's pea plants was likely to be true for most, if not all, living organisms and that this would have important implications in understanding evolution as well as heredity.

The young Vavilov, meeting Bateson in Cambridge for the first time, was inspired by all this heady stuff. He was also impressed because Bateson had visited Russia. He had been fully prepared to 'rough it' by undertaking arduous field trips to Vavilov's own homeland to study plants growing in the Russian steppes as part of his exploration of the relationship between the environment of plants and their genetic features – what they inherited. Bateson was by now a leader in what was an emerging scientific field of great importance. The principles of what would become known as 'genetics' – indeed, it was Bateson who coined the word – had been of increasing relevance to biologists ever since Darwin, and not just in relation to plants. The idea that certain individuals were endowed with desirable characteristics was a growing focus in science and society. But good basic laboratory work in the field and in the laboratory was only just beginning, and the tools available were still quite limited.

Among other studies, scientists were starting to do research on an animal species: the fruit fly, *Drosophila melanogaster*. Such flies turned out to be particularly valuable to experimental geneticists, partly because they are easy to keep and feed on rotten fruit which is widely available, but also because they reach sexual maturity within a week and therefore breed quickly. Moreover, they have only four very large chromosomes (on which the genes are carried), which are easy to stain and identify under a simple microscope. Indeed, Thomas Morgan later won the Nobel Prize in Physiology or Medicine for his work on *Drosophila* and these flies are still a key model for geneticists even today. Morgan's elegant work was another key influence on Vavilov: the young Russian realized that studies of this kind were going to produce answers to some of the most important questions in biology. Vavilov was a young man with a mission. He was strongly altruistic and, as in so many scientists, his pursuit of science was undoubtedly inspired by the idea that he would be doing good. He saw clearly that, in the field of botany, a better understanding of genetics was likely to be hugely valuable in improving crop yields and that this could mean improving the lot of many humans. So he decided to devote his scientific life to research into plant breeding.

After spending nearly two years with Bateson in Cambridge, Vavilov returned to Russia bubbling with enthusiasm. He was now aged twenty-nine, but the authorities had plans for him which made it difficult for him to get down to work in a university setting. The Russian army was fighting in the First World War, and Vavilov was sent on a mission to Persia to investigate why soldiers on the front there became ill after eating bread. Vavilov soon found the answer: the bread was contaminated with a poisonous species of rye. The problem was therefore easy to solve – considerably easier than Vavilov's expedition to find it, which was made under the most severe climatic conditions at high altitude in hostile territory. Vavilov, accompanied only by two bearers, an overweight guide and six pack-animals, had to cross the remote Demri-Shaug glacier – an expedition which the Governor of Bukhara had warned was the utmost folly. As has been pointed out by my friend Jan Witkowski,[2] a geneticist from Cold Spring Harbor, nowadays nobody would dream of making such a journey without survival clothing, satellite telephone and global positioning technology. But Vavilov did it wearing his three-piece suit.

Having accomplished the official part of his mission, Vavilov saw a wonderful opportunity to spend some time collecting and studying the remarkably hardy plants of the region. His painstaking and diligent work, now and later, was not merely driven by academic interest. He hoped that if he could gain an understanding of how these species had evolved in this harsh environment over generations, they might be cultivated or cross-bred to feed people who were dying of starvation in the northern parts of Russia.

When he got back home in 1916, the whole country was in turmoil, and by March 1917 the Tsar had abdicated. In October the Bolsheviks seized government and Russia faced civil war. With the descent into political chaos, there were severe food shortages by 1920; water and electricity supplies in many big cities had failed and many factories, shops, universities and hospitals were abandoned. In Petrograd, where Vavilov was now working, many people were suffering from starvation. By 1921 a terrible famine was stretching across Russia and Vavilov was determined to use his botanical researches to help feed his compatriots.

Although still a young man, Vavilov soon gained a senior academic position. Unlike many of the Russian intelligentsia – men such as Stravinsky, Prokoviev, Chagall, Kandinsky, Nabokov and Gorky – he refused to leave Mother Russia for the more plentiful life in Western Europe. Dedicating himself to what he believed was a truly important human goal, over the next few years he led numerous field trips in the search for cereals and basic vegetables which might be grown quickly to alleviate the hardships in his homeland. His many expeditions eventually led to his being credited with identifying the birthplace of more plants than anyone else in history. Flying in the face of received wisdom about the origins of agriculture, Vavilov argued that farming had originated not in fertile valleys like those of the Euphrates and the Nile, but in the most obscure mountain ranges, where there was the greatest variety of species.

Vavilov soon built up a formidable reputation. He began a systematic study of cultivated plants from all over the world, concentrating particularly on those growing in countries with mountainous regions. His expeditions took him to North Africa, the Middle East, Persia, Afghanistan, Mexico, China, Mongolia, Japan and India. During trips to Peru and Bolivia he identified twelve new species of potato. In all, he collected and archived the seeds of over 250,000 plant species – a collection never equalled by any other scientist. Thus, within a few years of the Russian Revolution, Vavilov had become one of the most respected scientists in the Soviet Union. He was elected a member of the prestigious Soviet Academy of Sciences and appointed director of the highly important Institute of Plant Breeding. He won numerous medals and in 1926 received the highest honour of all, the Order of Lenin, for his work on the genetics of cereal crops.

And what of the apple? It was in one of those inhospitable places he visited in the course of his researches – the Tien Shan mountain range between Kazakhstan and China – that Vavilov was to pinpoint its origins. Some sixty years later, research by Barrie Juniper of Oxford University's Department of Plant Sciences confirmed Vavilov's findings. Today, there are some 7,500 different varieties of apples in the world, all of them descended from a single small, sweet type, *Malus sieversii*, found in the Ili valley in Kazakhstan, on the northern slopes of the Tien Shan. Not for nothing is the largest city of the region called Almaty or Alma-Ata, 'The Father of Apples'.

With his knowledge of genetics, inspired by the principles of Mendel and the teaching of Bateson, and his vast collection of seeds painstakingly gleaned from many different parts of the world, Vavilov seemed to be in a prime position to do what human farmers had tried to do since the beginning of recorded history. He should be able to hybridize and breed prolific and fertile crops that would inherit hardy characteristics. Here was an opportunity to improve the lot of farmers throughout the Soviet Union, and alleviate the hunger that was ravaging his country.

LYSENKO: POLITICS TRUMPS SCIENCE

It is at this point that the sinister figure of Trofim Denisovitch Lysenko enters our narrative. Lysenko is, in some ways, the Rasputin of this story. He may well have been quite unlike Rasputin in physical appearance, with his slim ascetic-looking visage, but he was utterly ruthless and deeply ambitious. And, as he grew more successful, he appeared increasingly dishonest and profoundly vindictive. In all these respects he was the very antithesis of the highly idealistic Vavilov. His family background was also very different. Lysenko's family came from peasant stock and he had had a fairly limited education; he never, for example, had the opportunity to attend a university. Once he was trained in plant husbandry, he took a job as a junior technician in an agricultural station in Azerbaijan. There he was responsible for selecting pea seedlings to provide the best cattle feed and compost for spring planting. He worked in this obscure position for some time, and it was only a chance article by a local journalist that won him wider attention. Quite out of the blue, in 1927 the official daily newspaper, *Pravda*, published a glowing report of Lysenko's work. Apparently his peas had survived the harsh winter to provide a valuable crop. Here was good political capital for *Pravda*: a 'barefoot scientist' who was not dithering around with esoteric academic matters, but addressing practical problems. As *Pravda* reported, this 'admirable young man', aged just twenty-nine, didn't trouble with 'irrelevant studies on the legs of fruit flies' like the academic geneticists, but could apply his skills and technology to making desert lands fertile. This report was classic Marxist thinking, ridiculing the intelligentsia (in this case people like Vavilov and his colleagues) who were assumed to be too often absorbed in high-flown study of no practical use to the proletariat. In this political environment, it was perhaps inevitable that Vavilov's command of scientific resources, remarkable success and international acclaim would have inspired jealousy among some of his colleagues.

Lysenko, building on the publicity he had been given by making ambitious statements about horticulture, soon found his ideas were being studied and then promoted in Kremlin circles. His 'new' technology seemed like a brilliant solution to the problem of how to produce useful food supplies for the masses quickly. Lysenko proposed that manipulation of the environment of a seed could change not only the plant it produced after germination, but also subsequent generations of its offspring. Perhaps a communist state was a particularly receptive environment for this proposal, which fitted in with Marxist ideals: the perceived 'elitist' advantages given by heredity could be superseded by the socialist drive to provide a better environment. Lysenko claimed that the temperature at which seeds were stored had a profound influence on their later development. In particular, he maintained he had discovered the process he termed 'vernalization' – effectively, the recreation of vernal (or spring) conditions – which he asserted was highly beneficial in producing better crops. As it turns out, this was not a new idea at all. A quite similar phenomenon had already been reported some seventy years earlier and found to be of severely limited value.

When he heard about Lysenko and his work, Vavilov did not reject these ideas automatically – even though his entire scientific training would have made him naturally very sceptical. Indeed, he was admirably open-minded, as a good scientist should be. He sought to understand Lysenko's claims, however unlikely they seemed. Rather than reject them out of hand, he tested the hypothesis that seed temperature might play a part in later development. But all his experiments led to the same negative conclusions; all the evidence he carefully collected argued that the laws of genetic inheritance were the real basis for breeding better plants. Crossbreeding, or hybridization, of different strains turned out to be a much better bet than relying on some manipulation of the environment in which plants germinated or grew. However, Vavilov readily agreed that expert 'gardeners' like Lysenko, who had worked close to the soil and had observed how to make plants thrive, could make valuable contributions to improving agricultural yields. Although Lysenko obviously knew little of scientific method and despised the need for mathematical evaluation and statistical analysis, Vavilov thought of him as an excellent technician. Feeling that Lysenko and others like him should be encouraged, he kept in contact and was helpful and protective, treating Lysenko like a young protégé. Whenever the possibility arose, he offered him valuable opportunities to present his work to scientific audiences and to become better known among the scientific community.

But, increasingly, what Lysenko was proposing seemed scientifically discredited. Reputable biologists had shown long before that the principles of Lamarckism, the idea that acquired characteristics could be inherited, were flawed. Jean-Baptiste Lamarck, a distinguished French botanist and an early evolutionary theorist, had argued well over a century earlier, at around the time of the French Revolution, that a plant or animal could change as a result of its environment and that these adaptations could be passed on to subsequent generations. An often-quoted example is that of the giraffe, which, it was suggested, grew a long neck so that it might feed off the greenest shoots at the tops of tall trees and then passed this useful attribute on to its offspring to save them from starving. Similarly, according to Lamarck, disused structures in an organism's body would be expected to shrink, and such atrophy, too, would be exhibited in its offspring. So Lamarck theorized that a successful species was the result of its continuous, gradual change as it became adapted to its environment. Although these ideas gradually fell into disfavour and were increasingly dismissed in the twentieth century by nearly all respected mainstream scientists, at the time they were an important step towards acceptance of evolution as a key force in biology.

But Lysenko was fortunate in that the state-controlled press, not known for its scientific literacy, continued to publicize his work in admiring terms. In growing favour within the Kremlin, he eventually gained access to Stalin's inner circle at a critical moment in Soviet history. At that time agricultural policy was a high political priority. Stalin was in the process of collectivizing the Soviet Union's agriculture, and needed all the help science could offer him. By the end of 1929 the world was in financial chaos. Wall Street had crashed and Stalin had ambitions to make the USSR the truly pre-eminent global power. He saw science as the driver for his new economy – in his utopia, nature would be bent to the will of the technologists. One key part of his project was to modernize Soviet agriculture, which was still highly inefficient and run on almost medieval lines. Stalin's plan was to replace nomadic peasant farming with intensive grain cultivation, converting huge tracts of the open plains of Central

Asia into large areas of cereal-growing fields. The advice of experts like
Vavilov, who maintained that most of this land was unsuitable for
anything other than animal herding, were ignored. So the nomads
of Kazakhstan were driven into huge collective farms on substandard
soil, often without any tools or housing, and told to produce grain.
Understandably, the Kazakhs put up some resistance, many
slaughtering their livestock rather than hand them over to the state.
Stalin responded by sending in troops. By the time his collectivization
plans were completed, more than a million Kazakhs – a quarter of
the population – had either been executed or died of starvation.
In the Ukraine, the situation was even more horrific. It is calculated
that Stalin's reforms there resulted in the death by starvation of at least
5 million land-holding peasant farmers, the so-called kulaks.

  Lysenko was increasingly given resources. He was offered a free
rein to carry out numerous costly and pointless agricultural experiments, with the goal of turning unfarmable eastern steppeland into
golden seas of cereal crops. One such experiment called for 800,000
pairs of tweezers – and proved no more useful, in the long run, than
any of his others. But failures were explained away as the result of
interference by counter-revolutionaries. Faced with the growing
resistance of the kulaks, Stalin demanded a socialist offensive
'against capitalist elements in town and country'. There followed
mass arrests and the forced deportation of around 6 million
peasants. The scientists who disagreed with this policy were also
persecuted. Those who dared to find fault with Lysenko's work were
treated as enemies of Stalin, indeed of Socialism itself. In 1929
around 700 senior academicians were dismissed or arrested, and
many other agricultural scientists were purged. For the time being
Vavilov seemed safe personally, but several of his colleagues were
accused of 'sabotage' and arrested. And it was inevitable that
Vavilov would begin to feel uneasy about Lysenko's experiments and
the line he was pursuing. Fairly soon he found himself seen as an
outsider. By March 1930 OGPU, Stalin's secret police, had opened a
file (no. 006854) on Vavilov. An informer had reported that there
was a counter-revolutionary cell at work inside his Leningrad team;
soon one or two previous colleagues who bore a grudge or disagreed
with him over research strategy were making secret accusations.
Quite slight differences of opinion started to corrupt relationships

within the team. So eventually it was easy for the inquisitors to 'prove' that there was 'evidence' that Vavilov was the leader of this dangerous group. More junior colleagues were arrested and, subjected to massive psychological pressure, sleep deprivation, threats and indeed physical violence, produced 'confessions', incriminating Vavilov in the process.

By 1936 there were effectively two camps in Soviet biology. Lysenko became increasingly influential and was able to publish unscientific criticisms of genetics.[3] It wasn't merely that people such as Vavilov had 'failed to recognize' the importance of the environment; Lysenko rejected the very notion that genes are inherited structures that pass unchanged from generation to generation. This open espousal of Lamarckism was a source of increasing ridicule outside the Soviet Union and bewildering to orthodox Soviet scientists. Meanwhile party policy increasingly favoured Lysenko – dooming Soviet agriculture and causing more starvation in the process. 'Isn't it strange', wrote Stalin's right-hand man Molotov in 1935, 'that a number of scientists have still not found it essential to show Lysenko active support?' Lysenko was asked to denounce those who argued against his methods in public. He named Vavilov with three of his colleagues. The following day *Pravda* announced that Trofim Lysenko had been given the highest honour in the USSR, the Order of Lenin.

That year, 700 people were due to attend the party's agricultural congress. Vavilov had decided to use this event to challenge Lysenko's total ignorance of genetics by holding a scientific demonstration in the lobby of the congress centre. He had his people set up microscopes with stained slides showing plant chromosomes in varies stages of development during cell division. To give the whole display complete clarity, the microscopic preparations were accompanied by explanatory diagrams and labels. Lysenko spent a few minutes peering down one microscope, looked at the diagrams and repudiated what he had seen contemptuously. He dismissed the chromosomes he saw down the eyepiece as mere artefacts caused by the way Vavilov's team prepared specimens. Scientific controversies often attract huge public attention and this conflict was no exception. The reports in *Pravda* aroused such interest that the whole congress had to move to a bigger hall as over 3,000 people flocked in, swamping the expected 700 participants.

A confrontation between the two men was increasingly inevitable. While Vavilov remained conciliatory and tried to steer a diplomatic and restrained course, Lysenko was vicious and arrogant. He continued to assert that genes did not exist and that the chromosomes seen down the microscope were some kind of fabrication by the geneticists. He had no serious experimental data to support his neo-Lamarckian views, but the geneticists were unwilling or unable to demolish the mumbo-jumbo he was peddling. Towards the end of the congress, at the height of the scientific discussions, there was the most unfortunate intervention. The American geneticist Hermann Muller rose to defend the principles involved in genetics. But he did more than that – he refused to let Lysenko's earlier nonsense go unchallenged and called him a 'quack', suggesting he was practising mere 'astrology'. If Lamarckism were true, Muller argued, it implied the inherent inferiority of peoples and classes who had lived under conditions which gave less opportunity for mental development. Such views, he stated, were fascist and racist. The effect was disastrous for Vavilov. By this time, Hitler had already introduced eugenic manipulation and sterilization of those 'unfit to reproduce', and in response the Kremlin had banned even mere discussion of the topic of eugenics. Muller, perhaps naïvely, had not understood the import of all this: that he was effectively accusing the supporters of neo-Lamarckism – and with them Joseph Stalin – of being both racists and fascists.

THE NET CLOSES

Vavilov continued to hope that he would be safe because of his international reputation and his many connections around the world. After all, his work was one of the reasons why Soviet science was still respected. But Stalin's bloodletting had started and by 1937 the Terror was in full swing. Stalin demanded absolute loyalty to the party line and even senior figures who expressed any dissent were arrested and shot. Vavilov was attacked mercilessly in the Soviet press for his 'religious view of the gene and heredity' and for the anti-Marxist view that certain people – for example, wealthy or aristocratic families – had the 'most precious genes': a travesty of Vavilov's real arguments. For the present, although the purges were gaining momentum, Vavilov remained at liberty, for the public arrest of an internationally respected figure would be an embarrassment to Stalin. But now, whenever he spoke at the big scientific conferences, Stalin either did not attend or ostentatiously left the hall.

By 1939 Lysenko had been appointed head of the Lenin Academy. He was now the most powerful man in Soviet biology. Bent on Vavilov's total humiliation, he called him to Moscow in the spring to account for his views. A transcript of this dreadful interview was recorded by another party thug, a man named Lukyanenko. The interrogation that took place is an example of a blight which infects scientists from time to time: fierce aggression rooted in professional jealousy. Lysenko argued that although it was claimed that potatoes had been imported into Russia from America, new varieties could be produced in Moscow, Kiev or Leningrad by merely manipulating their environment. At various stages during the interview Lysenko's hatchet man, Lukyanenko, made bizarre interjections which are still on the record – such as questioning the fact that our species, *Homo sapiens*, could have originated outside the Soviet Union. He asserted that 'Marxism is the only science' and that 'Darwin is only a part and the real knowledge of the world was given by Marx, Engels and Lenin'. It seems that at first Vavilov was able to keep his temper and answer in a rational fashion, but finally he seems to have given up defending his scientific view of biology. The interview concluded with threats from Lysenko that 'measures' would have to be taken against Vavilov.

By the autumn of 1939 Vavilov, though still in post, was virtually destroyed. After he had been refused permission to exhibit some of his work at a national conference, he arrived in Moscow and bearded Lysenko in his office. There are eye-witness accounts of Vavilov grabbing a terrified Lysenko by the collar and screaming at him for ruining Soviet science. Although the regime found it difficult to dispose of a man who commanded such respect internationally, a subtle way of finishing his career was eventually found. In 1940 he was sent on a fact-finding mission in an obscure part of western

Ukraine. Once he was travelling and out of communication with friends, family and the rest of the world, four secret service agents accosted him in the obscure town of Chernovtsy. They told him he was required urgently in Moscow on government business. He was bundled into a car and driven to Lvov and Kiev before finally arriving in Moscow. His team with whom he had travelled, and his family, were left totally unaware of his abduction by the state.

Once in the Lubyanka for questioning, Vavilov had effectively been 'disappeared'. Over the next few months he was subjected to sleep deprivation, relentless interrogations and increasing tortures. Descriptions of the sufferings of political prisoners under interrogation in Stalin's gaols do not make pretty reading, and perhaps are best avoided here. Suffice it to say that Vavilov was not spared pain and indignity. On the basis of the 'evidence' given by Lysenko and other supporters of Stalin's machine, he was charged with espionage and other counter-revolutionary activities. Over the next eleven months he was to be interrogated over 400 times. Usually questioning went on for at least four hours, sometimes all night. Pictures taken of him by the secret police show an emaciated, prematurely aged, vacantly staring individual. Fellow prisoners later testified how, after a night of interrogation by the senior inquisitor, a monstrous man called Kvat, he could no longer stand and was able only to crawl very slowly on all fours to his corner of the cell. There, sympathetic inmates prised his tattered boots off his swollen legs to give him a brief respite.

Although a subsequent show trial found Vavilov guilty and condemned him to death, the authorities did not wish to risk the adverse publicity which might have attended his execution. So they simply left him to rot. By a cruel irony he was transferred to a prison in Saratov, in a relatively remote part of Russia – just fifteen minutes' walk away from where his wife and son, who had been desperately trying to find his whereabouts and who had spent most of the past two years searching in vain for him, were lodged, not knowing even whether he was alive or dead.[4] Within a few months of being incarcerated at Saratov, this good, honest man, who had hoped to feed the malnourished millions, died of starvation and dysentery in his cell.

LESSONS FROM A TRAGEDY

This sad story touches upon many themes that run through this book. We shall return several times to the issue of genetic manipulation and the potential value and huge threats associated with the practical application of research in this field. Another theme that will recur is that of government control of science and how politicians are likely to use or misuse the knowledge that the scientists have produced. It would be foolish to pretend that the political environment in the USSR during the time of Stalin was, or is, unique. There are a number of totalitarian regimes in the world today that are still prepared to use technology in a totally unscrupulous manner. And even 'benign' elected governments such as those of our own relatively liberal democracy are quite capable of misinterpreting scientific data, or using the information which comes from science for their own ends.

A further issue concerns the consequences of the failure of complex technology. When agriculture failed in the USSR, the result was catastrophic. Estimates of the total number of people who starved to death during Stalin's dictatorship vary, but the total breakdown of Soviet agriculture, in which the misuse of technology was a prominent contributory factor, was certainly responsible for the deaths of many millions of Soviet citizens. In the Soviet Union, as elsewhere, people had become totally dependent on farming technology. When farming failed, there was no other practicable way to feed most of the population adequately.

There is another curious theme to which we shall also return: the nature of scientific 'truth'. As the twentieth century progressed, no reputable scientist was prepared to entertain any idea that there might be something in Lysenko's claims after all. Today, Lamarckism is pretty well anathema to serious biologists. Scientific 'truth' emphasizes that acquired characteristics are not inherited. If they were – even occasionally – Darwin's notion of 'natural selection', which is widely agreed to be the very keystone of the theory of evolution, would be challenged. Yet as we shall show in chapter 11, the relatively new science of epigenetics suggests that organisms may change in response to their environment, and that these changes can be occasionally inherited and are therefore transgenerational. That is to say, some traits exhibited by offspring are present because of parental experience – even though the actual 'letters' of the DNA have not been changed. Through the study of epigenetics we are beginning to understand some of the mechanisms which make this possible. It is ironic that Lysenko, who through his ignorance and prejudice did much to destroy or damage much of Soviet agriculture – and hence encompass the starvation of millions of his fellow citizens – may have not been so totally mistaken about heredity after all.

The reason for telling the story about Lysenko and Vavilov is not only to honour a great scientist and to show how scientific information can be manipulated by unscrupulous people in a position of power but also, primarily, to introduce the subject of farming – one of the very earliest of all human technologies. At a certain point in our history, in very specific locations, we began cultivating plants and domesticating animals, and spreading this method, very slowly, around the planet. Our consummate skill at doing so set off further changes: rising populations, increasingly complex societies, new skills and professions. It is easy to see the advent of agriculture as the point at which we became masters of our environment. But Vavilov's downfall and the collapse of Stalin's plans should act as warning signs. There can be no question that we have gained immeasurably from tilling and tending the earth, instead of just taking from it. But what might we have lost?

Chapter 3

A Grinding Existence

AN IMPORTANT SHIFT in human affairs occurred around 15,000 years ago, as global warming started again in earnest after a prolonged cold spell. The ice-sheets and glaciers that had covered a substantial part of the northern hemisphere – including, for example, Scandinavia – melted, causing sea levels to rise dramatically. The bridge of land between what are now Alaska and Russia was swamped and the two continents were separated. In South-East Asia, the islands became parted from the mainland. Temperatures rose too, on average by around 7 degrees Celsius. Within about 3,000 years, at the end of the Pleistocene age, many large mammals of North and South America had been wiped out. These species included the huge sabre-toothed tiger (a rather flea-bitten fur-and-plaster model of which can be seen snarling unconvincingly in a glass case in the Natural History Museum in London), mastodons and most woolly mammoths, the woolly rhinoceros, giant skunks, giant rabbits and, in North America, horses and camels. Scientists have wondered whether some kind of bacterial pandemic may have been responsible, but little evidence of a deadly infection has been found. Certainly no bacterial or viral DNA has been recovered giving irrefutable evidence of any global plague. And it seems unlikely that the growing human population was sufficiently adept at killing on this scale to account for the extermination of so many creatures.

The probability is that this particular climate change was unusually sudden, and that these large mammals simply did not have time to adapt. Although there had been previous ice ages, their onset and their cessation had been much slower. There is an important message here for us, because the general consensus today is that global warming is again happening at great speed, with the likelihood of severe adverse effects on life on earth, including on our own species.

There were perhaps some 5 million humans scattered across the planet when climate change occurred at the end of the Pleistocene age. Our hunter-gatherer ancestors adapted with more specialized tools – such as bows and arrows – and intensified the quest for sustenance. As it happened, weather patterns stabilized fairly quickly, creating the necessary conditions for agriculture. Using archaeological evidence and genetic studies, we can talk with some certainty about where this change took place, and when. We can even take a reasonable stab at explaining how. It is not – as we shall see – quite so easy to say why. Whatever the cause, it certainly was not human activity, which appears to be behind our predicament today.

In five areas of the globe, the changing climate assisted people in domesticating plants and then animals. In the Near East – an area called the 'Fertile Crescent', stretching from the Jordan valley into what is now Iraq – people began to select and cultivate cereal crops around 8500 BCE and then to breed small species of animals with desirable characteristics. Considerably later, around 3500 BCE, similar activity happened in Mesoamerica (the region lying between Panama and northern Mexico); and there is evidence for similar independent developments in what are now China (7500 BCE), the Andes and the Amazon basin (both around 3500 BCE), and the eastern United States (2500 BCE). This is, indeed, a prime instance of a theme to which we shall repeatedly return – namely, that human innovations often occur simultaneously in many different parts of the world not obviously connected by any form of communication.

These early agriculturalists lived in very different ways. In Jericho
– the world's oldest known continuous human settlement, and
incidentally the lowest point below sea level where humans have
lived over long periods – the early farmers lived in mud-brick
houses, arranged in haphazard fashion within a stone-walled
cordon. They cultivated wheat and barley, and hunted wild animals.
Some 5,000 years later, in the Tihuacan valley at the southern tip of
modern-day Mexico, people cultivated tiny corn cobs, beans, squash
and chilli peppers. They adapted to the rugged terrain by cutting
fields from the hillside and living in small, semi-permanent pit-house
villages, each dwelling comprising a wooden roof placed over a
shallow excavation in the ground. Though they invested much effort
in their farming activities – also making use of baskets and nets to
harvest crops – cultivated foods only contributed about a fifth of
their diet. Across the planet, at Jiahu in China, inhabitants lived
in a village of round and square sunken-floored houses surrounded
by a moat. They cultivated rice along the Yellow River, and seem to
have decreased their reliance on wild foodstuffs at an early stage.
There is disputed evidence that these early Chinese farmers also
developed alcohol and crude forms of writing. The archaeological
site there is remarkable for its cultural and artistic remains.
Researchers found pottery and turquoise carvings and, most
remarkably, the fragments of about thirty flutes made from the
carved wing-bone of the red-crowned crane. According to Laura
Tedesco of New York's Metropolitan Museum of Art, some of these
flutes are still playable, with five to eight holes capable of producing
tones in a nearly accurate octave.

But how do we establish whether farming took place in a particular area? This question is best addressed by considering what our
ancestors wanted from the process. If we contrast wild wheat with
einkorn wheat – a strain of wheat which begins to appear in the
archaeological record of the Near East from 8500 BCE – then things
become clearer. Wild wheat uses the wind to propagate. When its
pods become full and heavy, their brittle casings burst, spreading the
seed far and wide upon the breeze and in the coats of animals – and
of human beings, too. Einkorn wheat is tougher: its pods yield their
precious cargo only when deliberately threshed with a tool.
(Incidentally, einkorn wheat may have been particularly helpful in

early human settlement because it thrives in soil with poorer
nutrients.) The appearance of a type of wheat that cannot spread its
seed on its own is an aberration, a variation that would have come
about by accident. But humans would have seized upon it, realizing
that a wheat that kept its seeds was a useful plant. In the terminology favoured by biologists, they 'selected' its desirable
characteristics. They eliminated the competition by stripping away
weeds and those types of wild wheat that did as wheat plants are
supposed to, leaving only the tougher stuff that required harvesting.
When this was ready for threshing, the seeds from these 'mutant'
seeds were gathered and consumed; some found their way back to
the soil through human faeces and some were replanted.

So, when archaeologists look for evidence of crop domestication,
they look for signs of plants being selected for characteristics con-
venient or pleasurable to human life. That can include aspects like
soft flesh, large, sweet fruit, edible seeds and the absence of bitter-
ness. When these markers are found in one variant of plant or fruit
and not another, there is a strong indication that the former has been
domesticated. It is the reason why we are able to enjoy treats like
almonds, potatoes and cabbages, whose wild counterparts are far
less pleasurable on the palate.

But, in a book that celebrates human inventiveness, we should ask
if farming was really an invention at all. What is the evidence that it
was 'discovered', in the way that knowledge of bacteria or the
power of steam was discovered?

AGRICULTURE: A PAIN IN THE NECK

The story of man's expulsion from the Garden of Eden – as related
in the Book of Genesis – needs no retelling. But key elements
warrant attention. In the beginning God works and, after six days
of toil, he rests. In addition to explaining the origins of the Sabbath,
this is significant for another reason. Because Adam – God's creation
– does nothing. He is furnished with animals, birds, fish and fruit for
food, but no mention is made of anything being required of him.
Certainly he does not have to cultivate or farm the flora and fauna

of his paradise. The biblical account refers to God bringing the various species of created life to Adam, so that he can name them – hardly a very onerous job. Beyond this, the daily life of our first human seems pretty cushy. He has one just rule to observe – the simple one of not eating the fruit of the Tree of Knowledge.

This happy situation, of course, changes after Eve breaks that single rule. God admonishes his sinful creations, casting them out of the garden. 'Cursed is the ground for thy sake; in sorrow shalt thou eat of it all the days of thy life; thorns also and thistles shall it bring forth to thee; and thou shalt eat the herb of the field; in the sweat of thy face shalt thou eat bread, till thou return unto the ground.'[1]

God's curse upon humankind is farming. No longer shall his creations live an idyllic hunter-gatherer existence, serendipitously plucking copious ripe fruit from the trees for their daily food and scooping water from clear streams to drink. Sustenance shall be won henceforth 'in the sweat of thy face', that is, through back-breaking toil. With this as one of the founding myths of Western civilization, it may seem odd that we have seen farming as any kind of progress at all.

It seems even odder when we look at some detailed archaeological evidence. When examining bones at the Natural History Museum, Theya Molleson found a strikingly unique pattern of arthritis – one causing severe deformities of the lower back and feet.[2] She found the root of this puzzling situation when she compared this evidence of traumatic disease to contemporary painted murals. These depicted people grinding corn. Two workers would sit on opposite sides of a saddle-quern – that is, two heavy round stones, one on top of the other – arduously pushing the upper stone back and forth to grind the corn. These stones are massive and the grinders' backs would have been under immense strain, while their toes were bent under the feet for hours at a time to provide extra purchase. It is also interesting that other studies have shown rows of abscesses along the roots of the teeth – caused by dental wear after eating ground grain contaminated with sand. In the sweat of their faces, indeed.

And problems persisted, even as agriculture became more sophisticated. At the end of the ice age the average height for the hunter-gatherer peoples living around Greece and Turkey was about 1.78 metres for men, 1.65 metres for women. But the height of the average human plummeted as agriculture was adopted. It seems almost incredible, but modern-day Greeks and Turks are still not quite as tall as their hunter-gatherer ancestors.

Similar findings have emerged from the study of thousands of American Indian skeletons excavated from burial mounds in the Illinois and Ohio river valleys.[3] With the adoption of farmed corn as a staple foodstuff – which occurred around 1000 CE – the health of these humans became noticeably worse. The number of cavities in the average mouth jumped from one to seven, along with an increase in abscesses and tooth loss. Enamel defects in children's teeth suggest their mothers were severely undernourished when breast-feeding. Conditions such as anaemia, tuberculosis, yaws and syphilis became more widespread, and around two-thirds of the population seem to have suffered from osteoarthritis. Whereas 5 per cent of the hunter-gatherer population had lived past the age of fifty, now only one in a hundred made it to that age. Around 20 per cent of all children died before the age of five. By comparison, Kim Hill's studies of the Hiwi people in Venezuela (a group of hunter-gatherers often regarded as a good model for our Palaeolithic ancestors) show that about 9 per cent of children died before the age of five before these peoples had contact with 'civilization' – and, interestingly, many of those died as a result of accident or infanticide rather than disease.[4] Viewed in the light of such statistics, cultivating corn may seem more of a blight than we might expect.

Alongside the detrimental effects to health, agriculture is, plainly speaking, a right drag. In the 1960s Richard Lee and Irven DeVore started to study the !Kung Bushmen of the Kalahari desert – a group subsisting in some of the harshest conditions on the planet.[5] We might expect these tribesmen to endure a grim existence – rising with the sun to spend all day scratching out a few meagre calories from the dust. In practice, the average gentleman of the !Kung spent around six hours of his week hunting, and the rest of the time yarning with his pals under the trees. These trees included the mongongo, whose richly nutritious fruit and nuts supplied the !Kung with around half their calorific needs. The kernel is 57 per cent by weight fat and 26 per cent protein, rather similar to cashews and many legumes. With the sugars contained in the fruit surrounding the nut, the average adult would meet 71 per cent of his daily energy requirement by eating just 100 fruits. Indigenous people are reported as eating around 100–300 fruits a day in parts of Namibia. 'Why should we farm,' one Bushman eloquently put it to researchers, 'when we have so many mongongo nuts?' Compared to the pursuits of the average twenty-first-century Londoner, hunting and gathering sounds more like being a leisured member of the Garrick Club – so leisured, in fact, that we might well wonder why anyone chose agriculture.

ANT FARMERS

Among various species of ant found in America, there is a type of behaviour that warrants only one description. These ants, of the genus *Atta*, cultivate fungi in little gardens within their nests. Rather than using whatever soil is to hand, they seek out particular kinds of media – one species growing their crop upon caterpillar faeces, another on insect cadavers and others upon fresh leaves. The efforts undertaken by this latter group are impressive – they clip off leaves from favoured types of plants, slice them into handy chunks, scrape off foreign fungi and bacteria, and carry them into their nests. At this stage, the leaf fragments are pounded into a paste, fertilized with ant saliva and faeces, and seeded with the ants' particular choice of fungus. While this is growing, the ants guard their gardens, removing any spores of rival fungus species that they find. Whenever a queen ant leaves to start a new colony, she takes one thing with her: some spores of her favourite fungus. The best word to describe this behaviour – in layman's terms, at least – is farming.

Ants also purchase a concentrated sugary substance, so-called honeydew, from various of their fellow insects – such as aphids and caterpillars – giving them protection from parasites and predators in return. Some aphids have evolved to fit perfectly with this arrangement, having no means of defending themselves, and a specialized anus which allows the ant to suck out the droplet of honeydew with maximum ease. In return for this biological convenience, the ants will look after their aphids in the warmth of their nests during the winter, then transport them to the most useful site on plants above ground when spring comes.

So agriculture is not just a human occupation. It is a practice that developed out of systems already in place – among our own kind, and even many, many links back from us in the food chain. People didn't 'choose' farming – certainly not in those fertile valleys of the Near East and Mesoamerica – because they couldn't have known what farming was. It was a strategy for survival, one among many, but one which eventually came to dominate over the others.

The biblical account means us to assume that we were thrust out of hunter-gatherer happiness and into agricultural misery on one specific, fateful day – or at least, it takes that approach for the purposes of telling a good story. In reality, though, the lines between the two modes of existence are very blurred. Archaeological evidence from southern Africa suggests that, between the seventh and twelfth centuries CE, the Bushmen themselves herded goats, and occasionally farmed crops of sorghum, millet, melons and cowpeas.

Until it was flooded by the artificially created Lake Assad in 1973, Tel Abu Hureyra, in Syria, was a magnificent example of an early Near Eastern village. Dating back some 11,500 years, it consisted initially of a group of temporary circular dwellings housing a few hundred people, who were dependent on wild crops and upon hunting conveniently located species like gazelle, sheep, cattle and birds. Around 11,000 years ago people began to cultivate crops there – until the site was abandoned, possibly because of an increasingly dry climate. It was then reoccupied around 8,000 years ago, at which point the inhabitants lived in more permanent mud-brick houses, cultivating a wide variety of crops and herding animals.

Agriculture implies a settled existence: crops are planted in the place where they will thrive best, and the farmer remains reasonably close by to tend, harvest and plant them. But at Tel Abu Hureyra, as elsewhere, people were living in village-like settings centuries before they adopted farming. Some societies – like the Coast Salish Indians along Canada's Pacific coast – established permanent villages, but did not depend on agriculture at all. From areas as diverse as Palestine, coastal Peru and Japan we have evidence of peoples settling down, usually next to abundant stands of wild crops, yet not practising any cultivation for centuries. Conversely, our world still harbours human groups who move around and yet still practise quite complex proto-agriculture. In Papua New Guinea and

Australia, mobile bands 'manage' the flora and fauna they depend upon. They pull out weeds, they burn scrub to encourage the growth of new shoots, they leave the stems of yams in the ground to ensure further crops; they see off the rival predators of their own prey, and they corral their quarry on to the hunting grounds with fire.

Farming involves planning. Staple foods take time to grow, and require considerable human effort along the way. For that reason, no right-thinking human would abandon his other sources of food – his hunting and gathering – until his agriculture was long-established. Even then, he would be ill-advised to ditch his older way of life completely. Agriculture was one of a number of alternatives for our ancestors when deciding how to spend their time and resources. If a recent hunt had brought in a good supply of fresh meat, then this might be a suitable opportunity to do some planting. If the crops had inexplicably succumbed to a certain type of pest, on the other hand, then it was time for a people to sharpen up their spears. So why did one strategy – agriculture – come to dominate?

THE GREEN LANES

When I was a seven-year-old living near Palmers Green, I was quite excited when my mother told me we were going to visit a house in Green Lanes. In my mind's eye I saw some delightful village area like those I had discovered in my children's books. Nothing remotely prepared me for that dingy stretch of the A105 which even then ran through what seemed like one of the shabbiest parts of north London. Green Lanes is the longest street in London, the remnant of an old path which was once used to drive cattle down from the pastures of Hertfordshire to their demise in the slaughterhouses of Smithfield. These days, the road is filled with roadworks and lines of slow-moving buses rather than cow-dung and the shouts of drovers. But now a drive along it provides a fascinating glimpse of London's ethnic make-up: Turks, Kurds, Greeks, Chinese and now newcomers from Eastern Europe have established enclaves along its length. Among the grime and the traffic, the colourful displays of fruit and vegetables outside the Turkish and Kurdish grocers' shops

are a rather encouraging sight. Silky black aubergines nestle alongside plump red tomatoes, bunches of lush parsley next to okra, peppers
and beans. It is somewhat paradoxical that, in this age of convenience shops, supermarkets and overprocessed food, this scruffy
street still retains a remote connection to its agricultural origins. But
these elaborate displays don't merely exist to entice the customer to
buy something. They are to some extent a symbol of national pride
– the grocers' way of reminding us that they hail from what was
once one of the world's richest, most fertile and varied gardens.

It is interesting to consider why it was that certain areas
developed such a vigorous agricultural tradition, while others never
acquired it at all. According to Andrew Sherratt of Oxford's
Ashmolean Museum, the Near East and Mesoamerica were global
'hot-spots' – areas of immense geological pressures and constant
change in both landscape and climate. These conditions gave rise to
a specific kind of terrain – the sharp juxtaposition of hills with
deserts, and the creation of narrow strips of land. These led in turn
to a build-up of population, so that traditional methods of hunting
and gathering exhausted the available resources and compelled
people to seek other strategies to survive.

Theories about the origins of agriculture often focus on one basic
idea: that rising populations led to an extinction of mammals which
could be hunted for food, forcing people into a new way of life. The
problem is that, very observably, agriculture wouldn't have provided
the answer to the problem. Some have suggested that things
happened in a different order. Mark Nathan Cohen, for example,
Professor of Anthropology at Plattsburgh State University of New
York, points out that nomadic groups tend to breast-feed children
for at least two or three years.[6] Successful breast-feeding is
dependent on production of the hormone prolactin, and while
prolactin levels in the mother's bloodstream are high, ovarian
activity – and therefore ovulation – tends to be suppressed. So
breast-feeding is a form of natural contraception. No more children
are likely to be born until the nursing infant is able to walk – an
important consideration for a group of humans on the move.
Settling in one place for long periods of time, in Cohen's view, may
possibly have led to people abandoning breast-feeding as the only
way of nourishing their infants. If breast milk is supplemented, the

contraceptive effect of lactation is reduced. So this might have resulted in more frequent births and a rise in population. Such growing communities may then have depleted herds of mammals suitable for the hunt, thus necessitating a new way of living.

This interesting theory paints agriculture, with all its hardship, uncertainties and potential unhealthiness, as a measure taken in response to a crisis, rather than as a leap forward in human progress. But it seems a poor explanation why, when plentiful game was available, these hunter-gatherers adopted the sedentary way of life in the first place. Second, as a response to overpopulation, depending on agriculture is observably a bad move. Just possibly our ancestors would have noticed this increase in pregnancy rates rapidly enough to be concerned.

Archaeologist Les Groube, formerly of the University of Papua New Guinea's Department of Anthropology, speculated that warming of the climate led to the spread of diseases from Africa into the Mediterranean and Near East. A nomadic existence, with its spaced births, couldn't respond adequately to the consequent devastation wrought upon human populations who had little natural resistance to these new infections. So people settled down in order to have more children – a mode of existence favouring the adoption of more and more agricultural means of subsistence. His argument is impressive, because it recognizes that settling in a static community might occur before, and independently of, farming. But if humankind had succumbed to such a crisis, we would expect to find more telling evidence for epidemics in the archaeological record. Moreover, such a theory does not account for why agriculture arose in other parts of the world, for example on the other side of the Atlantic Ocean.

There may well have been several circumstances that acted as triggers for the development of farming. It seems highly likely that people in areas like the Fertile Crescent and Mesoamerica depended increasingly on agriculture as their reserves of wild food – animal and vegetable – dwindled. Perhaps this explains why the archaeological record of the Near East shows that the numbers of wild gazelles decreased as humans in this region gradually adopted cultivation.

It seems that more wild, but potentially domesticable, crops became available as animals suitable for hunting became more scarce. There are around 200,000 different plant species in the world – a figure that emphasizes how remarkable it was that Nikolai Vavilov collected nearly 250,000 varieties of seed during his global travels. Only a few thousand of all these plants are suitable for human consumption and, of these, only a few hundred have been domesticated. Today, around a dozen species account for more than 80 per cent of the world's crop production: wheat, barley, maize, sorghum, soya beans, potatoes, manioc, sweet potato, sugar cane, sugar beet and banana. And it seems our ancestors knew exactly where to put their greatest efforts. As a result of the climatic changes occurring around 12,000 years ago, areas like the Jordan valley were rich in stands of crops whose biology favoured domestication. They grew quickly and abundantly, they had large seeds, they were easy to harvest and they could be stored.

Studies of this region by Ofer Bar-Yosef and Mordechai Kislev showed that these proto-farmers had some twenty-three different types of wild grass available to them.[7] Of these, they appear to have plumped for emmer wheat and barley – the two that were most prolific and most easily grown for food.

Before the advent of proper agriculture, the availability of these wild crops spurred people to develop suitable technology to harvest them. Tools such as the sickle blade to reap the grain, baskets to gather it, the stone quern for grinding and structures such as the plastered store-house to prevent the stored grain from going damp were all important advances. These innovations began to appear at hunter-gatherer sites from 12,000 BCE onwards, after the big shift in climate; their users could not have managed adequate cultivation without them. Their development was probably stimulated by the environmental changes that occurred almost simultaneously in different parts of the world.

Hunting and gathering may be a relatively healthy way of life, but it yields fewer calories per acre than agriculture. Better food supplies would have encouraged both an increase in population and increased growth of humans, a trend which is still often true in the modern world. And a warmer, less harsh climate would also be conducive to population increase, thus aiding the development of better technology for gathering wild foods and the gradual adoption of a sedentary way of life.

So when we consider the 'spread' of an idea like agriculture, we don't really mean that it just took off, like some form of contagion. Almost certainly whole populations did not simply drop their spears and start hoeing the earth. The practice of agriculture spread because the numbers of people farming spread. Those following the hunter-gatherer lifestyle either died out alongside the wild resources on which they depended, or adopted agricultural modes of survival themselves. A third and smaller group, of course, found a niche for themselves in the world's least hospitable and least farmable regions, marginal territories like deserts and jungles, where they remain today.

GRUESOME INGREDIENTS

Friedrich Christian Accum, who was born in 1769, had a father called Markus Herz. Herr Herz was part of the eighteenth-century epidemic of German Jewish baptism which is now remembered with some shame. Many Jews in Germany at this time sought emancipation through conversion to Christianity, and Herz was presumably one such. He took as his new names Christian, 'follower of Christ', and Accum, derived from the Hebrew word *akum*, meaning 'not-Jewish', and became a full citizen some years later. Whether this helped his son isn't clear, but after an apprenticeship as an apothecary in Hanover, Friedrich left Germany to come to London. England's capital city was rapidly becoming a centre for people who were minded to follow the sciences, and Accum attended lectures at the School of Anatomy in Great Windmill Street. This private establishment had been founded by the great Scottish anatomist William Hunter in 1745. Its run-down red-brick façade can still be seen next to the Windmill Theatre, now rather more famous for a different kind of interest in human anatomy. Anatomy at the school was often taught by dissection of hanged criminals, and visitors such as Accum would have been allowed to attend lectures – although not those on 'the organs of generation and the gravid uterus'.

In 1800 Accum and his family moved from Haymarket to Old Compton Street. Friedrich had become increasingly interested in

chemistry and foodstuffs, and his house now served not only as a
dwelling place, but also as a school, a laboratory and a shop dealing in scientific instruments. There are no records of the details of
his conversations with his wife. Accum had business cards printed
which described him as follows:

> Mr Accum acquaints the Patrons and Amateurs of Chemistry that he
> continues to give private Courses of Lectures on Operative and
> Philosophical Chemistry, Practical Pharmacy and the Art of Analysis,
> as well as to take Resident Pupils in his House, and that he keeps
> constantly on sale in as pure a state as possible, all the Re-Agents and
> Articles of Research made use of in Experimental Chemistry, together
> with a complete Collection of Chemical Apparatus and Instruments
> calculated to Suit the conveniences of Different Purchasers.

Accum was an entrepreneur with a talent for self-promotion, and
soon his shop and laboratory had become the leading institution in
Britain where people could learn chemistry and do practical experi-
ments. His seminars were attended by students from well beyond
London, some from as far away as the United States. He also
attracted a pretty fashionable audience which included the Dukes of
Bedford and Northumberland, and the Prime Minister, Lord
Palmerston.

As interest in chemistry grew and new laboratory apparatus was
developed, Accum manufactured and sold mechanisms designed to
enable farmers to analyse the content of the soil which they used for
cultivation. For an investment of between £3 and £80 (no mean sum
in those days) a farmer could be provided with a fully portable
laboratory to increase the quantity of crops he could produce. In
1801 Accum joined the recently founded Royal Institution
in Albemarle Street, working there as an assistant to Humphry
Davy, appointed that year as a lecturer in chemistry. Accum continued to develop wide-ranging interests in chemistry, becoming a
board member of the Gaslight and Coke Company.

Accum is of particular interest to us because of a remarkable
book he published in 1820, *A Treatise on Adulterations of Food,
and Culinary Poisons Exhibiting the Fraudulent Sophistications of
Bread, Beer, Wine, Spiritous Liquors, Tea, Coffee, Cream,*

Confectionery, Vinegar, Mustard, Pepper, Cheese, Olive Oil, Pickles, and Other Articles Employed in Domestic Economy. It never topped the bestseller list, but it made waves. Its readers were appalled to discover that beer – that most treasured resource of Englishmen – was routinely mixed with sulphate of iron to give it a frothy head. Bread flour was mixed with alum, to bulk it up and improve whiteness. Accum also revealed some alarming facts about colouring agents, having uncovered the use of copper in pickles, red lead in Gloucester cheese and pepper spiced up with floor sweepings. One suspects that had a similar treatise been published in France, the Bastille would have been promptly repopulated with provisions merchants, but in England Accum's whistle-blowing prompted no immediate changes in law or practice. It did, however, commence a trend towards greater scrutiny of what people were eating and drinking.

In the early 1850s Thomas Wakley, editor of the *Lancet*, published reports on the adulteration of food and called for an official inquiry. This was set up in 1855, with Dr Arthur Hassall, physician and lecturer at London's Royal Free Hospital, as its chief witness – revealing some distressing facts. Orphan children were dying because unscrupulous tradesmen were bulking up their daily gruel with barley, which gave them diarrhoea on top of their existing ailments. Publishing the names and addresses of the guilty parties, the inquiry revealed that around eight tea factories in London were actually peddling recycled leaves – paying household servants to collect the lees from their employers' teapots and return them. To satisfy the vogue for green tea, one enterprising soul had simply added verdigris, or copper rust, to it. Nearly a decade after Wakley began his series of exposés in the *Lancet* the government passed the Adulteration of Food and Drugs Act. The principle behind the Food Standards Agency is not so new.

These horrifying findings reflected changes that were taking place in the nature of British society. The landlord who brewed his own ales, on his own premises, had become an anachronism by the latter half of the nineteenth century. The industrial revolution, as Karl Marx observed, resulted in an increased separation between labourers and the products of their labour, and between those involved in the different stages of production. Food producers, processors and consumers were alienated from one another, with activities like brewing, milling and baking taking place in centralized factories, away from the production of the raw materials, away from the places where they would be sold. A traditional miller, who sold his flour to his neighbours and fed it to his own children, wouldn't tamper with it. Flour from a factory in some far-off city location, sold by some intermediary somewhere else, was much easier to adulterate. Food production became more complicated and more anonymous; there were more opportunities to interfere with it, and less likelihood of that interference being detected.

This had not been so until the nineteenth century. From 1266 until 1815 the quality and the price of staple foodstuffs and beverages like bread and beer had been controlled by the local courts (the assizes). Even as the Black Death ravaged the country-side, local inspectors oversaw the production of foodstuffs, fining offenders heavily. And the reason these careful measures were lifted was grubby: money. In the nineteenth century, free trade was the watchword. Regulations and restrictions alienated the free-wheeling adventurer from his right to profit. Competition, it was argued, would improve the quality of bread and beer far more than a meddling local magistrate.

The removal of such restrictions did indeed result in greater com-petition. By 1850 there were some 50,000 bakers in business, and three-quarters of them were selling their wares for less than the cost of producing it. They could only do this through a combination of slave labour and adulteration. Competition meant contamination. It also meant that, for the first time, food had become the raw material of an industrial, profit-led process.

This trend intensified in the latter half of the twentieth century. In Britain, where the German blockade of the Atlantic and the Channel led to food shortages throughout the Second World War, post-war leaders won elections by promising plentiful, cheap food to their voters. In countries like France and Germany, devastated by the same war, leaders had to placate a generation who had known genuine starvation. From now on, maximum food for minimum cost became the order of the day. Governments made conditions for farmers as favourable as possible with tax breaks and subsidies, once again fuelling the potential for agriculture to function as a business concern. British citizens have recently recovered from outbreaks of *E. coli* bacteria in supermarket meat, a resurgence of foot-and-mouth disease costing £3 billion, and daily reports of an antibiotic-resistant bacterium, MRSA, sweeping our hospital wards. These, too, are consequences of a process that began, innocently enough, with our ancestors selecting certain favourable plants for consumption almost 10,000 years ago.

SILENT SPRING

In the 1960s marine biologist Rachel Carson published a haunting book entitled *Silent Spring*. The title says it all, conjuring up an eerie world in which the lively, vibrant, colourful bustle of nature has all but died away. Carson was describing what she'd witnessed at first hand in rural America as a result of the use of pesticides. The hedgerows no longer teemed with insects and birds. The lakes were emptying of their fish. The only sound was that of the wind whistling through the intimidatingly vast stands of intensively farmed cereal crops. Carson's arguments were rubbished by an industry that now had a very serious influence in the upper echelons of finance and government. Ripostes were issued, depicting her as an obscure tree-hugger, sick with cancer and making hysterical claims about an industry upon which she was unqualified to comment. We had, her opponents pointed out, been using pesticides for thousands of years.

True, we know that as far back as the fourth century the Chinese protected their rice crops with silkworm droppings. In the early years of the nineteenth century, with chemistry emerging as a formal scientific discipline, salts of copper began to be used to treat types of wheat fungus. Unfortunately, our nascent understanding of biochemistry was not much use when it came to dealing with the Irish potato famine. In the benighted circumstances of mid-nineteenth-century rural Ireland, farmers were powerless when their crops turned black and slimy within a few days of harvesting. We now know this black substance on the potatoes was the culprit – an organism called *Phytophora infestans* – but in those days it was assumed merely to be a fungus feeding on potatoes which had already succumbed to some other, unknown blight. In 1845 the naturalist clergyman the Revd M. J. Berkeley proposed that the phytophora might be the cause rather than the effect, but he was ignored. Fungus, it was widely assumed, could feed only on dead organisms, not upon living tissue. This error led to the deaths of around a million Irish people, and to the resettlement of 2 million more in America, England and Australia.

It was research by Charles Darwin and his son Frank that set us on the path to the silent spring. In 1880 Darwin conducted an elaborate experiment which involved fitting plants with hoods, or painting their surfaces, in order to understand what made them turn towards the light. Lacking sufficient knowledge of chemistry, he was able to conclude only that some 'influence' was passed from the upper to the lower parts of the plant. But this observation, in the hands of a Belgian chemist called Frits W. Went, led to the isolation in 1926 of the plant hormone indoleacetic acid, or IAA. It subsequently became clear that IAA could disrupt growth in certain plants, but not others – potentially giving it a future as a weedkiller.

IAA, being isolated from substances within plants themselves, proved effective and benign. But when scientists began to synthesize other products, the end results were far more toxic. DDT, as well as being an excellent deterrent of the pests that feed on grain and wool, proved effective in preventing malaria and typhus among human populations. Between 1945 – when post-war populations in southern Europe were literally drenched with it to control body lice and fleas – and 1953, a further twenty-five synthesized products joined the ranks of the new pesticides. DDT, in particular, remained in the environment. It entered the food chain, concentrated increasingly in the insects that fed on the crops and then in the fish that fed on the insects, and finally proving toxic to the birds that fed on the fish. In the 1960s, as Rachael Carson poignantly observed, species like owls and sparrowhawks were being wiped out in the areas where crops were sprayed.

Nature is a complex of interdependent species, and by eradicating pests we had set in motion the eradication of swathes of other wildlife. Since intensive spraying of crops with synthesized products began in Britain, we have seen drastic declines in the number and variety not only of weeds and pests, but also of birds: the bullfinch and song thrush, skylark and linnet are all under threat. Equally at risk are the small mammals that feed upon the crops and the insects: the vole, shrew and fieldmouse are all disappearing from our countryside.

And what about us – the consumers of these crops, of the animals that are fed on them? Sadly, the history of crop-spraying is one of successive prohibitions and permissions. DDT and the family of chemicals to which it belonged, the organochlorines, were made illegal in the 1970s, to be replaced by the organophosphates, chemicals which break down more quickly and were thus considered safer. Sheep farmers were obliged by law to use these chemicals in sheep dip between 1976 and 1992, during which time a number of concerns arose about the risk they might pose to humans. There is some evidence, albeit inconclusive, that people exposed to organophosphates can suffer mental disorders, chronic muscular paralysis and a range of unpleasant symptoms sometimes colloquially dubbed 'dipper's flu'. Endless questions on the subject are asked by the Countess of Mar in her crusade for more information in the House of Lords, but successive governments have consistently maintained that there is no link. Nevertheless, the potential risks have been acknowledged to the extent that use of these chemicals on sheep is now optional – and meanwhile organophosphates have been replaced by pyrethroids. Once again, the new family of compounds is billed as far safer than the old, because they break down so quickly. But they have to be applied in conjunction with other chemicals – potentiators – to make them work. These are just a handful of the many different substances added to pesticides to make them more effective. While they are not active ingredients, they do have side-effects. About 20 of around 200 inert chemicals classified by the United States' Environmental Protection Agency are possibly carcinogenic; 14 of them have been assessed as 'extremely hazardous'. Every active ingredient comes packaged to farmers in a variety of different mixtures and formulations: one powerful pyrethroid is available in twenty-six different products. It is extremely difficult to police this multitude of differing uses and therefore unsurprising that the British Medical Association, in its *Guide to Pesticides, Chemicals and Health*, concludes that 'it is impossible for any member of the population to avoid daily exposure to very low levels of different pesticides in food and water.

Consequently, there is concern about possible adverse effects on human health.'

Although a number of compounds have been flagged up as potentially harmful, the review process undertaken by governments is ponderously slow. The Israeli government restricted use of the pesticide lindane in 1982 because of suspected links to breast cancer. In 1988 and 1990 the governments of Sweden and New Zealand followed suit. In Britain, however, the matter was placed 'under review' in 1992. The chemical remained licensed for use until July 2000, when the UK and other European countries finally imposed a continent-wide ban across the EU.

There may also be inherent deficiencies in the testing procedures which allow dangerous chemicals to escape a ban. Toxicologist Jane Axelrad of the US Food and Drug Administration examined the testing protocols used by industry scientists, and found them unchanged, in many respects, since the advent of mandatory testing in the 1980s.[8] All studies, she noted, relied on standardized groups of animals, and as such failed to address the issue of how individuals might respond. The same few species end up being used time and time again, because scientists already have huge banks of data on them. In many cases, the animals used are albinos, because skin pigment makes some indicators difficult to spot. During the studies they are housed together under controlled conditions of temperature and humidity, and fed and watered in exactly the same way. And if they are genetically very similar they may all react to these substances in a similar way.

Many of these elements are unavoidable features of the testing procedure. But together, as Axelrad says, they generate an unrealistic indicator of how a substance is likely to affect various segments of the human population. Some people are overweight. Some people smoke, or only eat vegetables. We now know, of course, that an individual's physical and genetic make-up can also have a profound effect on his or her response to toxins. Standardized tests on a small range of animals bear little relationship to the highly textured realities of the human world.

THE CHANGING COUNTRYSIDE

If you drive through the British countryside these days, you may well be impressed by two aspects of the scenery. First, many fields are given over to one type of crop – miles of bright yellow oilseed rape, for example, or maize. Second, dotted in between these vast plains of production you may see villages without post offices or schools – or, in many cases, pubs. A few delectably shabby farm-workers' cottages may catch your eye, but much of the living accommodation will now be smartened up, gated off and occupied either by city-dwellers as second homes or by commuters.

When farming became a business, the key criterion of achievement shifted from the quality of the end product to the efficiency of its production. The rule of the factory became the rule of the fields: maximum output, minimum cost. The greatest cost in traditional agriculture has always been the labour required to produce it. By cutting labour, it became possible to boost profits. And the subsequent loss of rural jobs has utterly changed the shape of our countryside. It has resulted in ghost villages and the spread of monoculture – the vast, bland stretches of single crops which now dominate our horizons. On this model a single worker can be responsible for thousands of hectares, whereas a mixed farm – comprising smaller quantities of livestock and arable crops – requires more manpower, and therefore more wages.

Politicians tend to keep telling us that all this is progress. Agriculture in Britain is increasingly more efficient than anywhere else in Europe. Our citizens are healthy and our well-fed children taller than their parents. If the countryside no longer has the bucolic charms of an H. E. Bates novel, we are better off this way. But the rise of food production as a business concern may actually be undermining our well-being. Although government subsidies have made farming a source of profit, agribusiness faces one basic problem: the limits on how much of its products we can eat. Most of us consume around 3,000 calories a day – although we can get away with considerably less. A Channel swimmer in training may consume around 7,000, and an Olympic rower perhaps 11,000, but most of us would rapidly become ill on such an intake. Hence the problem for the food business: there's a ceiling on how much food we can eat, which puts a limit on how much profit can be made from us.

This industry has approached this problem in several ways, none of which are particularly good for the health of the planet, including the people on it. First, they have increased the cost of the produce with a great deal of superfluous packaging. Sometimes the food has been subjected to quite unnecessary processes, such as the ready-peeled fruit offered by one supermarket chain. Sometimes it has been flown thousands of miles in refrigerated containers – most of this consists of fruits and vegetables either foreign to our climate zone or out of season. And the addictive components of certain foods, such as salt and fat in crisps, or sugar in biscuits, have been boosted to encourage us to consume more than we want. Finally – and this is perhaps the most cunning tactic of all – they have boosted the output of the cheapest foodstuffs, namely cereals, by feeding them to animals, and then selling us the most expensive bits of those animals' flesh.

Together, these techniques have had a serious impact on our environment and our physical health. In the long term, the effects could be devastating. Our government anticipates that in fifteen years' time over 2 million of its citizens will be suffering from diabetes. Given that diabetes affects virtually every system in the body, increasing the risk of stroke, high blood pressure, heart disease, dementia, blindness and kidney failure, and causes chronic ill health for a multitude of other reasons, perhaps we need to take what we eat rather more seriously.

THE DESERT ONCE BLOSSOMED LIKE A ROSE . . .

It is in pursuing the subject of crop cultivation that we see most clearly how environment and the growth of agriculture were linked. It is not enough to conclude that some types of plants were simply worth the effort of cultivation. That is true – but why, of that handful, did some become domesticated in some places, and not in others? One part of the world, the Sahel, is of particular interest in this context. This is the sub-Saharan belt stretching right across the African continent from Mauritania in the west to Ethiopia in the east. This whole vast area is currently afflicted by one of the world's most shocking ecological disasters, for changes in the climate have resulted in it being stricken with severe drought. Most of the land is now arid and starving; most of the inhabitants who have not yet left or died are destitute. Yet only a few centuries ago, this was the proud country of the horsemen who ruled the ancient empires of Ghana, Mali and Songhai. The most famous town in the region is, of course, Timbuktu, which was a major centre for trading gold and salt. It became very important for its Islamic learning and was the seat of what is generally regarded as one of the oldest universities in the world. Now, like Ozymandias, it is surrounded by the sands of the desert which threaten to envelop it. When Mansa Musa, the king of Mali, went on the hajj – his pilgrimage to Mecca in 1324 – he visited Cairo en route. People there were astonished at his wealth, which, even allowing for the wide-eyed exaggeration of onlookers, was undoubtedly impressive. He was accompanied by 100 camels each carrying 300lb of gold, and each of his 500 slaves carried a gold staff weighing 4lb. Apparently 500 attendants accompanied his senior wife. His lavish spending in Egypt is said to have caused a substantial fall in the value of gold in that country – an effect echoed in our own time following the printing of new money in the wake of the global banking collapse. The staple crop on which the people of this region had always relied was sorghum, which used to grow there readily until a few hundred years ago – and in sufficient quantity not only to nourish the population but to bring in substantial wealth via trade.

More puzzles are posed by the olives, figs and grapes we think of as so inherently Mediterranean. In their wild forms, these were found right across western Europe, but they were not domesticated until a whole assortment of non-native cultivated crops arrived from the Near East.

The answer to these puzzles almost certainly lies in the phenomenon of climate change. Westerners working in the tropics often complain that they long for 'a bit of weather' – referring to the sharp and visible changes of season in their native parts of the world. This annual cycle of climate shift favours plants with an annual life cycle. As they have to do all their growing, germinating and dying in the space of a year, they do not grow too large; more of their own resources go into the creation of seeds and seed pods than into trunks and branches, and in consequence they are more likely to be edible. Because the ancient Fertile Crescent abounded in these types of plants, the hunter-gatherer peoples of the area had already begun to settle down and rely upon the large stands of these crops before they started to domesticate them.

But another side of the question remains. Other areas around the world possessed quite similar climates, but did not domesticate crops or animals until much later. Domestication occurred, for example, some 5,000 years later in Mesoamerica and 6,000 years later in the eastern part of the United States. So why did farming seemingly begin in the Fertile Crescent?

When he pinpointed the origins of domestication in the world's mountain ranges, the Russian geneticist Nikolai Vavilov was somewhat off the mark. But he was right to maintain that it occurred in areas with the greatest diversity. This may be why the trend occurred first in the Near East: this region possessed the greatest range of altitudes and types of terrain, making it possible to harvest different crops or to pasture different animals at different times of the year. It also possessed the greatest diversity of plant and animal species and therefore an increased probability of some of these species being amenable to domestication.

This was of great importance when it came to the goat, sheep, pig, cow and horse. There are approximately 4,500 different mammalian species on the planet, but very few have been domesticated. Those that were, it seems, were all brought within the human ambit during the relatively short period of time between 8,000 and 2,500 years BCE. This, more than anything else, suggests that our ancestors domesticated all the likely candidates as swiftly as possible. The animals that remained did not justify further efforts.

Surprisingly, few mammalian species are really suitable for agricultural use. Domesticated animals need to be relatively small, so that they can be managed by humans and will not consume much more food than them. They must be capable of breeding in captivity, and ideally should reach reproductive age and full growth fairly early, so that the farmer receives a swift return on his investment, in terms of wool, milk, meat, skins or traction. Most suitable animals would have lived in their wild state in herds or packs with a 'pecking order': the hierarchical instinct can be exploited by a human owner who can put himself or herself in the place of the dominant animal. And it is helpful if they don't bite the children: so domesticated species should be reasonably docile and not dangerous.

In addition to the five animals already listed, apart from the dog only another nine were domesticated before the twentieth century – and only by peoples in particular unique areas of the planet. These include the Arabian and Bactrian camels, the donkey and the yak. In spite of all the advances made in reproductive science and animal husbandry, we still can't induce endangered species like the rhino and panda to breed in captivity, and efforts to domesticate species like the elk, moose and bison have proved commercially unrewarding. Domesticating animals may seem one way we became masters of nature. But, as we shall see, it left our ancestors more unstable than ever.

Chapter 4

Animal Farm

IN 1913 RUSSIAN archaeologists dug into a burial mound at Solokha, on the banks of the Dniepr river in Ukraine. Some 100 metres in diameter and 19 metres high, it contained two separate tombs, dated to around 400 BCE. The first was of a woman, accompanied by some ornaments, bronze and silver vessels, and the remains of two horses. In the second burial chamber, N. I. Velesovski discovered the body of a male ruler, covered from head to foot in gold. He had been interred with his weapon-bearer, a servant and five horses. The figure of this man was armed with a bronze helmet, a sheathed sword covered in gold sheets and a silver quiver containing eighty bronze arrowheads. The grave-goods also included a golden comb, featuring an exquisitely detailed gold casting of three fighting warriors wearing finely worked armour.

This remarkable burial mound has been found to have many counterparts across an area stretching from Romania to Siberia. Known as kurgans, they are most prevalent in the steppelands of Russia and Ukraine, and are associated with the Scythians, a nomadic group who originated in the Altai region of Siberia around 3000 BCE and began moving westwards around 900 BCE. The permafrost prevalent in the region has ensured the excellent preservation of their remains, an invaluable source of information.

The Kurgan people of the mountainous north-west Caucasus also practised metallurgy 5,000 years ago. They were expert in crafting copper axes and knives, and possessed gold and silver vases, beads and rings, also making figurines of bulls, lions and goats. Their gold, silver and copper was mostly alluvial, obtained from streams in the region. Oddly, no bronze objects have been found; perhaps they had either no knowledge of alloying or no access to tin. Jewellery, chariots, horse harnesses – even a miraculously preserved woven carpet – attest to the high levels of artistic skill they developed.

HORROR ON HORSEBACK

The contrast between the lavish 'royal' kurgans of the Scythian chiefs and the ordinary, simple graves of commoners suggests that this was a society that was rigidly ordered and stratified into distinct classes. Such a rich culture with a complex social structure might be expected to indicate a people who lived in villages, who had settled and farmed the land. But the Scythians were nomadic horsemen, making use of the animal first domesticated in the steppes around 5000 BCE. Ruled by small élites, they were fierce warriors – sometimes hiring out their services to settled kingdoms in return for agricultural produce and gold, sometimes simply taking what they wanted. Women seem to have fought alongside men – in some burial sites they are dressed the same and interred with weapons such as bows, arrowheads and axes, and indeed with their horses – and may have been the inspiration for the Greek legend of the Amazon. The classical world was certainly fascinated by the Scythians. Homer spoke of their habit of drinking mare's milk. Herodotus, the Greek writer of the fifth century BCE who is sometimes described as the world's first historian and, in his many travels, almost certainly visited Scythian country, describes their costume in detail: padded leather trousers tucked into boots, worn with an open tunic. Of their women he said, 'No Scythian woman may marry until she has killed a man of the enemy.' He also refers to their use of cannabis, as a

fibre for clothing and as a sacramental smoke – which is backed up
by archaeological finds. Perhaps it is not surprising that the modern
Cossacks claim descent from the Scythian people.

Although the Scythian philosopher Anarcharsis impressed the
Athenians with his wisdom six centuries before Christ, the general
reaction to the mere name of this nomad tribe was one of terror.
Herodotus claimed their diet consisted of horse-blood, and saw
them as the epitome of savagery. This fed into later accounts, so that
by the time Shakespeare was penning the words of *King Lear* it was
perfectly reasonable to speak of the 'barbarous Scythian . . . he that
makes his generation messes / to gorge his appetite'.

The Scythians were probably among the first humans to
domesticate horses and possibly the first to use them as mounts for
archers. The most elaborate Scythian kurgans often contain many
horses, seemingly of two breeds. The commonest are the smaller, so-
called Yabou horses, which seem to have been pack animals; several
of these are often interred together. But there is also a larger
thoroughbred, the Turanian, which was highly prized. These horses
were of great value and are invariably buried singly. Quite often they
were either strangled or killed with a single blow, or in at least one
case several blows, to the head at the very end of their working lives.

These horse-borne nomads must have terrified the settled farmers.
Even in modern parlance, words for brutality and fierceness are
synonymous with that lifestyle, passed down from the Scythians to
later peoples – the Huns and the Mongols (also called 'Tartars'). My
Jewish ancestors in Ukraine and Russia would have trembled as the
Cossacks galloped through their villages raping and pillaging, and
probably the ancient Scythians were even more cruel. An unpopular
boss or mother-in-law may still be referred to as 'Genghis', or a bit
of a 'Tartar'. During the two world wars of the twentieth century,
Britons described Germans as 'the Hun'. This kind of gallows
humour may have softened the terror our ancestors would have felt
as the earth began to rumble with the sound of distant galloping,
and on the tilled horizon they glimpsed a dark, seething mass.
Horsemen are taller than unmounted men, they move faster and can
jump obstacles, and the hooves of their horses are terrible and
intimidating. Settled people are both vulnerable to nomad attack
and, by virtue of the crops they produce, an attractive target. Their

vulnerability would have been even greater after 3300 BCE when these warrior tribes developed another technology, the wheel. Chariots drawn by horses are formidable weapons. Throughout central Europe in the latter half of the first millennium BCE we find evidence of farming villages being built in tight huddles, highly fortified against the ever-present threat of nomad invasion.

This conflict between nomads and farmers may partly explain the origins of European languages. Common grammatical elements and similar-sounding words across the various languages of Europe have led scholars to propose the existence of an original tongue called Proto-Indo-European (PIE). By comparing words across a variety of different languages, linguists have isolated some 2,000 root terms – including, perhaps unsurprisingly, universals like 'brother', 'sky' and 'mother'. Although scholarly opinion remains hotly divided about where these PIE speakers came from, certain of these words offer some useful ideas. For instance, similarities in the word for 'snow' across a number of different languages suggest that the Proto-Indo-Europeans came from somewhere rather chilly. If we look for words for items which were available to people in 2000 BCE but not 4000 BCE, we gain some idea of when these PIE speakers were on the move. Similarities in the words for 'wheel', for example, suggest that 3300 BCE was within their timeframe. We don't find the same for 'battle chariots', though, suggesting that by 2000 BCE, when these deadly machines came on the scene, they were gone.

Scythians – and peoples like them – may have been the carriers of this Proto-Indo-European tongue. Omissions are as important as inclusions, and it is interesting that PIE is rich in those terms that would have been on Scythian horsemen's tongues all day long – wheel, axle, shaft, harness, hub – but not in agricultural terminology. We know their culture included wheels and horses, words for which we have been able to reconstruct; but though there are words for 'plough' and 'sickle', there is only one word for an unspecified type of grain. They practised little agriculture, and this is reflected in the vocabulary. Yet, if the speakers of this language were on the move between 4000 and 2000 BCE, this was exactly the time when Europe was adopting agriculture. What are we to make of this?

Kurgan burials of the Scythians, or any other relic from their culture, are not found west of Hungary. Instead, after 3000 BCE, central and western Europe became dominated by a kaleidoscope of cultures named after their artefacts – like the Beaker Folk and the Corded Ware People. These communities mixed certain elements of the steppe culture with that of the settled one. They practised agriculture, but they rode horses, worked metal and waged war on each other. How did the nomads' language survive, if all traces of their way of life vanished?

It could be that, having reached the limits of their natural habitat, these nomads had to adapt. Hordes of horsemen could sweep across the open steppes, but would have encountered a problem in the densely forested regions of central Europe. It is even possible that some of them dismounted and mingled with the groups they encountered there, passing on certain elements of their own culture, taking on elements of others'. But their language spread onwards, through Western Europe, even to the shores of Ireland.

If the domesticated horse gave us a common language and culture, it eventually also gave us something far more deadly. When our primate ancestors descended from the trees, they were vulnerable to attack from predators. But we were never more vulnerable than when we first began to tame nature.

SPREADING THE BAD NEWS

To watch a Big Man of Malakula tending his pigs is to behold a scene of surprising tenderness. Britain's class system may be still alive and kicking, but that of Malakula – an island in the Pacific archipelago-state of Vanuatu – pre-dates it by thousands of years, and is far more complex.[1] For men, life is a lengthy ladder of social grades known as *nimangki*. To progress up it, and to gain entrance to the secret societies associated with each grade, he must obtain, slaughter and donate pigs to men in the appropriate group, a custom found in many islands in the region. The sacrifice increases in value depending upon the type of pig being offered. In some societies, hermaphrodite pigs are the ultimate gift. But on Malakula, the men of the very highest grade are entitled to sport circular pig's tusks on their arms – the most precious items spiralling two or three times around the biceps, and worth a fortune in white man's money.

The effort a man goes to in order to obtain these tusks seems almost incredible. His wife may suckle the piglet at her own breast. Later, he performs elaborate dentistry, removing the upper canines to ensure that the tusks grow perfectly. The pig must never be at risk of getting into a fight with others of its kind, so it is castrated. Nor must it damage its precious tusks while foraging for food, so it will be hand-fed on mash and vegetables and kept tethered close to its owner's quarters. For the ten to twelve years necessary to grow a double-spiral tusk, the pig will be treated like a member of the family.

When Western missionaries arrived in these islands at the end of the nineteenth century, and saw this degree of closeness between man and pig, they were revolted. To them, pigs were synonymous with dirt and ignorance. To wean the people of the Pacific off their traditional ways, missionaries ordered them to rebuild their villages along the neat lines of an army camp, with the livestock kept at a distance from the houses. Then the islanders kept succumbing to epidemics. Dysentery, measles and influenza devastated island populations, sometimes killing as many as a third. The missionaries worked tirelessly to cure their flocks, but this only increased the suspicions of the locals. Believing themselves to be victims of some deadly form of magic, the islanders turned violent, killing and expelling the invaders. And they were right to assume these foreigners had brought the sickness. But it wasn't through witch-craft; it was through contact with the outside world and through farming.

Bacteria – including those that spread diseases – may be the lowest form of life, but they are better than us at reproducing to propagate their species. We accomplish this task in one enjoyable way (or two ways, if you count IVF), but bacteria and other small, virulent organisms do so via a multitude of methods and very rapidly. Mostly their reproduction is asexual – that is to say, each bacterium divides into two when it is mature. This is simple and efficient, and works well if the life cycle is short. Asexual reproduction does not encourage much genetic diversity, of course, because there is no mixing of genetic material from two parents. But if the reproduction cycle is short and quick, genetic change or diversity will eventually occur by occasional mutation. *Escherichia coli*, which lives in our intestine and is arguably the most studied of all living creatures (except humans), produces a new generation every twenty minutes. Most of the time *E. coli* is totally harmless, but some strains can produce severe food poisoning or even fatal infections. *Salmonella*, one variant of which causes typhoid, more commonly produces a form of food poisoning. But it can lie around inside the intestinal tract of some infected individuals causing no problems at all, until that person infects somebody else – perhaps through failing to wash their hands after defecation. Cholera, caused by the bacterium *Vibrio*, induces bouts of severe watery diarrhoea, with an afflicted person expelling copious amounts of dangerously infected fluid into his or her surroundings. Other infective agents may be carried by a third party, or vector. *Yersinia pestis*, the bacterium responsible for deadly bubonic plague, is carried by fleas which reach their human hosts in the fur of rodents, particularly rats. Other micro-organisms, such as the malaria parasite and the trypanosome which cause sleeping-sickness, cadge a lift inside the salivary glands and gut of insects. The viruses which cause flu and colds, on the other hand, force bouts of sneezing and coughing, which then propel them outwards on a fine spray towards fresh victims.

Humans have developed important responses to bacterial and viral infection. One of the most interesting is fever. The ability to produce the fever associated with infections may be the result of evolution. Perhaps it is an attempt to exterminate the culprits through heat, and expel them through sweat. Matthew Kluger of the University of Michigan pointed out that fever is a useful response not only in mammals, but in birds, reptiles and even fish.[2] Readers who have kept tropical fish in a tank will no doubt have noticed that when they are ill, they tend to congregate near the fish-tank light or other source of heat. If this is a protective mechanism produced over millennia of evolution, the modern approach of keeping patients comfortable by cooling their surroundings may be counter-productive. Another important response involves our immune system. The white blood cells form a protective army against unwanted microbes and can actively ingest them and then destroy them. And if microbes escape that process, they can be neutralized by the antibodies we produce. But microbes are 'clever', and because they reproduce at a very rapid rate and can mutate, they may change their structure to evade interference from our antibodies. And if we evolve more antibodies, more resistance, to catch up with them? Well, then some micro-organisms will spread to hosts who have not got the antibodies to resist them. It is a war – and a war that we humans are always likely to lose.

This is the way of life that has been in place since we adopted agriculture. The shift to farming resulted in human populations enlarging between ten and a hundred times. This itself increases the potential for the spread of diseases, but two other factors really matter. With agriculture comes a more settled life. Hunter-gathers leave or bury their excrement when they move camp. But agriculturalists remain close to their own bodily waste, sometimes even using it to fertilize the crops they intend to eat. This is not particularly intended as a comment on organic farming, but perhaps this fad is not quite so healthy as it has been presented even when waste products from humans are not involved.

The problem of waste is increasingly serious, and it is one that magnifies with the growth of human populations. When we come to consider the next big technological advance – the city – it is easy to see that the cramped conditions of settlements like Jericho would have been ideal breeding grounds for bacteria and for the animals which carry potentially deadly human diseases. The development of city-states was followed by improved techniques for navigation and transport, facilitating trade across long distances – and giving further scope for the spread of diseases. Rome succumbed to small-pox in the first century CE, probably as a result of trade and imperial connections with Africa. In medieval Europe, the Black Death came along the Silk Road, probably in flea-infested furs from Central Asia.

But agriculture also carries blame for this unhappy trend in human evolution. Many of the microbes that have done us the most harm come from the animals we have domesticated. Measles is a close relative of the virus that causes rinderpest in cattle. An ancestor of tuberculosis is brucellosis, a micro-organism that can be spread through infected cow's milk. A variant of whooping cough affects dogs and pigs. In each case, what has happened is that bacteria and viruses on the brink of dying out – either because they have killed all the available hosts, or because they have been baulked by antibodies – have evolved mechanisms for shifting between species. For example, *Rickettsia*, the typhus bacterium, initially spread between rats on their fleas. Occasionally, rat fleas could infect humans with typhus, too. But this was an inefficient means of spreading. Faced with the imperative to evolve or die out, typhus made the leap to human body lice, ensuring it a long and flourishing career in cramped, crowded, dirty conditions. This is why typhus was so lethal in the Nazi concentration camps. It has also been prevalent in crowded cities where rats and humans coexist closely. Although there is some doubt about this, typhus was most likely to have been responsible for the plague of Athens in 430 BCE, during the Peloponnesian War, which is thought to have killed one-third of the population.

A more recent threat to humans has been posed by so-called 'bird flu'. So far, human deaths from the H5N1 strain of avian flu have occurred only in people connected closely with poultry farming. The H5N1 strain appears to have started ravaging poultry stocks in Asia in around 2003, and to have spread to Europe with migratory birds. The World Health Organization states that this strain of the bird flu virus had infected 385 people by 2008, killing some 243. Although it originated in domesticated birds and has so far infected only people in direct contact with them, there is a real fear that the virus might mutate. If it does, and humans infected with the disease become infectious to other humans, a massive epidemic could follow. Liam Donaldson, the Chief Medical Officer for the UK, initially said that 50,000 people could die in the UK as a result, but when cross-questioned admitted that the death toll might be as high as 750,000. The truth is that nobody knows how serious such a pandemic might be in Britain and, at the time of writing, our government is still not really well prepared for it, despite the fact that it would be likely to cause massive disruption to industry, education, and social and medical services across the country. Similar concerns have not gone unrecognized in parts of Asia, where infection of birds with H5N1 has forced the slaughter of 300 million chickens. This in itself is an economic catastrophe in Asian communities where local farmers depend for their livelihood on producing these birds and where chickens are one of the cheapest ways of getting animal protein – in parts of China, indeed, they are effectively a staple diet.

According to the WHO there are hundreds of strains of avian influenza viruses, but until very recently only four – H5N1, H7N3, H7N7 and H9N2 – were known to have infected humans. H5N1 was a preoccupation of the world's business leaders at the Davos economic summit in 2005. A 22-page risk study presented at that conference suggested that this infection was a bigger threat than international terrorism, warning that 'If the avian flu H5N1 virus mutates to enable human-to-human transmission, it may disrupt our global society and economy in an unprecedented way.' Some experts estimated that the number of deaths could be worse than that from the 1918–19 Spanish flu pandemic, which is estimated to have killed between 40 million and 50 million people worldwide.

Vaccines are being developed which may mitigate these effects. But viruses are 'clever' and the risk of mutation is genuine. More recently, the new strain called H9N2 was also shown to pose a serious threat. Tests on this strain, seen mostly in birds, now show it is capable of infecting and spreading among humans, and it has infected several children in Hong Kong. Although it seems to cause only a relatively mild illness in humans, it can also cause illnesses in pigs and ferrets. The biggest concern is that a human patient might become infected with two strains simultaneously. If a human or animal is infected with two strains of flu at the same time, there is the possibility that the DNA of the virus might undergo 'reassortment', which could increase its infectivity across species. Until now, we have been fortunate in one major respect. Unlike the virus causing the common cold, bird flu virus has not seemed to be transmitted in droplets in the air. So sneezing and snuffling are not currently thought to pose a risk. But nobody knows what further mutations might produce; if droplet spread did occur, the effect could be truly devastating.

The modern farmer has a handy solution to the spread of disease among his flocks and herds. While feeding them low doses of anti-biotics does not kill viruses, it does control bacterial infection, and also boosts growth. This was discovered by accident on a farm in the United States in the 1950s. Chickens were being fed the waste product left over from the manufacture of the antibiotic aureomycin, and these birds grew up to 50 per cent larger as a result. This may have been because of the absence of infections, or perhaps because the drugs enabled them to absorb more nutrients from their feed; the precise mechanism is unclear, but farmers recognized that by whatever means this additive resulted in bigger chickens – and therefore more profit. Governments – trying to please undernourished populations after the Second World War – also understood the benefits. With little resistance, the Therapeutic Substances Bill was passed by the UK parliament in 1953, permitting the addition of small quantities of penicillin and tetracycline to animal feeds.

At that time, only one voice was raised in protest. It was that of Tayside North's Conservative MP, Colonel Sir Alan Gomme Duncan. 'May I ask whether we have all gone mad to give penicillin to pigs in order to fatten them?' he demanded of the House. 'Why not give them good food, as God meant them to have?' His protest went ignored.

We have considered, in the foregoing pages, some of the problems that the farmer's plough left in its wake. And right now, in the age of the supermarket, the genetically modified crop and the year-round strawberry, we might be facing the greatest one yet.

THE MEAT MYTH, THE PLIGHT OF PIGS AND A NEW FLU

Most of us in the developed world eat far more meat than we need, and it's interesting to reflect on how this came about. In 1932 Dr Cicely Williams, first woman medical officer in Africa's Gold Coast (now Ghana), reported an unpleasant condition in the local children. It was known locally as kwashiorkor: the victims had dry hair and skin, reddened hair and swollen bellies. Blood tests revealed malnutrition, despite the fact that their parents farmed rich crops of cassava root. Dr Williams proposed protein deficiency as a cause of the disease, and although this was only a tentative hypothesis the idea rapidly became accepted as gospel in a scientific community scattered around the globe and lacking the more efficient communication systems we have today.

So entrenched had the idea become that by 1953, when the recently established WHO was formulating its policies, kwashiorkor was considered to be the most serious nutritional disorder in the world and its single cause to be a lack of protein. This thinking fed into the food policies of the developed nations, too, so that it became the accepted wisdom that a diet rich in meat and milk was the route out of malnutrition. When I was a teenager at school, pictures of African children with their swollen bellies were used as an example of the need for protein in a proper diet. Not until the 1970s did scientists begin to suggest that malnutrition resulted from a lack of all kinds of foodstuffs, and that the protein needs of humans were relatively modest – modest enough to be obtained from cereals and beans. By then the damage had already been done, in terms of the switch agriculture had made in order to provide us with, and profit from, the foodstuffs we had been told we needed. Cereal crops – a perfectly adequate source of protein in their natural state – were boosted to provide the enormous quantities required by livestock, while meat and dairy animals became subject to intensive production.

Today there are grim reports about the battery farming of pigs. Many accounts relate how these animals on factory farms are forced to live in their own faeces, among their vomit, and even surrounded by the corpses of other pigs.[3] In these overcrowded, poorly ventilated and filthy conditions disease is rampant. Respiratory problems are common because of high levels of humidity and toxic gases from the manure pits – in fact, 70 per cent of pigs on factory farms have pneumonia by the time they are sent to the slaughterhouse. Conditions are so filthy that at any given time more than one-quarter of pigs suffer from mange. In spite of the protestations of people like Alan Gomme Duncan, intensively farmed pigs are fed massive doses of antibiotics to keep them alive in these conditions because of the importance of the market. Today, humanity is more carnivorous than ever before, thanks to intensive breeding of animals, low world grain prices, global distribution networks and, until very recently, the Chinese economic boom. In 1965 the Chinese ate just 4 kilograms of meat per head per annum:

today the average Chinese citizen consumes 54 kilograms a year.

The Center for Disease Control and Prevention in Atlanta, Georgia, the most respected public health organization in the world, has repeatedly emphasized that these appalling conditions are extremely risky for human health. Overcrowded, intensive pig farming has led to MRSA infections in farm workers and their families, outbreaks of deadly Japanese encephalitis, fatal attacks of the Nipah virus and the frightening SARS virus, among other serious human illnesses.[4] Stephen Morse, Director of the Center for Public Health Preparedness at Columbia University, writing fourteen years ago, emphasized in a clear and graphic scientific article the viral risks to human populations.[5] Since then burgeoning pig farms have continued to expand and seem to be perfect incubators for infectious disease of all kinds. In damp, warm pens where weakened animals are tightly confined in barely enough space to move at all, with ordure spread all around, new strains of rapidly changing bacteria and viruses seem all too likely to emerge, and it is inevitable that eventually some of these strains will be infectious for humans as well as pigs. Although it is not certain that so-called 'swine flu' definitely originated in pigs, there is a wide body of responsible scientific opinion which considers that farmed pigs are the most likely source. So when dense populations of factory-farmed animals are kept in unhygienic conditions near human villages, there is potential for a catastrophe – as has recently been seen in La Gloria in Mexico, the possible epicentre of the recent outbreak of swine flu.

But if you turn to the website of the largest pork producer in the world, Smithfield Foods, an American company with major interests in pig farms in Mexico, you will be greeted by video clips of a buxom platinum blonde, who calls herself (I think – the Southern drawl is so pronounced her English is difficult to interpret) Paula Deen, assuring 'y'all' that 'we still make bacon in the same old-fashioned way'.[6] She goes on: 'Genuine Smithfield Hams have crowned America's table for generations. They are as distinctive as they are delicious, as elegant as they are hearty. They are the flavor of Southern Hospitality' – and, an astonishing revelation from this elegant lady: 'Queen Victoria had such a love for Smithfield ham, that she had half-a-dozen delivered to her court, every week.' This is cue for an exhortation that 'You don't have be a royal to eat like one.'

Smithfield Foods seems important because, whether or not they 'still make bacon in the same old-fashioned way', and whether or not our royal House of Hanover really did patronize this establishment in its former colony, as I write there has been the most frightening outbreak of porcine influenza in Mexico. Almost two-thirds of the country's population have recently had upper respiratory tract infections with flu-like symptoms. The La Gloria plant, part-owned by Smithfield Foods, is in the state of Veracruz, where it is reported that 950,000 pigs are intensively reared each year. So far over thirty people are known to have died across Mexico, and the disease is reaching pandemic proportions, with reports of human infections (and numerous deaths) in the United States, New Zealand, Spain, Britain, Israel, Canada, Germany, France, South Korea, Costa Rica, Ireland, Italy, Switzerland, Austria, Hong Kong, Denmark and the Netherlands. At this stage it is difficult to be sure what started the infections, or how widespread this particular mutation of the flu virus really is. This new strain appears to be H1N1, and it is alarming that this virus does seem capable of being transmitted from human to human. Moreover, according to reports from Italy it may be somewhat vaccine- and Tamiflu-resistant. Given that the UK government has stockpiled large amounts of Tamiflu, a drug which knocks out the virus, one wonders if the failure of the UK press to mention this feature of the pathogen is merely coincidence. Moreover, treatment with Tamiflu and a similar drug, Relenza, has made many flu victims feel much more ill and it is not clear how effective it is – particularly in young children, a number of whom have died as a result of flu. Carl Heneghan and Matthew Thompson of the University of Oxford have just published data showing that there is little convincing evidence of the value of these drugs in children.[7] These authors argue that the government should have demanded much more data about the efficacy and side-effects of these drugs before stockpiling them. The response from the government's Chief Medical Officer, that 'this [the Oxford] study is limited in scope', is interesting.[8] If his judgement is correct, one wonders why, three or more years after a pandemic was predicted, the government has not done more to promote 'better' studies that are less 'limited'.

Whatever the connection with the strain recently reported from

Italy, the evidence points to a need for concern about the pig farms in Mexico. People in La Gloria were repeatedly complaining about a massive number of upper respiratory tract infections for over three months before this news about pig flu hit the world's press. In view of repeated complaints about filth, liquid manure, flies and smells around the pig farm there, the response of Smithfield's chief executive, Larry Pope, seems somewhat questionable. In a recent interview on CNBC he said: 'But I tell people when you visit our farms, I'm not concerned about you. I'm concerned about the pigs. I'm concerned about you contaminating the pigs. Not the pigs contaminating you.' He then went on to reassure the audience: 'And again, it [the virus] doesn't transmit through the meat.'[9] The Luter family of Smithfield, Virginia, has been curing and selling hams since the turn of the century. Joseph W. Luter Senior's first job was at a local meat-packing plant. His young son, Joseph W. Luter Junior, followed in his father's footsteps, learning every aspect of the meat-packing industry. What would the upright Southern family which founded this business make of the current situation?

A SUPPLEMENT TOO FAR

One day in 2005 war veteran Tom Mason went into a north London hospital for a routine hip operation. The next day he was gravely ill, a ferocious infection ravaging his body. Over the next few weeks he yo-yoed back and forth between intensive care and the general ward, succumbing repeatedly to the infection, until finally his exhausted system gave up.

Tom Mason had a daughter who is a journalist, Angie. She was puzzled to see that his death certificate mentioned only pneumonia as the main cause of his demise, so she paid for a private post-mortem. The findings indicated that methicillin-resistant *Staphylococcus aureus* (MRSA) had been behind her father's rapid decline. Initially the hospital agreed to amend the certificate, but then reversed its decision, a change of heart which may have been related to their realization that Angie Mason was making a television documentary about the case. Confronted on camera, the previously helpful consultant had the rabbit-in-headlights look of some junior minister caught by John Humphrys. Why hadn't the bugs that really killed Tom Mason gone on to the death certificate? 'I can't fit them all on,' he stammered, an excuse so flimsy it could almost have been amusing, were the point at issue not involved in a man's unnecessary death.

When I was a doctor in training it was relatively routine practice for some hospitals to list 'pneumonia' as a general term for the cause of death; the various hospital-dwelling bugs that might have led to the infection were hardly ever recorded on a death or cremation certificate. Even now, hospitals forced to compete in much-publicized league tables have a strong incentive not to mention the possible presence of MRSA in their wards. Hospital practices in the NHS – staffing and hygiene procedures as well as those un-helpful league tables – may have had a prominent part to play in the spread and persistence of this frightening infection, often dubbed 'the hospital super-bug'. Simply blaming dirty wards in the past – as the Department of Health commonly does – is not the answer. A key problem is that moribund patients, or patients with indwelling catheters or drip lines, are at most risk because of the tubes inserted into veins and other parts of their bodies, which function as con-venient points for bacteria to enter and gather. It is often forgotten that the hospitals treating the greatest numbers of most gravely ill patients inevitably will be those where there is most risk of MRSA – a fundamental reason why, in my view, league tables are an almost total nonsense. But the chief cause of the MRSA problem, I would argue, lies squarely in a food industry which has for decades adulterated cattle, sheep and poultry with low doses of antibiotics.

Antibiotics have brought enormous benefits – preserving the lives of children who would have died in former times, and controlling many serious infections such as tuberculosis. But there are problems associated with them – or rather, with the bacteria they are designed to combat. In the first place, some kinds of bacteria, and most viruses, cannot easily be defeated with these drugs. Second, bacter-ial mutation gives rise to resistant varieties – like MRSA – which flourish in many antibiotic-rich environments. We now know, too, that bacteria can exchange genetic information, within and across different species. So they can not only acquire immunity to antibiotics, they can pass it on to subsequent generations.

This means that constantly exposing bacteria to antibiotics is a bad strategy, as it gives them the maximum opportunity to develop resistance and to pass on that trait. This is what happens with intensive farming, when animal feed is adulterated with small doses of antibiotics. And the problem is compounded by housing animals in cramped conditions, lowering their natural resistance to infections and increasing the capacity for those infections to spread from one animal to another. It would be hard to think of better conditions for the creation of new, antibiotic-resistant superbugs.

It is now illegal throughout the EU to supply antibiotics used to fight human diseases to animals for non-therapeutic purposes. But that legislation – resisted at every stage by the British government – still permits the inclusion in animal feed of antibiotics not used in human medicine. Also, the decline in rural livelihoods means increased potential for flouting rules. As agriculture has become dominated by big business, vets and farmers equally have felt the pinch – increasing the chances of people being prepared to supply and to use banned substances. With the internet as a new source of pharmaceuticals, some food producers no longer have to rely on prescriptions from their local vet. And the global economy, of course, is just one facet of a world in which people and products can move about with greater ease. Legislating against antibiotics in animal feed is of limited value when aircraft land in our country every day from regions where similar laws are not in place.

GONE FISHING

It is often said that your goldfish will never get bored swimming around its bowl because its attention span is so short and its memory lasts only eight seconds. This urban myth was ignored by scientists in Woods Hole, Massachusetts, where 5,000 sea bass were raised in a geodesic aquadome. Here, every time the fish were served dinner, a low tone was sounded electronically. The expectation was that when the bass were eventually released into the ocean they would remember the tone and, when it was sounded, return to base for a protein-rich meal. The idea was that farmed fish, having been trained in this Pavlovian way, could be set free to feed naturally and then persuaded to return when fully grown to be caught for the market. But this research was halted recently by a lawsuit initiated by an environmental organization, Food and Water Watch, which is protesting about the release of farmed fish into the ocean.

The lawsuit may seem unlikely to succeed, but it illustrates some important concerns. Uncontrolled hunting was clearly responsible for the depletion of many mammalian species in many parts of the world in Neolithic times. Now it contributes to a global problem with one of our best sources of protein, fish. Overfishing is an important reason why so many fish that were once so plentiful, such as Atlantic cod, are now comparatively rare. Overfishing threatens a catastrophe. In 2000, Professor Jeffrey Hutchings from Dalhousie University, Nova Scotia, warned that in parts of the world up to 90 per cent of the ocean's supply of cod, halibut, flounder and sword-fish had been destroyed by fishing,[10] and that fish cannot reproduce fast enough to keep up with the devastation caused by humans.

Aquaculture is seen to be one solution. Fish farming is hardly a new idea:

And the rivers shall become foul, and the canals of Mazor shall be diminished and dried up. The meadows by the river by the mouth of the River and everything sown by the river shall wither, be driven away and be no more. And the fishers also shall mourn and all that cast angle into the River shall lament and all they that spread their nets shall be wretched . . . and all that make sluices and ponds for the fish.

So says Isaiah, lamenting the fate of fish farming in Mizraim (Egypt) some 3,500 years ago.[11] Humans have farmed fish, a rich source of protein, in ponds and dammed rivers since prehistory. The Chinese farmed carp in lakes from very early times. But it is only now, since we have greatly overfished most of our oceans, that the world's fish are seriously at risk of dying out.

So a boom in fish farming has been almost inevitable. Since the late 1960s commercial fish farming has expanded rapidly, to the point where it is now responsible for nearly one-half of all fish and shellfish eaten anywhere in the world. Humans consume roughly 52 million tonnes of fish annually, and fish farming is valuable business, worth some £55 billion a year. About 8 per cent of the world's population depend on fisheries for their protein or their livelihood. The United Nations Food and Agriculture Organization (FAO) argues that this is not enough; that we need to double the output of aquaculture to prevent the burgeoning world population depleting more of our dwindling supply of wild fish. But fish farming raises major environmental concerns.

For one thing, farmed fish need to be fed. Whether they are grown in freshwater ponds or ranched in enclosures in the oceans, fish are voracious feeders. There is a haunting drawing by Pieter Brueghel the Elder in the Albertina Gallery in Vienna. Dated 1556, *Die großen Fische fressen die kleinen* ('Big fish eat little fish') is a treatment of a famous proverb. In the centre of the picture is an enormous beached fish. Climbing up a ladder to scale the back of this sea monster is a small helmeted figure with a trident with which he is about to stab his prey. Another figure with a massive knife is slicing open the big fish's belly, revealing a profusion of swallowed marine creatures. The great fish has vomited still more fish on to the beach, and each of these in turn has attempted to devour other fish. In the foreground another fisherman points towards this bizarre scene for the benefit of his son, to warn him (as a Flemish inscription emphasizes) that we live in a senseless world where the powerful instinctively and continuously prey on the weak.

Aquaculture exploits the cannibalistic*nature of many fish. Vast quantities of anchovies, whiting, sand eels and sardines are hoovered up to provide the diet of farmed fish. It can take 5 kilograms of small fry to produce one kilogram of good salmon for the table; some predatory farmed fish, such as tuna, require even more food. Removing small fry from the ocean to maintain this industry is certainly not preventing the continued depletion of the world's stocks of fish. Krill are also at risk – around 200,000 tonnes of these little crustaceans are taken annually to feed stocked fish, which means 200,000 tonnes less of the staple food of many whales, some sea mammals and many sea birds. So it is not surprising that fish farming is threatening the lives not only of wild fish, but of birds and coastal mammals as well.

One solution is to farm more vegetarian fish. But species like

*If fish eating fish is cannibalistic, then so are we for eating other mammals.

tilapia, carp and catfish are not particularly popular, at least in part because they are not generally as tasty as salmon, trout or tuna. The aquaculture industry is managing to replace fish meal with vegetable-derived proteins in feeds for some carnivorous fishes, but vegetable-derived oils have not successfully been incorporated into the diets of carnivores.

Farmed fish, vegetarian or not, are often reared in extremely crowded conditions. Commonly, there may be over 62,000 fish in a one-hectare area. This means that each fish, which may measure nearly a metre in length, has less water than the amount held in an average bath. Trout are stocked at even higher densities. Generally reared in freshwater raceways or earth ponds, they are sometimes stocked at a density of 60 kilograms of fish per cubic metre of water, which is equivalent to twenty-seven trout, each one foot long, in one bathtub. In these conditions it is inevitable that injury and disease are common. Packed so tightly, the fish rub against each other and against their nets and cages, resulting in trauma to their fins, scales and tails. This undoubtedly causes stress but it also literally lays the fish open to parasites and infections. Antibiotics have increasingly been used to control these infections, but even so, up to 30 per cent of farmed fish may die prematurely. There is also some evidence that the drugs used to counter infection have entered the environment, as well as concern that, as through the farming of land mammals for meat, the residual presence of these drugs in human food products has contributed to human antibiotic resistance.

In spite of overcrowding, freshwater fish farms also use vast quantities of water. In the UK, fish farming, cress growing and amenity ponds together use the same volume of water as all industrial and agricultural uses combined, except the public water and electricity supply. Although most of this water is cleaned and returned to the rivers from which it was extracted, in some parts of the world – for example, India, parts of China, and Africa – the exclusive use of scarce water for aquaculture is a serious issue. The situation in Thailand, where good quality clean water is in short supply, is typical. Approximately half of Thailand's shrimp ponds were formerly rice paddies. Now, aquifers and groundwater levels there have decreased significantly, exacerbating what was already a serious water shortage. Shrimp ponds in Thailand are a serious environmental hazard. Nearly 20 per cent of Thailand's mangrove forests have been destroyed to increase the output of the shrimp farming industry, leaving coastal areas more prone to erosion and flooding, so that in many places there is increasing contamination of the soil with salt. Moreover, the natural habitat for both land and water species is at serious risk through the polluted water that is flushed into coastal and river waters near the shrimp farms. This water contains heavy concentrations of fish faeces, some uneaten food and other organic debris, leading to oxygen depletion and excessive growth of damaging algae. Some 1.3 billion cubic metres of effluent is discharged from Thailand's shrimp ponds into coastal waters each year.

There is a further environmental risk with saltwater farming and farming in the open sea.[12] When fish are encircled by nets or cages, or reared closely together, infestation with parasites such as lice can cause serious disease. Sea lice are little crustaceans which eat into the skin of the fish. Damage to the skin means that the fish cannot protect themselves against the absorption of salt from seawater and are vulnerable to a variety of infections. The female sea louse produces a thousand larvae or more during her life, most of which, under normal conditions, drift in the ocean where they die. Only a few will ever drift close enough to a fish to attach to it. But in the confines of a fish farm, larvae have an excellent chance of finding a fishy host. Consequently, more larvae survive to become lice, and those lice put more larvae into the water. This has been a particular issue for the health of Atlantic salmon, as the wild fish near farms have also become infected. In the Broughton Archipelago in British Columbia, juvenile wild salmon on their journey out to sea have to 'run a gauntlet' of large fish farms located offshore near river outlets. Some experts maintain that these fish farms are responsible for such serious lice infestations that there is likely to be a huge loss of wild salmon. Claims have been made that well over 90 per cent of the wild population is at risk, though this high figure is disputed.

Another problem emerges when farmed fish escape from their confines, posing further risks to the environment and health of wild fish. Philip Lymbery, an animal welfare consultant, claims that 411,000 salmon escaped from fish farms during 2000.[13] This 'leakage', coupled with a decline in wild salmon, has become so serious that the numbers of escaped farmed salmon are said to out-number catches of wild salmon by seven to one in some parts of the world.

A further concern is the question of animal suffering. The distress caused to battery-farmed birds is well recognized, but there should probably be much more concern about fish. Under certain conditions, farmed fish seem to be suffering. Philip Lymbery points out that, as about 70 million fish are produced and slaughtered on British farms each year, farmed fish now represent the UK's second largest livestock sector after broiler chickens. More farmed fish are slaughtered annually than all the pigs, sheep, cattle and turkeys put together. Yet, he points out, farmed fish have tended to receive second-class attention from legislators. He claims that some legally permissible methods of slaughter cause intolerable suffering from a stressful and prolonged death. Of particular concern to him are methods of allowing fish to suffocate, in air or on ice. It is claimed that these may cause immense pain to fish and would not be tolerated in respect of terrestrial farm animals. Animal welfare agencies also raise concerns about the stress of slaughter, which starts when fish are harvested. Salmon, for example, cannot adapt quickly to being hauled to the surface from cages which may be 20 metres deep, where the pressure is three times that at the surface. Unable to release excessive air from their swim bladders quickly enough, they may suffer severe distress. While assessing the painful experience of any animal is extraordinarily difficult, there should be more concern about how fish are treated.

Finally, the marine environment is under grave threat from climate change, a theme which is always in the background in any discussion about the consequences of human technology. There is very good evidence that global warming and its associated environmental changes are having a major effect on the marine environment and that the natural food supplies of fish are being seriously affected by the increasing amount of carbon dioxide in the atmosphere which is acidifying the earth's oceans.

About one-quarter of all CO_2 released into the atmosphere by human activities is absorbed in the sea. The best estimates show that over the period 1750–2009 the atmospheric concentration of CO_2 increased from 280 to 387 parts per million, and the surface water of the oceans has become increasingly acid, its pH value changing from 8.2 to 8.1.[14] This represents a virtual doubling of acidity – making the sea more acid, we think, than it has been for many millions of years. If CO_2 emissions continue to rise unchecked, it is calculated that the pH of the seas will be between 7.6 and 7.8 by the year 2100.[15] It seems that the polar oceans will be first to suffer the impact of acidification, and some shelled sea-animals will suffer corrosive effects. In a statement issued in June 2009 by the Academy of Sciences of the Developing World (an umbrella organization which represents some seventy scientific academies, including the UK's own Royal Society), it is argued that even if CO_2 emissions are stopped entirely, it will take thousands of years for marine pH values to return to the levels that prevailed before the industrial revolution.[16]

Ocean acidification is already resulting in loss of coral reefs, and the wildlife which the reefs support. And as CO_2 builds up, many other marine species will be threatened. In the UK, the trade in scampi, scallops, crabs and lobsters, which was worth around £230 million in 2007, is likely to be severely affected. And it is not just those animals with a shell of calcium carbonate (which will be weakened or dissolve in the more acid environment) that may be in jeopardy. We do not know how fish will respond to changes in acidity or whether their food sources will be seriously depleted, but their larval and developmental stages are likely to be at serious risk. One problem which makes prediction particularly difficult is that the build-up of CO_2 appears to be occurring at a time when the earth's temperature is increasing. As the oceans of the world heat up, the deleterious effect of any increase in acidity is likely to be much more severe.

A DANGEROUS DELICACY

In 2003 the BBC programme *Watchdog* highlighted an illegal trade believed to be controlled by criminal gangs. Small organized groups, often obtaining their deadly produce in the deprived farmlands of Wales, were shipping it to south London, where it was popular among that area's sizeable African population. The product in question was not a drug, but a food. 'Smokie' is the carcasses of sheep and cows, blowtorched with the skin on, to, in some people's view, improve the taste. Regarded as a delicacy in some African countries, smokie has usually been sold with the spinal cord and brain intact – a practice which poses considerable risks to human health, because of the risk of transmission of bovine spongiform encephalopathy (BSE or 'mad cow disease').

It is possible that BSE was first seen around 1983, but nobody took much notice. The first real concerns arose in 1987, when a cow died a peculiar death after being attacked by fits of trembling; this cow had lost weight and was seen staggering in a rather 'drunken' manner. Soon other cows followed suit. At first it was assumed that this was a bovine form of scrapie, a disease of sheep that has been known about since 1732. It was called scrapie because infected sheep and goats scrape their fleece against trees and stones as if they were suffering from severe itching. Like cows with BSE, sheep with scrapie become emaciated and walk in a very uncoordinated fashion. Scrapie is a highly contagious disease but seems to be infectious only between sheep or goats, and apparently never affects humans. Because it makes flocks very ill, the standard practice has been to kill all affected sheep after putting the flock in quarantine.

Research revealed that BSE was caused by a hitherto unknown kind of pathogen. This was an abnormal protein very similar to another protein found in the normal nervous system, but somewhat different in its physical structure. This altered structure seems to prevent nerve cells from functioning properly, and prompts other normal proteins of the same kind in the brain to adopt a similar structure – consequently the protein propagates itself by contact. The discovery that this protein was 'infectious', that it could spread from animal to animal, was worrying, to say the least.

The first person to understand that this kind of brain disease was likely to be caused by a protein was a very extraordinary colleague of mine at Hammersmith Hospital, Dr Tikvah Alper. A South African, she went to Berlin in 1930 to work on a doctorate in nuclear fission, but never completed her degree; being Jewish, she was lucky to escape from Nazi Germany and to be able to return safely to Johannesburg. After the war she worked on the biological effects of radiation, but became *persona non grata* in South Africa because of her antagonism towards apartheid. At Hammersmith, where she eventually settled, she headed the Medical Research Council Radiopathology Unit. Although this indomitable woman retired in 1974, she kept up her scientific work, arguing that the transmissible agent in scrapie and mad cow disease could not be either a virus or a bacterium because, as she demonstrated, it did not contain any DNA or RNA. Her theory was greeted with disbelief, not to say derision, in some circles.

It was not until Dr Stanley Prusiner, from the University of California in San Francisco, did his research that the true nature of the infective agent was documented.[17] In identifying it as a specific protein – he coined the word 'prion' – he eventually won the Nobel Prize for Medicine in 1997. It had become clear that prions were the cause not only of scrapie in sheep, but of severe brain disorders in humans – such as Creutzfeldt–Jakob Disease (CJD) and the 'kuru' sickness which had afflicted New Guinea tribal groups up until the 1950s. Prions, fortunately, are not easily passed on, except through ingestion of affected tissue. However, prions are not destroyed by the usual measures which kill most bacteria, such as heat – so special care needs to be taken when cleaning or sterilizing surgical instruments which might be contaminated. Sheep pass scrapie to one another by eating placentae left in fields. The Fore tribe of New Guinea used to eat human brains, causing a condition the locals called 'laughing death'. This progressive, fatal brain malady robbed its victims of the ability to walk, talk, and even eat. Examination of their brains after death showed massive holes in the cerebral cortex – it looked like a sponge or a Swiss cheese.

In Britain in the 1980s theories abounded as to the cause of these prions in Britain's cattle. Perhaps, it was suggested, the use of organophosphate pesticides had affected animal feed in some way, causing the mutation. This hasn't been entirely ruled out, but a far more likely cause is the fact that the infected cattle were fed animal produce. Once again, this boils down to the fact that farming is a profit-oriented concern. It is virtually impossible to obtain commercially profitable quantities of milk by feeding cows grass alone, so cattle are also fed high-energy pellets. These are usually made from oily and fatty seeds like linseed and soya, but during the

1980s it became commonplace to include the leftovers from abattoirs – a rich but very cheap source of fat and protein. Few regulations controlled the conversion of bits of otherwise unusable animal carcasses into cow feed. Nobody knew this might cause a problem, although in retrospect the practice of forcing a naturally herbivorous species to eat animals, including its own kind, should have rung some alarm bells.

In 1989 the government conducted an inquiry into BSE, and concluded that it was unlikely the disease could spread from cattle to humans. In 1990, in an effort to show that it was completely safe to eat British beef, the public were treated to the unedifying sight of Mr John Selwyn Gummer, the Minister for Agriculture, trying to persuade his daughter to eat a beefburger on College Green outside the Houses of Parliament, for the benefit of the cameras at a photo-shoot. This sensible little girl refused, and her father's credibility suffered a massive blow. Six years later, the first human deaths were reported from not BSE but a new, variant type of CJD. In March 1996 the UK Health Secretary admitted there was a 'probable link' between BSE and vCJD, and the European Union immediately responded by banning imports into EU countries of all British beef (while completely denying that there might be a similar problem in other EU countries where similar cattle feeds were employed). The British government reluctantly ordered the slaughter of all cattle more than thirty months old – reckoning that no beasts younger than that would have been fed the infected meal. But it then emerged that the disease could be passed from mothers to their offspring, forcing a rethink. The government subsequently banned all sales of beef on the bone, on the ground that bones and, in particular, the spinal cord contained the highest proportion of nerve tissue, which was where the affected prions resided. This rule remained in place until 1999.

Meanwhile, people continue to die of vCJD: the surveillance unit in Edinburgh reckons there have been 161 deaths definitely or probably caused by the condition since 1995. Although the much-dreaded epidemic never happened, this crisis, and the measures taken to shut the barn door after the cow had bolted, are estimated to have cost Britain at least £5 billion. And the cost in terms of people's confidence in the food they eat is immeasurable.

MALARIAL GOATS AND MUTATOES

According to the WHO, malaria infects 300–500 million people a year, and kills 1 million. An overwhelming proportion of these deaths is concentrated in the poorest parts of the world and among the poorest people of those regions, who are often coping with HIV and malnutrition alongside this mosquito-borne menace. Mass vaccinations and mosquito nets are the only solution, but the costs of providing them are crippling.

Help may be at hand, though, from a peculiar source. In 2001, a team led by Dr Anthony Stowers of the US National Institute of Allergies and Infectious Diseases succeeded in implanting a gene from a deadly malaria parasite into a goat embryo. The gene in question is designed to be activated by cells in the goat's mammary gland, enabling goats so treated to produce a cheap and plentiful vaccine in their milk. It may be some time before this is a reality, but earlier trials that involved producing the vaccine in mice and then giving it to monkeys provided favourable rates of success.

In April 2006 the European Medicines Agency gave the green light to the use of the world's first drug made from a transgenic animal – in this case a genetically engineered goat. ATryn is an anti-clotting agent, and it has been cleared for use in surgery for people with the rare congenital disease antithrombin deficiency. Unlike malaria, a condition estimated to kill a child every thirty seconds, antithrombin deficiency affects just one in every 3,000–5,000 people. The first transgenic animal – made by transferring the gene of one animal into the embryo of another – was created in 1980. Twenty-four years later, the first transgenically produced drug became available, for a condition affecting a tiny minority of people, and the subsequent approval of ATryn is a watershed for those involved in the so-called 'pharming' industry – but it also indicates the degree of caution with which licensing authorities are approaching the issue. Meanwhile our newspapers abound with lurid but fascinating reports from biotech laboratories around the world. Goats implanted with a spider gene to produce large skeins of suture silk. Cows which can produce human antibodies to a range of deadly diseases: botulism, anthrax, smallpox. And of course, the notorious 'fishberry' – half strawberry, half-fish.

Perhaps rather sadly, that last product never actually existed. Back in 1991 some researchers at a biotech laboratory developed a genetically engineered tomato, carrying a gene from the Arctic flounder fish which enabled it to keep warm in freezing waters. It was hoped that this gene would enable the tomato to thrive in frosty climates, but in fact the experiment failed. This hasn't bothered the numerous sources – including the BBC – who have cited the success of this experiment as a reality, and used it to bolster their arguments against the genetic engineering of crops and animals. In fact, the said tomato was never injected with a fish gene – just an engineered copy of one, translated, as it were, into plant terms. Furthermore, no one has, to my knowledge, attempted the same thing with strawberries, or even announced their intention to do so. It may be that scientists from the biotech laboratory informally speculated about the future uses of their experiment in strawberries. It may also be that the timing of the venture corresponded roughly with research into Frostban – a genetically engineered bacterium used to protect strawberries from frost damage.

There has been a considerable amount of alarm about anything that involves the words 'genetic' and 'modification'. But GM is definitely here to stay. In 2006, some 102 million hectares of GM crops were planted in twenty-two different countries. The majority of these crops were herbicide- and insecticide-resistant soya beans, corn, cotton and alfalfa. The United States is the biggest producer, followed by Argentina and Brazil. Almost half of all the soya beans grown in the world are genetically modified.

The advantages of modifying a crop or animal are plain to see. By cultivating plants with an inbuilt resistance to pests and disease, farmers can boost their output and, potentially, apply fewer toxic chemicals. Engineering can also be performed on an organism's susceptibility to climate, creating plants that can grow in inhospitable conditions. And crops can also be boosted, to provide extra quantities of the nutrients lacking in the diet of a given area.

A good example of the last kind of GM crop is golden rice – engineered with special daffodil and bacteria genes to contain an extra quantity of A-vitamins. This might be of particular benefit to people in the developing world, where vitamin A deficiency lurks behind a range of conditions, not least blindness. This is especially so in parts of India and South-East Asia, where many people eat little but rice, three times a day.

But golden rice has its opponents. The product in question – and others like it – can only be manufactured by large, affluent Western companies. There is a definite perception that genetically modified crops are being manufactured simply to increase the profits of big business, with little regard to the hazards of such foods and the risks of propagating 'new' species in a vulnerable environment. There are also those who argue that the future for the developing world should rest on increasing self-reliance, not dependence on and exploitation by the wealthy West. Engineered golden rice plants are sterile, meaning that each year farmers are forced to purchase more seeds from the Western companies – a process typifying how, in some people's view, the poor become enslaved to the rich.

There are clearly problems associated with the creation of pest-resistant crops. Concerns have been raised over the use of Bt-corn, a genetically engineered maize which produces its own insect repellent. This repellent substance is a poison, and because it is generated inside the plant, the insects are exposed to it for a longer period than they would have been if it had simply been sprayed on by the farmers. As a result, the insect has more potential to become resistant to the poison – rendering both crop-spraying and genetic modification useless, and necessitating the use of even stronger and more toxic compounds. This kind of problem could be exacerbated if genes from the engineered plants found their way into wild ones: farmers could face a generation of super-weeds, immune to all attempts to curb their spread.

A related issue is that plants engineered for specific purposes could get crossed or mixed up with those intended for general consumption. During the 1990s a biotech company produced a maize with a built-in insecticide called Cry9C. It was approved for use only as animal feed, but by 1999 it was discovered that some of it was being sold in American supermarkets in products intended for humans. The company in question was forced to buy back every bushel of maize grown across a third of a million acres. That was most probably a case of simple human error; but there is uncertainty whether some maize, fit for animal consumption, might cross-breed with the types we consume. And in an increasingly globalized economy, we can never be sure where all the separate ingredients of our foodstuffs have come from.

At present, because of objections such as these, and the need to conduct extensive, rigorous trials, only three GM crops are approved within the EU, and none are grown commercially in Britain. The situation in our country is due in part to widespread public resistance, fuelled by media reports on so-called 'Frankenfoods' and coming in the wake of reasonable fears about BSE and foot-and-mouth disease. The public resistance to GM crops is, in my view, completely understandable. When trials of GM crops were planned in Britain, there was no serious attempt to take the fears of the public seriously, and industry and government were prepared to ride roughshod over public opinion. This resulted in many people applauding when trial crops of GM rapeseed and potatoes were targeted by activists. Destruction of experimental fields of GM crops, and the accompanying outcry against the experiments, resulted in three of the major biotech companies being compelled to suspend their work in 2003 – although a fourth, BASF, began trials in 2007.

The activists have a right to express their views about the so-called 'mutatoes', but there is a risk that much good work will be undermined and ultimately blocked unless scientists are more prepared to recognize and respond to public concerns. It is not sufficient for scientists to condemn governments for 'pandering to gut reactions'. We need instead to encourage healthy debate, and a good first step would be to admit much more openly that there are some serious concerns about growing GM crops and damaging the local ecology, and about the potential risks involved in consuming GM foods. In the 1980s 37 people died and some 1,500 were left disabled after consuming an engineered food supplement, tryptophan. Normally developed from naturally occurring bacteria, the modified tryptophan contained a toxin, which caused a disease known as EMS. Did this toxin arise as a direct result of the genetic engineering procedure? Perhaps it is significant that, at the same time as it was trying to cut costs by using GM tryptophan, the company in question also cut corners in its filtration process – potentially allowing some impurities into the bacteria that way.

Unfortunately, further investigations are unlikely because the manufacturers destroyed all stocks of the engineered bacteria strain when the scandal came to light. Inevitably – and not unreasonably – some people see that move as significant in itself.

The British position on GM foods has been substantially influenced by Dr Arpad Pusztai of the Rowett Research Institute. In April 1998 this scientist told the nation that he had fed GM potatoes to rats, which had gone on to suffer significant intestinal changes. He reported alterations in the structure and function of their digestive systems, changes in the development of their internal organs and a slowing down of their immune response to injuries. In light of these observations, he said it would be 'unfair' to give such products to humans. Within two months of this announcement, seven supermarket chains – already sensitive to public concerns after the BSE and foot-and-mouth crises – had withdrawn GM food from their shelves. Dr Pusztai, meanwhile, was sacked from the Rowett Institute for publishing the results of his research before it had been completed or peer-reviewed. The Royal Society then stepped in and reviewed his work, declaring it to be flawed. The *Lancet*, meanwhile, a frequent sparring partner of the Royal Society, printed Pusztai's work – not so much in defence of his findings as in opposition to the way the Royal Society had judged Pusztai on the basis of partial and incomplete work.

Scientists remain divided over the case of Dr Pusztai, but some of us have an uneasy feeling about how this work was reviewed and accepted for publication. There are reasonable concerns about some of the wider issues involving GM foods. I admit to sharing with a large section of the populace a feeling of revulsion on seeing the featherless chicken, engineered so that it could be housed in intensive battery conditions in the fierce heat of Africa. But certainly, instinctual reaction to a few extreme cases should not necessarily colour our judgement about a whole technology. Most scientists are deeply disturbed by the persistent attacks of activists on properly conducted scientific trials, because we will never establish the safety or otherwise of GM foods if we cannot test them. And if the gut reaction is always allowed to win, what will we lose? Genetically engineered crops to feed people in benighted, drought-ridden areas? 'Pharmed' goats that may prevent thousands of deaths from malaria? We should look at the achievements as well as the failures and drawbacks of this technology. Above all else, we should be proud that we have developed this capability – something that would not have been possible if our ancestors hadn't started gently and tentatively to modify their own environments, ten millennia ago.

Chapter 5

Wild and Whirling Words

IN 1907, JOSEPH Conrad's thriller *The Secret Agent* depicted a failed bomb plot, its gory conclusion taking place in a smart south London park. But it was not known to many of Conrad's worldwide audience that his story described events which had actually occurred in London thirteen years earlier.

On a Thursday afternoon in February 1894, Messrs Thackeray and Hollis, two members of the Greenwich Royal Observatory staff, were 'working late'. It was only 4.45 p.m., but in those days of sparse street lighting and blinding smog it wasn't uncommon for many workers to leave for home before nightfall in the winter months. Thackeray and Hollis were in the Observatory's Lower Computing Room when they heard 'a sharp and clear detonation, followed by a noise like a shell going through the air'. Making for the windows, they saw the Observatory's doorman running across the courtyard. They followed him, reaching a point where they could see down the hill that sloped away to the north of the Observatory. They watched as the doorman, a park warden and some schoolboys ran towards a figure apparently crouched on the path below the Observatory.

Joining the group, the two men's first impression was that the man had shot himself. In fact, his injuries were far more horrific. His left hand was missing, he had a hole in his abdominal wall, the path on which he was lying was awash with blood, and fragments and splinters of bone were spread 60 metres around him. In spite of this, he was conscious and still just able to speak. He was carried by a stretcher to the nearby Seamen's Hospital, where he died half an hour later, having revealed nothing of his identity or the cause of his injuries.

The police soon established that the man's name was Martial Bourdin and that he had left his rooms in Fitzroy Street and taken a tram all the way to Greenwich, carrying a parcel under his arm. It seemed that this parcel contained an incendiary device that had detonated unexpectedly. Bourdin had a large amount of money on him, suggesting that his intention had been to leave the bomb somewhere and flee the country – probably for his native France. Subsequent investigation revealed he was a member of the 'Club Autonomie', a Soho-based gathering place for foreign anarchists. The following evening, 17 February, the club's premises near Great Titchfield Street were raided just after midnight. Eighty arrests were made, some shady characters (the newspapers refer to Frenchmen and Bohemians) were questioned, and a few were returned, by force, to the Continent, but no charges were ever made in connection with the incident. Bourdin's funeral attracted large crowds, many of them in sympathy with the anarchist cause.

While Conrad embellished the story considerably, some intriguing questions hang over the reality. Why would Bourdin have picked a target like the Observatory – out of the way, sparsely occupied – and with too small a bomb to inflict any notable damage to persons or property? By 'notable', I mean great enough to cause sufficient suffering, loss of life and destruction to ensure that the event would be reported by the press, thereby spreading the terrorists' gruesome message to the widest audience. Some believe that Bourdin may have been duped into carrying a bomb, or that he intended to dump it somewhere and flee to France. His brother-in-law was thought to have been a police informer, and many anarchists believed the whole event was a publicity stunt, set up to show the nation how ineffectual the terrorists were, and how swiftly the authorities could act.

Whatever the truth behind it, this story demonstrates how both the terrorists, and the governments they intended to topple, depended upon the same invention. Publicity – before the days of the radio, the satellite television broadcast, the internet and the text message – meant newspapers. These newspapers, along with novelists like Joseph Conrad, relied heavily upon the presence of a large, literate public. And none of that would ever have been possible had it not been for an innocuous little invention by accountants in Iraq some 5,000 years before. This invention has visited far more destruction upon humankind than the dynamite and gunpowder of the terrorists.

A POTTED HISTORY OF OLD WORLD WRITING

The earliest mention of Johannes Gutenberg – regarded, perhaps erroneously, as the father of printing – comes from court trans-actions in the city of Strasbourg. In 1439 a legal case was brought in which Gutenberg and three other men – Hans Riffe, Andres Heilman and Andres Dritzehen – disputed the rights to a mysterious industrial process. Gutenberg had taken substantial sums of money from his business partners in return for teaching them this un-specified process. It was to be kept secret and if one of the partners should die, Gutenberg maintained the right to buy back his 'share' in the idea. Matters came to a head when Dritzehen died, and Gutenberg gave orders that a piece of equipment lodged in the dead man's house be returned. The equipment was never found, and to this day we do not know what it was. But we do know that Gutenberg had spent some of his partners' money on lead, various other metals and a winepress. Given the form and content of the invention for which Gutenberg later became famous, it seems likely the mysterious item was a prototype printing press.

So the first mention of a printing press in European history comes hand in hand with disputes over property and money. And it is for this purpose – recording what was owned by whom, what was due by someone's hand to someone else – that writing first came into being. In its earliest forms, writing was about as similar to the lines you are reading now as the average human is to a fish. The ancestor of writing came in the form of two distinct series of tokens or counters, fashioned out of clay, and found in the Near East between the eighth and fourth millennia BCE. The two series – developed to keep track of goods – evolved into two distinct kinds of writing.

Plain tokens, the earliest kind of proto-writing, arrive in the archaeological record at the time of the beginnings of agriculture. Found right across the Near East, they have smooth surfaces and simple forms, and, judging by the nature of the small-scale societies among whose remains they are found, seem to have helped people keep track of harvests and herds.

Complex tokens, by contrast, are found only in the ruins of the first Near Eastern city-states, like Uruk, which became prominent midway through the fourth millennium BCE. With intricate stylus-made markings on their surfaces, and taking a wide array of forms, they have been most frequently found around temple complexes. These temples, presided over by a king and an élite class of priests, supervised a complicated economic system whereby food was brought in from the surrounding countryside and redistributed fairly to the populace. The king and his priests, in turn, ensured the continuing fertility of the land through the enactment of certain rituals – including a sacred act of coitus between the ruler and a goddess, embodied in the form of a prostitute-priestess.

Along with standardized bowls and jugs – presumably for the distribution of fair rations to the people – and mosaic tiling, the complex tokens suggest a new type of society, based on that most hated ancient practice, taxation. In this period we also find the first use of seals, some of them depicting brutal beatings – possibly the ancient taxman's way of extracting revenue. The imposition and maintenance of such a system required weighty administration, along with sophisticated systems of record-keeping, rules of conduct and penalties for non-compliance.

The two types of token, though used for similar purposes, were stored differently. The simple shapes were kept in sealed clay envelopes, in order to preserve the details of whatever had been counted or exchanged. Once the tokens had been placed inside, the clay envelopes – sometimes rectangular, sometimes shaped like hollow balls – were sealed and baked. If the contents were sub-sequently disputed, an official could break open the envelope and confirm what was inscribed. A typical rectangular envelope from ancient Turkey, made around 1850 BCE, is on display in the British Museum. A significant number of the complex shapes, however, were perforated, suggesting that they were kept on strings. What were they for?

According to archaeologist Denise Schmandt-Besserat, the answer lies in the cuneiform – from Latin *cuneus*, or wedge-shaped – writing that came immediately after the token stage.[1] In the writing of the third and second millennia BCE, simple shapes like cones and spheres refer to the staple products of the kind of agricultural settlements that had once used the simple tokens. Cones and spheres of varying sizes referred to different measures of grain. Cylinders denoted animals.

We can also see the descendants of the complex tokens in cuneiform script. More intricate symbols, like incised cones and rhomboids, represented 'made' goods – processed foods, like beer and oil and bread, and manufactured luxury items, like perfume and metal objects. These were made in the city, using the raw products from the countryside, but they were subjected to processes within the temple precincts and under the supervision of its personnel.

The leap from tokens to writing occurred via a simple short cut, taken by overworked temple accounting staff. To ease their access to the sealed clay envelopes containing simple tokens, clerks began to impress the tokens on the outside of the envelopes when they were wet, leaving a visual and tactile impression of what was inside. From there, it was a short step to realizing that the impression on the outside of the envelope conveyed all the same information as the actual tokens inside it, and abandoning the envelope and tokens in favour of the clay tablet alone. The same tactic would not work, though, for the complex tokens, because the intricate markings on their surfaces would not leave a readable impression on the clay. These were therefore rendered using a stick cut from a reed – the same instrument that had been used to adorn the tokens in the first place.

From the earliest examples of cuneiform – there are about 4,000 inscribed objects from Uruk, dating from around 3200 BCE – it is possible to trace a line passing into the scripts of the Assyrians,

Babylonians, Hittites and Canaanites, and diverging from there into Phoenician and Etruscan, and thence to the more familiar Greek and Roman alphabets of modern times. I sit at my computer now using a remote descendant of cuneiform to set out the history of writing. Earlier today, I sent a text message telling someone I'd be late and an email about a book I'd ordered. I, and billions of others, use writing every day as an adjunct of that very basic human function discussed in chapter 1, language. And because that function is so basic, so instinctual, it is easy to overlook what a leap writing was, and what a complex process led to its emergence.

The record-keepers of ancient Uruk did not use writing as a means of daily communication. Cuneiform was a strictly defined device used solely for accounting. The first signs they used were pictographic – that is, they represented not sounds, but objects and quantities. The sign for 'hand' was a drawing of a hand, the sign for 'water' a series of wavy ripples. But as usage changed, and these symbols began to be etched in clay with a manufactured stylus, it became necessary to simplify and abstract them. Curves became straight lines, fine details were eliminated. Over the course of three millennia, cuneiform changed from strings of recognizable pictures into a bristling plethora of etched lines and triangles.

As the script changed, so did the way it was used. Rather than invent a symbol for every word, the accountants of the Near East combined symbols from a repertoire of around 700. The sign for 'eat' was a combination of the signs for 'mouth' and 'food'. This flexibility set the stage for a further development, intimately related to the character of the Sumerian language. Composed of mainly monosyllabic words, with many homonyms – words with different meanings, but the same or similar pronunciation – the language lent itself to the introduction of a phonetic form of writing. For example, the word for 'arrow' in Sumerian was *ti*, which was also the word for 'life'. The early writers, rather than having two symbols, simply used the arrow sign, with extra markers when they wanted it to indicate the meaning 'life'. Straightforward as this seems, it marks a major shift: for the first time, symbols were being used to refer to a sound, rather than an object or idea. An equivalent from our own tongue might be the symbol of a pear to denote both the fruit and the word 'pair'.

Once symbols have been uncoupled from their original points of reference, and have come to be associated purely with sounds, then they can be combined to represent more complex and even abstract meanings. Again, an example using our tongue might be the juxta-position of the symbols for a bee and a leaf to convey the meaning 'belief' – pictorial representations of two simple object-nouns being used to express an abstract concept unrelated to either of them.

This, in turn, changed the way the early writers arranged their information. In the earliest period of cuneiform, the space on a writing tablet was divided in such a way as to convey meaning in itself. Specific locations upon one of these baked-clay records – many of which still bear the fingerprints of the scribes who held them – told readers the names of the supervising officials. Certain commodities were only ever written on the left or the right, so that the spatial location of words said as much as the words themselves. As cuneiform became more sophisticated, however, the tablets began to convey everything through words themselves. Columns were given headings, totals and subtotals were labelled as such. Terse phrases enter the record, such as 'Wullu received it' and 'from the hand of Nashwi'.

Some 90 per cent of the Uruk tablets are clerical records; the remainder are exercises in writing for the new class of officials being trained to create and maintain them. Mostly, scribes were given huge, dull lists of commodities and administrative titles to copy out – though if they were a little luckier they might get some verses, a proverb or a whimsical description of their own working day. In a few, fascinating examples we see not only the tortuous hand of the trainee scribe, but the patient corrections of his teacher.

As cuneiform became increasingly capable of conveying more complex information, trained scribes added their own signatures to the bottom of tablets, along with the positions of their fathers. In most cases, these were well-to-do palace officials, suggesting that the 'writing classes' were a privileged élite. A tablet from the second millennium includes a boast by the Sumerian king, Shulgi, who declares: 'No one could write a tablet as I could.' King Shulgi, who clearly attached importance to this skill, established a school for the training of scribes – institutions replicated throughout the Near East and along the Nile.

The office was particularly revered in Egypt, where scribes could become rich and influential officials, and writing was itself sacred. Thoth, the messenger of the sun-god Re and the patron of scribes, was also the keeper of order, the timekeeper of planetary movements and the weigher of souls on Judgement Day. Small wonder that, in the second millennium BCE, an Egyptian official left his son with the following message, after dropping him off at scribal school: 'It is to writings that you must set your mind. I do not see an office comparable [with the scribe's]. I shall make you love books more than you love your mother.'

A GREEK'S GIFT: WRITING IN THE CLASSICAL WORLD

The earliest examples of Mediterranean writing have been found on the island of Crete, within the palace of King Minos at Knossos. Minos and other rulers of comparable stature were figureheads of a distinct civilization, centred on elaborate palaces. Again, the earliest writing associated with these palaces is pictographic – simple images like a man's head or an axe carved into a variety of hard surfaces, including jasper, amethyst and gold. Impressions of these images in seals on fragments of clay have also been found, suggesting that they provided some form of signature on goods or accounts.

Early Minoan writing bears a passing resemblance to Egyptian hieroglyphics, leading some to suggest a direct line of inheritance from the Nile valley to the Mediterranean. In fact, it looks as if the Greeks inherited writing from a rather closer neighbour – a point we shall deal with later. But this putative connection raises an interesting side-issue in the story of writing. There seem to have been three places around the world where it emerged spontaneously, without any outside influence: in the Near East in the mid-third millennium BCE, in China in the second millennium BCE and in Mesoamerica in the third century CE. A range of other societies also developed their own methods for recording speech, ideas and quantities, but in these cases the question of influences is less clear. There seems, for example, a world of difference between the aesthetically perfect hieroglyphics of the Egyptians and the simple, sparse notation of the Sumerians. But the two societies were in contact. Alongside its complicated catalogue of pictographic signs, Egyptian writing also employed a parallel means of conveying sounds. It therefore seems possible that the Egyptians, whose society was in its ascendancy some three centuries later than the Sumerians', inherited their idea of writing, but not their script. By contrast, when the central Italian region of Latium came into the cultural orbit of its literate neighbour Etruria, it took on the Etruscan alphabet, adapting it to the specific needs of its own language. Between the writing systems of Sumer and Greece, as we shall see, there exists a chain of continuity, sometimes in ideas, sometimes in signs.

As Ronald S. Stroud, Professor of Classics at the University of California at Berkeley, points out, it was a natural step for these early Minoans, used to making impressions with their signet rings in soft clay, to progress to a script based on the same principles.[2] From a period slightly later than that of the pictographic seal-stones, we find a script written in straight lines on clay tablets, which was dubbed 'Linear A' by Sir Arthur Evans, the archaeologist who unearthed it at the start of the twentieth century. It passed through stages: in the earliest examples, the pictograms still occur, but stylized and simplified, and the prevalence of numerals suggests that they were used solely for accounting and book-keeping purposes. Later, Linear A evolved further, keeping only about a third of its original stock of signs, refining its numeral system and being inscribed consistently from left to right.

It is clear from the relative frequency of the various signs that Linear A represented the sounds of a language. Unfortunately, we don't yet know what language. It has been suggested that it represents a unique, indigenous tongue of the island, pushed out of existence by later waves of Indo-European invaders. Classical archaeologist Gareth Alun Evans believed it could be a type of Luwian, an Indo-European tongue spoken in the area now occupied by Turkey.[3] Much controversy rages over one of the signs that has been deciphered – KU-RO, used to convey the meaning 'whole'. This may be close to the Semitic root *kl* (pronounced 'kol' in Hebrew), which covers the same range of meanings as 'whole'. Alternatively, it could come from the Indo-European root *kwel*, meaning 'to move around a centre' or 'to dwell'. At present, too little of the script has been deciphered to yield more than a fascinating debate.

The bulk of the Linear A inscriptions found so far seem to be lists of objects, goods and personnel, all associated with the massive administrative framework of the Minoan palaces. There are only around 200 examples, but their distribution across more than twenty different Cretan sites, and a few locations within other islands and on the Greek mainland, suggests fairly widespread use over a period of about 200 years (from 1650 to around 1450 BCE).

Alongside the pictograms and Linear A fragments unearthed by Evans from Knossos were 4,000 clay tablets written in another script – which he called 'Linear B'. This script used short vertical strokes to divide words from each other, and also employed ideograms – a wheel, a vase, a man – to represent certain commodities or people mentioned in the text. Linear B has also been found at sites on the Greek mainland, and the sheer number of found objects assisted the outstanding scholar Michael Ventris in his efforts to decipher it. His thesis of 1952 argued that Linear B was an archaic form of ancient Greek, employing the same system of word endings to convey meaning.

Tragically, Ventris died while still young, only two years after the publication of his ideas, but they had enormous influence. The cracking of Linear B enabled classical archaeologists to suggest that Knossos had been subject to some form of a takeover. Literate palace officials, using Linear A to make records in their own language, were kept on when Greek invaders came across from the mainland and imposed their rule. Impressed by this strange innovation, the new masters demanded a similar system for their own language – and the result was Linear B.

Readers will notice I avoid the term 'alphabet'. Linear A and B were not alphabets, but syllabaries – a range of signs used to depict combinations of vowels and consonants. An example of deciphered Linear A, found on a stone ladle, runs:

*atai-*301waja osuqare jasasa rame unaka nasi ipina ma siru te*

I have not the foggiest idea what that means – and as far as I know,

neither has anyone else. But there is no evidence to suggest that this
script was used for anything other than record-keeping, associated
with the passage of goods and people into the palace, and their
distribution into the surrounding countryside. There are no kingly
letters, no laws, no poems. Analysis of scribal handwriting has
identified around 105 separate individuals, all of them working
within the palace on administrative matters. Even more strangely, at
a certain point in time, it all vanishes. With the destruction of
Knossos in 1380 BCE, and the burning of a related palace on the
mainland 180 years later, all Greek writing disappeared for four
centuries. The epics of Homer were created as oral poems, and it is
telling that, within the 27,000 verses of the *Iliad* and *Odyssey*, while
we get illuminating references to subjects as diverse as sail-making
and farting, there is only one, tiny, ambiguous reference to writing.
Nothing suggests that Homer – or those poets of the eighth and
seventh centuries BCE who operated under his franchise – had any
knowledge of writing for the purpose of conveying a detailed
narrative, and when the Greeks did begin to write again in the mid-
eighth century BCE, they did so with an alphabet nothing like the
ancient syllabaries of Knossos.

So where did this new alphabet – the ancestor of our own – come
from? The most convincing thesis is that the Greek letters we know
came via Phoenicia, the civilization based in the north of Canaan,
now Lebanon and Syria. Scholars think this for several reasons.
First, the earliest examples of Greek alphabetic writing look similar
to Phoenician – which was itself a descendant of the earliest scripts
of Sumer. And the order of the letters is similar. In picking names for
the separate letters of their new alphabet, the Greeks adapted the
Semitic sequence of 'aleph, bet, gimmel' to 'alpha, beta, gamma'.
And they wrote, as the Phoenicians and Hebrews did, from right to
left.

There is substantial evidence of contact. By the mid-ninth century
BCE the Phoenician kingdom of Tyre was entering its ascendancy,
exercising commercial and colonial influence as far away as Malta,
Sardinia, Spain and Sicily. Greeks, also trading and travelling
extensively around the Mediterranean region, would have had
contact with Phoenicians. Herodotus refers to their settlements on
the Greek mainland, and archaeological evidence suggests the

presence of Phoenician traders and artisans on the islands of Crete, Rhodes and Cyprus.

Fittingly for a civilization praising 'wine, women and song', one of the earliest examples of the 'new' Greek writing comes from an inscription on an amphora. Dated about 740 BCE, this vessel bears a single line suggesting that whoever is the best dancer will win this jug of wine. Dating from the same period is a drinking cup found in a tomb on the island of Ischia, bearing the message, 'I am the drinking cup of Nestor. Whoever drinks from this cup swiftly will the desire of fair-crowned Aphrodite seize him.'

These texts deliver a host of interesting material. For example, as well as adopting the Phoenician alphabet, the Greeks added their own adaptations. The Semitic alphabet had five consonants not used in Greek, but no vowel signs. When they modified it, the Greeks deftly adapted the useless consonant signs for their vowels – and these have persisted, through Greek into Etruscan, thence into Latin and into the Roman alphabet we now use.

In contrast to the Sumerians and the Minoans, who used writing to keep track of goods, the Greeks seem to have turned the new skill to more frivolous ends. We find no contracts or accounts from this first period of Greek civilization, no laws, decrees or kingly letters. The earliest texts are private – declarations from the owner or maker of the inscribed object, celebrations of drinking, dancing and love. Writing often forms part of the design of objects – such as a little wine vase from Corinth, on which painted letters wind around the dancers and a flute player, and the message conveys little except that the name of one of the dancers is Pyrrhias. The inclusion of an inscription on an object seems to have been a way of making an impression, demonstrating skill in using a highly prized, foreign invention.

The Greek adoption of writing had profound and enduring consequences for the rest of Europe, mainly as a result of the establishment of a small Greek colony at Cumae, on the Italian peninsula, close to Naples. From there the invention of writing, in its specific form as the Greek alphabet, passed into the hands of the neighbouring Etruscans, and then to the north-lying civilization of Latium. The Etruscans were a wealthy and sophisticated group, whose Greek-inspired material culture had a profound influence on surrounding areas. From the seventh century BCE, lavish goods found in graves at Praeneste and Tibur – in the central heartlands of Latium – suggest the presence of a powerful ruling élite, who were either Etruscans themselves or at least heavily immersed in their culture. Throughout that century there is increasing evidence of contact with Etruria in surrounding centres, the most significant of which was Rome. Here that influence was at its strongest – apparent not just in material objects, but in the architecture of its houses and public buildings, and in the organization of public spaces, such as the Roman Forum.

But the most significant Etruscan export was writing. In the Praeneste burial site, dating from about 650 BCE, we find a silver cup bearing the word *vetusia* just under the lip. Vetus was someone's name, the *–ia* a way of signifying possession; in other words, this means 'property of Vetus'. Cups with similar inscriptions have been found in Rome and at Satricum, some 40 miles to the south, such as '[given/made] by Laris Velchaina' and '[of/from] Rome'.

The purpose of this early writing was not book-keeping but bragging. Wealthy families seemed to use it to enhance their prestige and the value of their possessions. Some Etruscan inscriptions suggest that exchanging gifts was an important custom among the élite – perhaps, as in other societies around the world, a way of ensuring long-lasting relationships. Rex Wallace, Professor of Classics at the University of Massachusetts, suggests this could have been the means by which the alphabet itself was communicated.[4] Wealthy Latin families probably received inscribed gifts from Etruscan neighbours, and began to copy their alphabet. The earliest Latin inscriptions date from the last decade of the seventh century BCE, and are written upon wine containers, which would have made useful gifts. The first bears the cheery legend *salvetod Tita* – 'may Tita be in good health'. The second says: 'I am the urn of Tita Vendia. Mamar[cos had me made]'. Wallace points out that in both cases the recipient of the wine was a woman, suggesting that a shipment of booze may have been an integral part of a wedding gift from groom to bride.

The fact that I'm writing now in a direct, mostly recognizable, descendant of that script is down to two factors. The first is the flexibility of the system inherited by Latium from Etruria. This system had, as we have seen, already gone through a number of mutations in its passage down the line of continuity that links the ancient cuneiform of Sumer to the scripts in turn of the Canaanites, the Phoenicians and finally the ancient Greeks and Romans. The end product was a highly adaptable system – one well fitted to being taken over, not just in idea, but in basic form, by a wide range of different linguistic groups, and particularly those sharing an Indo-European heritage. By virtue of being an 'alphabet' rather than a 'syllabary', it conveys a more atomized and extensive range of possible sounds. *B, d, g, l* and so on are simple phonemes, common to a great number of languages, whereas the world's tongues differ enormously in terms of the typical syllables formed with those phonemes. A syllabary has to be associated with a specific language – as is the kana script of the Japanese – or else be radically re-invented when imposed on to another tongue. The alphabet, by contrast, could be easily adopted. Roman letters could be clearly adapted to convey specific sounds, like the *ü* of German or the *č* of Czech. Signs useless in any particular language could be coopted for another purpose, as in Turkish, which uses the symbol *c* to convey the sound *j*. With its catalogue of twenty-six basic signs, the Roman alphabet was long enough to provide material for a huge range of sounds, yet short enough to be mastered swiftly.

Second is the pattern of Roman foreign policy. Whenever Rome conquered territories, it followed the same *modus operandi*. Cities were confiscated and land redistributed to Roman citizens. A large cohort of Latin-speaking and Latin-writing Roman citizens was then drafted in to administer each captured city. The indigenous inhabitants had to accept the new rule, and along with it the new means of communication. In some cases, such as Umbria, the pre-literate language died out alongside the advent of written Latin. In others, such as Campania, indigenous speakers of the Oscan tongue started to write their language in the Roman alphabet. This, in itself, indicates how useful an innovation the alphabet must have been. Colonized, enslaved and disenfranchised Oscan-speakers would have had strong motives for not adopting the culture of their conquerors, yet the alphabet proved too splendid an invention to ignore.

POTS AND PROPHECY: WRITING IN THE ORIENT

Crack-making on chia-shen, Ch'ueh divined.
Fu Hao's childbearing will be good.
The king, reading the cracks, said, 'If it be a ting day
    childbearing, it will be good. If it be keng day childbearing it
    will be extremely auspicious.'
On the thirty-first day, a chia-yin, she gave birth.
It was not good. It was a girl.

A scribe of the Shang dynasty carved these lines on a bone some 3,500 years ago. They, and some 150,000 other pieces like them made upon animal shoulder-blades and turtle shells, provide the earliest known evidence for writing in the East, and they seem to have been employed in a form of divination. After a bone is subjected to extreme heat – by holding it over a fire, for example – a pattern of cracks can be observed across its surface, and with this diviners claim to be able to foretell the future. It is a practice found in old civilizations throughout much of Asia, and even across the Bering Straits among North American Indians, but the Chinese were unique in carving the subject matter and results of their divinations into the bones themselves.

David Keightley, Professor of History at the University of California at Berkeley, points out that most modern Chinese readers would find these oracle inscriptions baffling.[5] But there is a parallel between these early characters and the medieval Latin manuscripts in our museums. At first sight, the neat, stylized, flowing Latin script appears to be quite different from our own – then we recognize an *a* and an *e* and a *g,* and that helps us to recognize whole words, and to realize that our own handwriting is not so vastly different – if possibly a little worse. The modern Chinese symbol for 'bright' consists of the symbol for a sun or moon juxtaposed with the symbol for a window. The symbol for 'good' is made up of the symbol for a woman next to the symbol for a child. A 'multitude' is conveyed by the symbol of three men under a sun. In each case, there is no vast difference between the modern form and the way it appears on oracle bones inscribed 1,500 years earlier.

Although heating up bones may seem like a rather haphazard way of deciding your future course of action, the Shang dynasty élite approached the business with considerable bureaucracy. The cracks were numbered by scribes, who appear, from the precise dating of inscriptions, to have filed bones away, bringing them out again at measured intervals for further prophecy. Some bones seem to have been reserved for enquiry on certain topics, such as the likelihood of success in the hunting field, or the varying fortunes of separate ten-day periods in the Chinese calendar. In the prophecy above, the scribe was careful to protect the reputation of his king as a diviner – a skill upon which his enduring power would have rested. The ruler, referred to here as Ch'ueh, promised a good (i.e. male) birth if the baby was born on a 'keng' or 'ting' day, but said nothing about a 'chia' birth. By stressing the type of day on which a girl was born – a chia-yin – the scribe makes it clear that the king wasn't wrong . . . even if he wasn't exactly right, either.

The earliest examples of Chinese script are difficult to decipher. Like the scribes of the Near East, the first Eastern writers employed what is called the 'rebus principle' to convert the sounds of their language into a repertoire of symbols. It is often said that Chinese writing is ideographic – that each of its symbols conveys an idea, just as a sign of a car skidding means 'slippery road ahead', or the symbol of a crown implies royalty. But Chinese writing is better described as 'logographic', meaning that each symbol represents a whole word. Of course, this is different from our alphabet, in which single symbols convey single sounds. But the two systems share something in common: in both, symbols denote the sounds of the spoken language. And in both cases, that system evolved by means of the rebus principle.

I have mentioned the possibility of using a picture of a pear to render the meaning 'pair'. This was a principle employed in both ancient Sumer and in ancient China. The Chinese word lai, 'to come', was represented by a picture of a plant, because the word for growing grain had the same pronunciation. Unfortunately, whereas Near Eastern scribes added extra markings to tell their readers which word they were referring to, those in China were more inscrutable and did not. The cauldron symbol crops up in a number of Shang dynasty oracle-bone inscriptions, but it is impossible to tell whether it means a 'cauldron', 'to divine' or 'to regulate' – all words with similar pronunciation.

Unlike their Sumerian counterparts, the scribes of China never took what we consider to be the logical next step of making their symbols refer to sounds instead of whole words. But, as Professor Keightley points out, to think of Chinese culture as having been in some way blighted by inability to make the leap to put its form of writing on a par with our own, enlightened form is to take a prejudiced view. To understand Chinese writing and its survival, it is necessary to appreciate that culture's reverence for tradition, and the prestige which would have been – and remains – attached to the mastery of such a complex system. Something similar was probably at work in Egypt, where hieroglyphics became more beautiful and complex over time. Writing was not simplified because to do so would have opened the skill, and the privileges of the scribe, to a wider group of people – something the élite had no intention of doing.

THE BLACK BISHOP

The world owes a debt to that most beastly bishop, Diego de Landa. A Franciscan friar, de Landa was sent to bring the Mayan peoples of Mesoamerica to the Roman Catholic faith. Appalled by their idolatrous habits, de Landa set about his task with appalling brutality, regularly using unspeakable tortures. He organized an auto-da-fé on 12 July 1562 in which a number of Mayan manuscripts and some 5,000 cultic images were burned. Speaking of his own actions later, de Landa remarked casually: 'We found a large number of books . . . and as they contained nothing in which were not to be seen superstition and lies of the devil, we burned them all, which they [the Maya] regretted to an amazing degree, and which caused them much affliction.'

The severity and ferocity of de Landa's Inquisition raised eyebrows even among his fairly brutal contemporaries. An earlier ruling by the Spanish king had exempted indigenous peoples from the authority of the Inquisition. But in his zeal to save souls, de Landa dispensed with much of the formal bureaucracy that accompanied the legitimized sadism of the religious authorities. Seen as a renegade, he was recalled to Spain to face charges for operating an illegal inquisition. He was also accused of falsifying evidence after it emerged that one of the victims of an alleged Mayan human sacrifice cult was alive and well. Yet de Landa was exonerated and thereafter appointed bishop of Yucatan by Philip II.

De Landa's methods were deplorable but he unwittingly left later generations a great gift in the form of his *Relación de las cosas de Yucatán*, written in 1562. While seeking to eradicate Mayan culture altogether, de Landa went to some lengths to understand it – exploring and noting down key aspects of its mythology, ritual and writing system. While he got some bits spectacularly wrong, de Landa's account has served as a first-rate tool for deciphering Mayan writing. How curious that there is an echo here of the behaviour of some Nazis, intent on understanding Jewish culture and traditions – even language – before attempting to blot it out.

Mayan hieroglyphics are visually arresting. Their lively, block-shaped glyphs carry a flavour of modernity – almost humorous and cartoon-like – at the same time as they attest to an ancient, vibrant and complex civilization. Writing experts find them difficult to decipher. As they consist of intricate squares laid out on a grid, the temptation is to imagine that each one represents a word, as in Chinese. In fact, each little square can contain up to five different, interlocking symbols, and the squares have to be read in paired columns, rather than as a straight line.

Many of these complicated symbols relate to the Mayan calendar, which encompassed several interlocking cycles, some tracking astronomical movements, others following seemingly abstract intervals of time. In common with other societies in the area, the Maya employed a 365-day solar calendar, called the *jaab*, alongside a 260-day ritual cycle called the *tzolk'in*. The *jaab* was further divided into eighteen 'months' of twenty days, plus five 'unlucky' days at the end called the *wayeb*. These two systems were also combined into a single 52-year cycle called the Calendar Round, and into an even longer affair that covered time cycles of 5,000 years. It may be confusing (though probably no more so than trying to predict when

Easter is celebrated next), yet the Mayan obsession with the calendar has enabled us to give precise dates to the events commemorated on stone plaques.

Writings on stone monuments conveyed information about rulers' dates of birth, their marriages, their accession to the throne and their victories. This was part of the propagandist use to which Mayan writing was put. In a hierarchical society, where the élite competed for the top positions, writing enforced a ruler's military power and demonstrated his descent from a long line of semi-mythical ancestors.

In addition to signs of the calendar, the Mayans possessed an extensive syllabary and some logograms that represented whole words. Scribes seemed to be able to pick and choose how they wanted to represent individual words or phrases. These factors have made the script notoriously hard to decipher. To date, the earliest examples come from the site of San Bartolo, formerly the location of an ancient Mayan city. The meaning of the very earliest piece of text, dated 300 BCE, cannot be unravelled entirely and bears no dates of its own, but what can be unpicked suggests that it refers to the accession of a ruler called Chan Muan. Symbols representing the lower body and thighs of a man were used in later Mayan writing to indicate enthronement, so this text may record the ascension of Chan Muan to the throne of some unidentified kingdom.

We know this thanks to de Landa – who hoped to eradicate all Mayan writing. As part of his bid to prove that the Mayans were just like us but for the light of the gospels, he asked Mayan informants to help him understand their writing. Assuming that they employed an alphabetical system as Europeans did, he asked them how to write *a*, *b*, *c* and so on. Conditioned by their own writing system to imagine these as syllables, the Mayans heard these as 'ah', beh', 'seh' and so on and gave him the corresponding glyphs. Although his method was faulty, it enabled de Landa to record a small but significant portion of the Mayan syllabary, which was invaluable to later scholars.

PRINTERS, READERS AND REBELS

The name of Johannes Gutenberg, alleged inventor of the printing press, was never far from controversy. Following his problems in Strasbourg, Gutenberg did manage to create a wooden press, which in 1440 won him a contract from the Catholic Church to print indulgence slips. The sale of these pieces of paper, purchased by sinners wishing to escape damnation, became a trigger for the Protestant Reformation – and Gutenberg would play no small role in that movement too. This part of his story began in 1450 when, with funding from one Johann Fust, Gutenberg printed the Bible in the town of Mainz. Five years later, a two-volume Gutenberg-printed Bible was on sale, at a price roughly equivalent to three years' salary for a clerk. This may seem steep, but a handwritten Bible, which would take a monk twenty years to transcribe, would have cost much more.

The Bible project must have had potential for profit – because once it was on the market, Johann Fust sued Gutenberg, taking possession of the printing equipment and the nearly completed Bibles. He subsequently went into partnership with Peter Schoffer, a former assistant to Gutenberg, producing more Bibles and the high-quality Mainz Psalter, the first book to bear a printer's trademark and imprint. For his part, Gutenberg attracted further funding from Conrad Humery, a wealthy lawyer and politician, and produced books printed in an elaborate, cursive typeface that resembled handwriting. In 1465 the Elector of Mainz recognized Gutenberg with an ecclesiastical office which came with a comfortable income. Here again, Gutenberg's name intersects with the background to the Reformation, whose proponents vigorously attacked simony, or the sale of church offices.

Scuttling from one patron to another, continually falling out with his business partners, Gutenberg cannot have known what a part he would play in history, nor what an impact his invention would have. By 1500 there were printing presses in 250 European towns and cities that had together produced some 27,000 editions. Just decades previously, scholars might have travelled for months to visit a monastic library containing just twenty-five handwritten manuscripts; now there were an estimated 13 million books in circulation.

Before we examine the impact of this tidal wave of written material, it is interesting to consider a claim which precedes Gutenberg. Citing a book called *Batavia*, written by one Hadrianus Junius in 1575, citizens of the Dutch city of Haarlem maintain to this day that their ancestors invented the printing press. In his book, Junius claims that a householder in Haarlem by the name of Laurens Koster had invented printing 128 years earlier – well before Gutenberg. He had, it was said, covered letters carved from beech-bark with ink and pressed them to paper. Initially performing this as a trick to amuse his grandchildren, Koster adapted his crude typeface into lead and tin and proceeded to print books. Koster had numerous visitors from all over Europe, not all of them benign. One, whom he took as an apprentice, was named as Johann Fust, who later fled with stolen materials and set up his own press in Mainz. But Hadrianus Junius himself conflates the identity of 'Johann Fust' with 'Doctor Faustus', the semi-legendary European figure who sold his soul to the devil, suggesting this account is only folklore.

The true origins of printing were Chinese. In 175 CE a Chinese emperor ordered the six classic texts of Confucius to be carved on stone. Scholars, eager to own these vital texts, took 'rubbings' by laying paper on the engraved slabs, and when subsequent emperors engraved further texts, a substantial library began to be built up. In Korea and Japan, in the eighth century, printers began to circulate copies of religious texts made using woodcuts – in which the white areas of the page to be printed are laboriously cut away from a block of wood, until the remaining parts of the flat surface represent the reverse of the image to be made. This costly and time-consuming process resulted in numerous editions of Buddhist scriptures and the 'Standard Histories' of Chinese imperial affairs. So many woodcut copies of the 'Diamond Sutra' were printed in Japan that many households still own one.

A Chinaman called Bi Sheng invented the first movable typeface, fashioning it from clay. His discovery never caught on because Chinese characters are too numerous and complex, and clay letters cannot withstand repeated pressure. Nor, although Buddhist and

Confucian texts were popular throughout the East, was there a single standard text like the Bible which could guarantee a fledgling printer sufficient returns on his investment. Nevertheless, the idea seems to have influenced Korean printers, who not only began to make durable, movable type out of bronze in 1380, but went on to develop a national alphabet – the *han'gul* – to aid the process. This occurred in 1443, just as Gutenberg was printing his indulgence slips for the Catholic Church. It is not impossible that there might have been some cross-fertilization: if the bubonic plague could travel along the Silk Road from the Orient and into European towns, so could the concept of printing with movable type.

Half a century after Gutenberg fulfilled his contract with the Church, his mass-produced indulgences were at the heart of a new storm. In October 1517 a young monk called Martin Luther angrily nailed ninety-five theses, or propositions, concerning the regrettable practices of the Church to the door of Wittenberg Cathedral. His action stimulated one of the greatest cultural upheavals of the modern age – the Reformation. By the Middle Ages, nailing notices to a church or cathedral door had already become an accepted way to protest or to begin a debate. It was a form of dissent that presupposed a relatively high level of literacy among the general populace. It depended on reading and writing; it also stimulated them.

Luther's protest against the corruption and opulence of the religious establishment centred on the twin notions of *sola fide* and *scriptura sola*. To gain salvation, he argued, one needed only faith – not the bought indulgences or faked relics of a church that ran itself like an empire. The priesthood, he argued, was unnecessary because there was nothing magical about the communion ceremony or the scriptures. This view could never have been put forward without the increasing availability of cheap, printed Bibles and other religious reading matter, and a literate public to absorb them. It was a view which altered European culture for ever, separating church from state and individual from community.

Literacy, of course, was not the only factor in this revolution; but it was a prominent one, with roots and repercussions beyond as well as within the Church. Europe's new abundance of keen readers – who could purchase little devotional texts and chivalric romances

from travelling pedlars for roughly the price of a broomstick or a
night's boozing – were a by-product of a burgeoning urban
economy. This was stimulated by improved techniques in ship-
building and navigation, opening up new continents for trade. The
mass of available printed books and pamphlets required an
improved network of roads to transport them, and, indeed, played
a part in its creation. Among the earliest bestsellers was a 1553
place-map of Europe, and one of the richest families of the era were
the Tassis, who couriered goods – and hence ideas – across Europe,
giving us the word 'taxi'.

Writing was a driving force behind the changes which took place
in Renaissance Europe. The nature of monarchy changed: kings and
emperors tended to stay put, leading by correspondence instead of
moving around their territories in a continuous circuit. Philip II of
Spain spent most of his life within the court at Madrid, a very
different existence from that of his predecessor Charles V – and one
that earned him the contemptuous nickname 'the King of Paper'.
The same pattern became established among the feudal aristocracy:
rarely glimpsed, and now distanced from those they ruled, the
nobles of Europe became increasingly vulnerable to the protests
and mockery now so easily distributed as a result of printing and
literacy. Petitions became a new way of registering protest: notable
instances include the 'Root and Branch' petition, signed by over
10,000 people, that was among the primary causes of the English
Civil War in the 1640s. A century and a half later, the French
Revolution was stimulated by attacks on the monarchy in the
burgeoning print media, and heavy-handed attempts to silence
them.

Books threatened the status quo not merely because they allowed
individuals to soak up new ideas, but also because they preserved
those ideas. An act of criticism or rebellion was no longer limited to
the memories of those who had seen it; it could be transmitted
across great distances, perused, analysed and mulled over. Literacy,
according to anthropologist Jack Goody, who died in 2009, was at
the heart of Europe's dynamism. In non-literate cultures, he argued,
power went to the elderly, those with the longest memory.[6] The lack
of the written word created societies favouring continuity and
repetition. But books give anyone with reading skills the ability to

judge things for themselves. They allow for the accumulation of knowledge outside of anyone's control, and for the constant, critical perusal of that knowledge. If we view history as a long list of other people's mistakes, it is clear that history sometimes stimulates change because writing gives later generations a capacity to review events, replay them time and time again, and find new ways of doing things.

Goody's thesis, however, overlooks the dynamism of many non-literate societies, and overplays the plasticity of some of those that were literate. The Ottoman empire depended for its success upon a class of literate administrators, drawn from the provinces it captured. Yet when the Ottoman Sultan Selim II first heard about printing in 1515, he ordered the death penalty for anyone engaging in it, and it was not until 1726 that the first secular press was permitted. To this day, few societies could be so rigid and backward-looking, so keen to ban books, as Saudi Arabia – and yet Islam, Arab civilization and writing go hand in hand. Meanwhile, in a flurry of activity during the late nineteenth and early twentieth centuries, unlettered peoples in the South Pacific responded to colonialism with quite drastic reorganizations of their own societies. In Fiji and New Guinea, for example, new leaders and prophets within the so-called 'cargo cults' challenged the authority of the elders, outlawed traditional rituals and developed new customs – all without the mind-expanding stimulus of books.

HATED BOOKS, HOLY BOOKS

In every generation, some people have considered writing to be a dangerous technology. Indeed, from time to time – and not all that rarely – books have been regarded as so dangerous as to justify their suppression or destruction. The campaign led by Bishop de Landa is not an isolated example, and it was not only the Mayans who had their cultural heritage suppressed in this way. Writing probably started in Sumer, and Sumer possibly also saw the first book-burning. Military conflict ravaged the Chaldees between about 4100 and 3300 BCE, and there is plentiful archaeological evidence from the sites in and around the ancient city of Uruk of premeditated burning of tablets by soldiers. The hymn to Ishbi-Erra, Amorite king and despoiler of Ur in around 2004 BCE, relays the orders of the Mesopotamian god, Enlil, who possesses the Tablets of Destiny: 'the country to be reduced and the city ruined . . . he had fixed as their destiny the annihilation of their culture'. The 'annihilation of their culture' presumably must have included destruction of written tablets because all over Sumer archaeologists have uncovered pulverized and burnt tablets from various stages of its history. There is some evidence that certain tablets were recycled to make bricks, or hard core for street pavements. Some archaeologists think that over time as many as 100,000 books may have been destroyed in Sumer in these ways.

The earliest known attempt to suppress writing in China is that of the tyrannical ruler Qin Shi Huang (259–210 BCE), famous for building the first Great Wall and commissioning the Terracotta Army. An enigmatic and curious figure, he is regarded as the first Chinese emperor. He claimed he was the son of King Zhaoxiang of Qin, but as his mother's pregnancy lasted for over a year (she was a woman with what might be termed a bit of a history) there is a degree of dubiety about his genetics. Qin unified China, taking draconian measures to quell dissent. He outlawed Confucianism, burying alive 460 followers of this religion, and had all books (except those dealing with medicine, farming and divination) burned.

Jews have repeatedly suffered at the hands of those wishing to burn their books. Since their very earliest times Jews have prized literacy. The admonition about reading and writing the scriptures, 'You shall teach them to your children . . .' (Deut. 11: 19) was observed as a cardinal obligation – and, apart from being a religious necessity, was also seen as intellectually liberating and a democratizing influence. It is expected that every male Jew should write for himself a Torah scroll, and though this is a custom more honoured in the breach than in the observance – after all, it takes even an experienced scribe at least a year of work to complete a single copy of the Torah – the expectation accords a high priority to literacy. Such respect for learning leads to ambitions of equality which are threatening to autocratic governments. So in 168 BCE the Seleucid monarch Antiochus IV ordered all Hebrew books to be rent into pieces and burned in Jerusalem, thus precipitating the Maccabean revolt. And, as I have described in *Human Instinct*,[7] the Romans under Hadrian burned Rabbi Chaninah ben Teradyon, together with a scroll of the Torah wrapped around his body.

Early Christians, too, suffered similar persecutions; so it is a matter of shame that Christians such as Torquemada later destroyed Jewish books in medieval times. At the very beginning of the fourth century the Emperor Diocletian, in an attempt to suppress Christianity throughout the entire Roman empire, had all Christian books burned. It seems a little ironic, therefore, that just a couple of decades later, after the Council of Nicaea in 325, Christians destroyed all books relating to Arius, who argued that God the Father had existed before Jesus.

Perhaps the greatest and most irreplaceable losses of literature were the successive burnings of books in Alexandria. That great library was the main depository of classical writing and was effectively a university campus. Were the books stored there still in existence, they would be of incalculable academic value, because so much ancient knowledge was recorded in them. Because of their destruction, we have only a very partial knowledge of classical literature, particularly from Greece. The complex of buildings had a garden walk, reading rooms, lecture and seminar rooms, and a place to eat. Approximately 100 scholars lived there during their studies, and some even received stipends for their work in lecturing, translating or copying. Books were mainly written on papyrus scrolls and stacked on shelves, and a group of clerks were employed to catalogue the acquisitions. This was one of the world's first research institutions, containing reference books on physics, astronomy, mathematics and philosophy. 'The place of the cure of the soul' was inscribed on a wall above one shelf.[8] The main library at Alexandria was well funded by the Ptolemaic government and charged with collecting the world's knowledge. Being situated at a major port at the end of the Mediterranean, it was ideally placed to have access to many books coming from East and West. It is said that the librarians there were not above finding convincing reasons for confiscating books from ships for archiving in their repository.

Exactly why the library was destroyed by successive fires is unclear. Arson is one possibility; the first attack on it was during the war between Julius Caesar and Ptolemy XIII in 48 BCE and, given that the Ptolemy family had appropriated quite a number of manuscripts to augment the library, it is quite possible that the assault was premeditated. The Roman writer Seneca reports that, whether by accident or design, during the war for the Egyptian throne 40,000 books were burned there. This catastrophe was just the first of a number of attacks on this greatest of libraries over the next 350 years before it was totally destroyed. Alexandria became a lawless place. Almost the last terrible act there was the assassination of Hypatia, the first woman in history to be killed because she conducted scientific experiments. The extraordinarily brilliant daughter of the philosopher Theon, Hypatia taught mathematics and astronomy, was regarded as a distinguished mathematician like her father, conducted astronomical observations, and is said to have invented the hydrometer to measure the density of liquids. Seneca Scholasticus, a Christian historian, said of her: 'On account of the self-possession and ease of manner, which she had acquired in consequence of the cultivation of her mind, she not unfrequently appeared in public in presence of the magistrates. Neither did she feel abashed in going to an assembly of men. For all men on account of her extraordinary dignity and virtue admired her the more.' Sadly, 'all men' was something of an exaggeration. Her prowess and her very forward behaviour (in order to dissuade an ardent suitor, she is said to have held up cloths stained with her menstrual blood to show there was nothing beautiful about sexual desire) outraged the local Christians. Eventually the local monks abducted her from the street, beat her with roof tiles, then cut out her tongue and pulled out her eyes. Once she was dead, they dragged her body to a mound, cut it into pieces and burned it. Most of her writings were also destroyed. An amazingly sexy painting of her, produced by Charles William Mitchell in 1885, hangs in the Laing Gallery in Newcastle upon Tyne.

One of the greatest of all Jewish philosophers was the orthodox rabbi, teacher and physician Moses Maimonides, who died in 1204. He was a prolific author, publishing the most definitive commentaries on the Bible, the Hebrew liturgy and the Talmud. Much of the time he wrote in Arabic, the vernacular of scholarship, rather than Hebrew. One of his most controversial books was

Moreh Nebuchim, or *Guide to the Perplexed*. The book is difficult to read and at times almost seems to be in coded language; this may suggest a fear on the author's part of being considered a dangerous thinker, for the work is a critique of conventional Jewish thinking. In it Maimonides emphasized the value of scientific method and, as a doctor, reflected that Jewish law permits of everything that has been verified by experiment. It is shocking to recall that in 1233 his fellow rabbis burned this book in public in Montpellier, condemning it as heretical.

The history of literature is bedevilled with such instances. The third caliph, Uthman, ordered unauthorized copies of the Qur'an to be destroyed. Abelard was forced to burn his own books at Soissons. In the Languedoc in the thirteenth century all texts of the Cathars were destroyed during the Albigensian Crusade. John Wycliffe's books were burned in Prague by the bishop there. Boccaccio's *Decameron* was burned by Savonarola in 1497. In 1526 the bishop of London, Cuthbert Tunstall, had Tyndale's version of the New Testament burned, and numerous books were burned by the Parliamentary forces during and after the English Civil War in the 1640s and 1650s. In an act reminiscent of the fate which befell Rabbi Chaninah ben Teradyon, Servetus was burned as a heretic in 1553 in the Calvinist city of Geneva because his translation of Ptolemy's *Geographia* was regarded as heretical: 'Around his waist were tied a large bundle of manuscripts and a thick octavo printed book.' The unedifying history of the destruction of literature goes on into more recent times: the Tsarist government in Russia, Germans invading Leuven in the First World War, Stalin's regime, the Nazis, the Iranian government, Senator McCarthy, the US Food and Drug Administration, Muslim opponents of *The Satanic Verses*, fundamentalist Christians in North America, Georgian troops in Azerbaijan, the Chinese government pursuing the Falong Gong, churches in the United States protesting against Harry Potter – all have tried to destroy the printed word.

Particularly distressing, and still in recent memory, was the hatred shown by Serbs as Yugoslavia disintegrated in the wars of the early 1990s. Under the iron hand of Tito, the different cultures had learned to live together in reasonable harmony. But the confederation of states started to collapse after his death and ethnic conflict

surfaced, first between Serbs and Croatians, then between Serbs and
Bosnians. Echoing what had happened in Sumer 4,000 years earlier,
the Serbs embarked on a campaign of cultural annihilation.[9] In
Croatia the libraries of Vinkovci and Pakrac lost close to 100,000
books, and the University of Osijek some 30,000. Vukovar lost all
its rare manuscripts which were 500 years old, and the public library
there had 76,000 books destroyed. The deliberate, merciless
destruction continued with the Library of History and the library in
the Franciscan monastery in the same city. Nor do the horrors end
here. In Dalmatia, an estimated 200,000 books were looted and all
books not in Cyrillic were burned, together with many ancient
manuscripts. Zadar lost 60,000 books by deliberate bombing of
libraries, and Dubrovnik lost 200,000 science books and many more
other volumes besides. The destruction was even worse in the
Bosnian capital Sarajevo, a city rich in history with remarkable
libraries – probably 250,000 books, manuscripts and periodicals
were burned, the Serbs deliberately targeting the library with incendiary shells, leaving other buildings nearby undamaged. Needless to
say, the Muslim community suffered particularly badly.

Speaking as a Jew, I find it particularly shameful to have to record
that only last year, in May 2008, hundreds of copies of the New
Testament were burned by a handful of 'Orthodox' Jewish students
in the Israeli town of Or Yehuda. Perhaps I might be permitted to
observe that the destruction of any religious text is actually illegal
under current Israeli law. One wonders if the authorities could have
done more to prevent this disgusting episode.

Most, though by no means all, of these instances of book-burning
were driven by religious convictions. The notion of 'holy writ', or
'divine scriptures', pervades the world's religions. Orthodox Jews
regard the Torah, the Five Books of Moses, as written by God
personally, the Ten Commandments being engraved on the tablets
by the divine finger. So it seems especially shocking that Moses,
descending from Mount Sinai, broke the first set of tablets with its
holy writing when he saw the Children of Israel dancing around the
Golden Calf.

Writing a copy of the Torah is so holy an activity that a scribe
must spend years in training, and is required to use the utmost care
when writing. He is forbidden to use his memory and must copy

from a model text. Certain letters must have prescribed embellishments, and the spacing of letters and columns is laid down by tradition. Before the actual act of writing – with a quill pen in indelible ink on specially prepared parchment – the scribe must be pure. Most scribes will undertake ritual immersion in water in a *mikveh* before starting work. If, after completion of the scroll, a letter is found to be unclear, an eight-year-old child is asked to read the word in question – itself a testament to the priority attached to early literacy by Jews. This test will determine whether or not the scroll is fit for use. If the scroll is irreparably damaged, it must be buried in consecrated ground. Each letter is holy – when Chaninah ben Teradyon was burned with the scroll, the Talmud reports that the letters flew up to heaven.

Muslims view the Qu'ran in a very similar way. The Qu'ran is received by divine revelation, so cannot be altered in any way. The Arabic in which it is written is regarded as the most pure form of the holy language. There are strict rules to ensure each book containing the Qu'ran is treated with reverence. It may not be placed on the floor, or come into contact with dust; no volume may be placed on top of it. Consequently it is usual to store a copy of the Qu'ran on a top shelf for safety and cleanliness. During its recital, there are formal instructions as to when and where the reader may breathe, which syllables may be stressed, where the reader may stop, and where a congregant may prostrate himself. When the Qu'ran is read, people should not talk, eat or smoke, or make distracting noises.

THE PERILS OF EDUCATION

Some historians are tempted to see the Universal Postal Union, created at the Treaty of Berne in 1874, as the zenith of a trend begun with Gutenberg and Luther. The treaty was impressive, providing for the creation of a flat-rate postal service, uniting millions of citizens from Siberia to Spain by the written word. It is no co-incidence that this came in the same era as the electric telegraph and the steam-powered rotary printing press, both of which revolutionized the speed and quantity of written communication by books and newspapers. At the same time, the governments of Europe promoted mass literacy for their citizens, largely by means of primary-level schooling.

But though bold laws were enacted by a range of governments, they did not necessarily translate into swift action. When the Spanish government assessed its efforts in 1897, it discovered some 2.5 million school-age children were not attending, and the figure was still 1.5 million in 1930. The British government began to subsidize state schooling in 1823, and then to inspect its quality in 1846. Twenty-four years later, it made the right to primary schooling universal and in 1880 it made attendance compulsory, finally declaring it free in 1886. The government, like those of France and Belgium, under-invested, spending between five and eleven times more on law and order than on education.

When governments did finally put their combined might behind the idea of mass literacy, it required effort to gain public enthusiasm. Surveys in the 1870s suggested children were not learning to read and write for the simple reason that they were required to help at home. In Bologna, in 1874, inspectors returned the depressing news that many pupils were unable to attend schools because they lacked suitable footwear, or because the nearest schools were too far from their homes. The government-designed curricula were often too full of 'useless' facts, which angered many parents. To combat understandable opposition to the loss of children's labour during the busiest time of year, governments were forced to alter the school timetable to fit the agricultural calendar – giving the long summer break we still enjoy. In an effort to persuade people of the benefits of reading and writing, the curriculum was streamlined and the method of instruction in literacy changed to allow children to master whole words from the outset, instead of disjointed syllables.

The benefits of mass literacy were consistently over-emphasized. Enlightenment philosophers and historians painted book-reading as the springboard not just for democracy and social advancement, but also for prosperity. In fact, the slow public uptake of education suggests it was seen as a luxury, indulged in by people who were already prosperous. The cities may have contained growing numbers of literate people, but few of them, even until the twentieth century, would have needed to be literate to perform their jobs, unless they were connected with the printing, publishing and message-carrying trades. Furthermore, they were drawn to the cities not because they had had their horizons broadened by books, or because others had written to them of the city's delights, but because they needed work. Their literacy did them few favours in that search. Church registers – a valuable source of information – indicate that, at the end of the nineteenth century, many literate men were in the same occupations as their fathers. In Britain, a sixth of those who signed the register between 1869 and 1875 were actually in lower-paid work than their unlettered fathers.

But literacy and printing did disrupt the status quo; and, just as the Protestant revolution had been fuelled by books and pamphlets, so the religious authorities of those European states newly freed from Catholic sovereignty were quick to crack down on dissent from their own orthodoxy. In Calvin's Geneva, every manuscript had to be submitted to a committee for clearance. Printers deliberately copied the typeface of religious texts for more inflammatory works, hoping to escape close scrutiny by the censors. Attacks on the authorities were often printed under pseudonyms, and published by shadowy outfits claiming to be based in the then mythical city of 'Freetown' or even the Vatican. It wasn't uncommon for contentious material to be published in the form of a pack of cards.

This relationship between reading, writing and rebellion reached a peak in the later part of the nineteenth century, when improvements in communication and printing fuelled the newspaper industry and created the first 'media moguls' – people who'd become rich by nothing more than the sale of information. At the same time, with the burgeoning wealth of that period came increasing dissent. Throughout the second half of the nineteenth century, anarchist groups engaged in what the French activist Pierre Brousse called 'propaganda of the deed'. The prime minister of Spain, the empress of Austria and the king of Italy were all murdered by anarchists seeking to convey a powerful message to the people of Europe. Indiscriminate acts of violence – such as the bombs set off at the Gare St-Lazare in Paris in 1892, and at Barcelona's Teatro Liceo in 1893 – were also calculated to create a climate of panic.

But the total number of casualties from all these events was just twenty-six. Although there were more violent episodes, the list of casualties remained small. The bomb-throwers and assassins, like Conrad's secret agent, depended upon their deeds being known, discussed and fretted over by a group of people far larger than those affected by the immediate detonation of their bombs. They depended upon newspapers to communicate their violent message. The same was true, a little over a century later, when terrorists flew jet planes into the twin towers of the World Trade Center. On that occasion there were vast numbers of casualties, but the true impact was worldwide, flashed across the globe by satellite and internet, those technologies which evolved directly from the written word. If we live in a changed world today, one that is more fearful, less tolerant, one where freedoms are restricted, then our invention of writing has clearly contributed to this condition.

THE DAILY NEWS: POLITICS AND PRACTICAL HARLOTRY

The first publications we would probably recognize as newspapers were handwritten news-sheets published in Venice in about 1560. Somehow it seems appropriate that this romantic but mysterious city, full of light and darkness, a centre for intrigue and gossip, should be the city that spawned the modern journalist. Although news bulletins go back to Roman times, these Venetian bulletins – usually called *gazetti* or *avisi* – were really proper newspapers. Unlike any previous publications, they were produced under the same name on a regular schedule, with journalists writing the copy. They often reported accounts of wars and political events in other parts of the country, and sometimes items of commercial interest. With the spread of printing, newspapers became popular in many parts of Europe, particularly in Basel, Frankfurt, Amsterdam and Vienna. The first newspaper printed in England was published in 1621, and the *Weekly News*, our first regular periodical, was published continuously from 1622 to 1641. English newspaper publishers were more innovative than their counterparts on the Continent, catching the reader's eye with headlines, and sometimes illustrating their copy with woodcuts. Most importantly, they raised revenue by printing advertisements, and proprietors were often sufficiently free-thinking to hire female reporters occasionally. English newspapers were the first to encourage circulation by paying newsboys or newsgirls to peddle their papers in the streets.

There was a widely held opinion, at least among the ruling élites, that the press could destabilize proper government, and consequently censorship was usual. Most European countries regulated news publishing using a licence system. Any printer wishing to print news had to obtain a government licence, and a print shop could be closed down if it published matter which was considered subversive. During the English Civil War, with Royalist and Parliamentary armies roaming the country, the populace became very eager for news and so papers became increasingly important. Several people, notably the poet John Milton, wrote eloquent monographs in support of a free press. Newspaper publishing was tolerated by the Parliamentarians and after 1649, when the war was won, discussion of domestic politics – provided it favoured the Roundheads – was not completely forbidden. However, once the Commonwealth was established that arch-manipulator Oliver Cromwell was determined to ensure that his power would not be undermined and generally gagged the press, allowing only a few authorized newspapers to be printed. Even after the Restoration of the monarchy the press were not allowed to offer overt criticism of the government. In the early eighteenth century parliamentary debates were regarded as secret, and a ban on reporting them was generally enforced by Standing Orders.[10] Some limited accounts of debates were occasionally printed, but only during the parliamentary recess; these reports, taking advantage of some loophole in the prohibition on publishing parliamentary proceedings, seem to have been ignored by the authorities. But in 1738 this loophole was closed by a resolution of the House of Commons and parliamentary journalism was halted. In 1747 the House of Lords took action over unsanctioned reports of the treason trial of Lord Lovat, who had offered duplicitous support to the Jacobite cause in Scotland.[11] Two printers were reprimanded, but released on condition that their offence against the Upper House was not repeated. In February 1760 the Speaker of the Commons complained that four newspapers contained

'Accounts of the Proceedings of this House, in contempt of the Order, and in Breach of the Privilege of this House'. The printers were ordered to attend at the Bar of the House. The breach of privilege seems pretty mild, to say the least; it turned out that the offending section in the issue of the *London Chronicle* merely reported a formal vote of thanks on behalf of the House by the Speaker to Admiral Hawke for his naval services, and Hawke's reply. The printers attended at the Bar, confessed and were found guilty; they were lucky to get a mere reprimand from the Speaker.[12] Eventually it was commercial competition between newspapers which drove publishers to challenge the embargo on reporting parliamentary proceedings.

All this seems very odd, because nowadays most of the news-paper-reading public finds our parliamentary proceedings so deadly dull that such reporting as appears in print is for the most part flippant, facetious or factually inaccurate (sometimes all three).

London was always the centre for British newspaper publishing. By 1782, 18 newspapers were published in the capital city, and 61 others throughout the British Isles. Thereafter the growth was phenomenal. By 1809 there were 217 British newspapers, of which some 63 were published in London, and by 1840, according to stamp tax returns, London published no fewer than 109 papers out of a national total of 493.[13] The circulation of newspapers was equally impressive. By 1840, well over 4 million copies of news-papers were read in the county of Yorkshire and over 3 million in Lancashire. In Edinburgh 1.7 million newspapers were circulated, and – perhaps most remarkable of all, given the size of that city – in Dublin, with a population of around 380,000 people, well over 3 million. Printed news was becoming an increasingly influential medium, not merely in Britain but all over the globe. The country with the biggest newspaper industry by this time was the United States. The first state in which a newspaper was produced was prob-ably Massachusetts, in 1704; by 1840, an estimated 100 million copies of newspapers were being published across the country every year. The owner of a newspaper has a different status from that of the owner of any other business, with apparent power over the minds of men and women. The story of Vavilov may seem extreme by our standards, but it is not just Stalinist totalitarian governments that try to manipulate what people think. Because free countries pride themselves on having a free press, a newspaper owner in a democracy considers that he or she can decide not merely what people should think, but also what they should think about. And the more liberal the country, the harder it is to regulate this power over citizens' minds.

One of the first examples of this was the case of Alfred Harmsworth, who was able to use his newspaper empire to control political decisions of the day, and even to bring down a relatively stable government. Harmsworth was no great shakes at school, where the only promise he showed was in editing the school magazine. After leaving – he did not attend university – he had various junior jobs in journalism before, in 1888, joining his brother Harold to found a popular magazine called *Answers to Correspondents*. The editors promised that any question sent in would be answered by post, and the more titillating answers had the privilege of being published in the magazine. Within four years they were selling over a million copies a week. Publication of *Comic Cuts* and a woman's magazine, *Forget-Me-Nots*, followed. But Alfred's real breakthrough came with the purchase in 1894, for a knock-down price, of the virtually bankrupt *Evening News* – his first proper newspaper. Eye-catching or lurid stories were presented to the readership under headlines like 'Another Chelsea Murder', or 'Killed by a Grindstone'. Within a few months the circulation topped 400,000 copies and two years later this had been doubled.

Harmsworth's next triumph was the *Daily Mail*, selling for just one-halfpenny a copy. It was the first paper in Britain catering for a new public that wanted something simpler, shorter and more read-able than the boring papers which did not put news on the front page. It introduced the banner headline, and sport and human-interest stories were given particular prominence. Robert Cecil, Lord Salisbury, the Prime Minister of the day, sneered that it was 'written by office boys for office boys'. This hardly stung Harmsworth, who was laughing all the way to the bank – and further: for, with his paper enjoying a circulation of 500,000, his influence was growing. He took a patriotic stance (in which he probably genuinely believed), telling the public that the *Mail* stood 'for the power, the supremacy and the greatness of the British Empire'. Eventually he changed the size of the newspaper from a broadsheet, giving birth to Britain's first tabloid. The *Daily Mirror*, a newspaper aimed at female readership, followed in 1903. This was not a great commercial success at first; for one thing, women did not have the purchasing power of men, and the sales increased sevenfold when his managing editor made it into a picture paper for both men and women.

Harmsworth was always looking for new angles, and the cult of celebrity was not alien to him. He was the first proprietor to recognize that stories depicting the royal family sold copy. Having refused a knighthood as being beneath him, he accepted a baronetcy in 1904. A year later he was elevated to the House of Lords as a baron, ultimately becoming Viscount Northcliffe. The new peer relentlessly increased his influence, purchasing the Sunday *Observer* and then his crown jewel, *The Times*.

Harmsworth was deeply interested in technology, passionate about engineering, motor cars and aircraft. So enamoured of the automobile was he that he did not allow reporting of traffic accidents in his newspapers for fear of giving this new form of transport bad publicity. Very early in the history of flying machines he came to the view that foreign aircraft posed a threat to island Britain, and as early as 1909 he campaigned for defences against the possibility of bombs being dropped on London. He had been suspicious of German intentions as early as 1897, but failed to win support from the government of the day. His forthrightly expressed concerns about the rising might of Germany, against whom, he said in a *Mail* editorial in 1900, England would lose a war, resulted in his being accused of being a warmonger. Once war did indeed break out in 1914 a rival paper, the *Star*, proclaimed that, 'Next to the Kaiser, Lord Northcliffe has done more than any living man to bring about the war.' He continued to agitate about what he felt was the incompetent British conduct of the war, and wrote devastating criticisms of Lord Kitchener, the war minister. As Kitchener was very much a national hero the circulation of the *Mail* dropped dramatically and Northcliffe was reviled in many quarters; at the London Stock Exchange 1,500 members joined forces to burn copies of his newspaper in public. He was equally forthright in his justified criticism of the Gallipoli campaign – 'forty thousand killed, missing or drowned; three hundred millions of treasury money thrown away' – arguing that the real need for Britain was to concentrate on the war in France. After Kitchener's death at sea two years into the war, he focused on bringing down the Prime Minister, Herbert Asquith. When Asquith eventually resigned in December 1916, largely as a result of this media pressure, his successor Lloyd George felt it would be wiser to have Northcliffe inside the government tent and invited him to become a cabinet minister. Northcliffe refused – he preferred to retain his influence, and his freedom to criticize the government.

Lloyd George clearly hated Northcliffe, describing the press magnate in a strictly confidential letter to his parliamentary private secretary as 'one of the biggest intriguers and most unscrupulous people in the country'. It was only in 1918, when Northcliffe was wooed by Lord Beaverbrook – another powerful press magnate and owner of the *Daily Express* – that he agreed to join the cabinet and became Minister of Propaganda. Such was Northcliffe's skill in smearing the conduct of the Germans that German ships were deployed to shell his house on the Isle of Thanet in an assassination attempt. Sadly, Northcliffe's gardener's wife and daughter were killed in this attack. Even when the war was won Northcliffe still refused to support Lloyd George, and continued to wield considerable power over politicians until his death in 1922. Stanley Baldwin, who became Prime Minister in 1923, may have been thinking of Northcliffe when he accused newspaper proprietors of 'exercising the prerogative of the harlot through the ages: power without responsibility'.

Press barons at the turn of the twentieth century such as Northcliffe, Beaverbrook and, in America, Randolph William Hearst, and more recently Robert Maxwell, Rupert Murdoch, Silvio Berlusconi and Conrad Black, may well have exercised great influence in the politics of their respective societies. But how important is the press when a totalitarian regime is in power?

In January 1933 the German cabinet had only three Nazis in it, yet within three months Hitler had consolidated his political power by legal means and with the support of the German public. Clearly, people could have been too afraid to rebel against the Nazis at this stage, or they could have been convinced of the value of what they offered. There is no doubt that Nazi propaganda, particularly through the press, played an important role. But the extent to which newspaper reporting contributed to the consolidation of the National Socialist regime is still open to question.

The historian Sir Ian Kershaw, writing about the Nazis, points out that 'the regime would attach a high priority to the steering of opinion'.[14] The Nazis did little market research and there were few, if any, non-Gestapo polls to analyse whether or not their propaganda was having the anticipated impact, aiming as it did simply at what Goebbels, the Reichsminister for Public Enlightenment and Propaganda (which gave him total control of the press, publishing, radio, cinema and the arts), called 'a mobilization of mind and spirit in Germany'. To this end, Goebbels used the technique of 'total propaganda' to create the image of Hitler as a Messiah who would be Germany's saviour. Total propaganda meant government control not only of the press, but also of culture. Goebbels was clever enough to realize that without variation people would soon get tired of the same message. Therefore he used his control of newspaper reporting in a subtle fashion – the style of the reporting was un-altered, but now all the newspapers promoted fascism. Any person who produced, distributed, broadcast, published or sold any form of written work, whether journalism or literature, had to follow all the rules laid down by the Goebbels ministry. Of course, all Jewish newspapers were prohibited. Without a licence to practise, no writers, publishers or producers could get work. With the elimin-ation of all anti-Nazi publications, the public must have felt that the general mood of their fellow citizens was pro-Nazi. So by disagree-ing they would step out of line, and the terror of not conforming helped consolidate power in Nazi Germany.

David Welch, Professor of Modern History at the University of Kent, agrees that propaganda played an important part in mobiliz-ing support for the National Socialists in opposition and keeping them in power once they had gained it. But he also argues that propaganda alone could not have sustained the Nazi Party and its ideology over a period of twelve years.[15] There is now considerable evidence that Nazi policies and propaganda reflected many of the aspirations of large sections of the German populace. Propaganda, he asserts, is as much about confirming as converting public opinion and, if it is to be effective, must preach to those who are already partially converted.

So, if Professor Welch's analysis is correct, it is politicians who are most concerned with the press. Newspapers may have their biggest impact on the way politicians think and behave; their impact on the voter may be rather less. And certainly, in Britain at least, there is some evidence that their influence with the public is open to doubt.[16]

In the closing days of the general election campaign in 1992 it was widely expected that Labour would win. Britain, with John Major at the helm, was in recession, and pretty well all experts concurred: if Labour did not win outright, at least there would be a hung parliament. On 1 April, just eight days before the election, the opinion polls published evidence of a substantial lead by the Labour Party. Even though this fell somewhat towards polling day, Labour under Neil Kinnock's leadership remained confident that it would win. On the morning of polling day, Rupert Murdoch's *Sun* had a photograph of Neil Kinnock's head inside a domestic light bulb on the front page. Taking up virtually the whole of the rest of the page, a banner headline read: 'If Kinnock wins today will the last person to leave Britain please turn out the lights'. On page 3 – above the space normally reserved for something voluptuous – the headline read: 'Here's how page 3 will look under Kinnock!' and featured an overweight woman. In the event, there was a massive turnout (77 per cent) at the polls and Labour gained a small swing – not anything like enough to win, or even to produce a hung parliament. Subsequently the *Sun*'s famous analysis of the election results was headlined: 'It's the *Sun* wot won it'.

Whether Rupert Murdoch's campaign seriously influenced the result of the 1992 election is doubtful. But there is no doubt at all that the *Sun*'s headlines greatly influenced the Labour Party inner circle when it came to planning Labour's campaign for the 1997 general election. In the run-up, Labour did an enormous amount to curry favour with the press. It was very concerned about its image – even to the extent of changing its name to 'New Labour' in 1994–5. It also made a concerted effort to ensure that members of the party were 'on message' when talking to the press. To help orchestrate the presentation of an attractive face to the press, and in attempts to get more positive press coverage for the party, in 1994 Tony Blair invited Alastair Campbell to run his press and communications office. Campbell had been a close adviser of Neil Kinnock and had worked closely with Robert Maxwell. He had worked at the *Mirror*, and been political editor of *Today*.

In March 1997 John Major, the Tory Prime Minister, announced that a general election would be held on 1 May. He hoped that some favourable press coverage might improve his apparently hopeless position in the opinion polls. But soon after the announcement Blair's and Campbell's careful work on the editorial staff of the *Sun*, and their wooing of Rupert Murdoch, bore fruit. The *Sun*, previously always thought of as a Tory newspaper, announced that it was backing Labour.

The *Sun*'s proclamation was but one of a number of press endorsements that New Labour obtained. The *Star* had been strongly pro-Conservative in 1992, and the *Independent* neutral. Both now declared for Labour. It has been calculated that in the run-up to the 1997 election more than twice as many people were reading a newspaper that backed Labour as were reading one that supported the Conservatives. Whether or not this was a key reason for the massive Labour victory is open to question. I think that it probably was not. But what it did do was to establish Alastair Campbell and his team of spin doctors in a very strong position, and in the ensuing years of the Labour government, management of communication with the press was strictly enforced by the press staff in Downing Street wherever possible. I remember being at the wrong end of a lash when I – taking the Labour whip in the House of Lords – broke silence and gave a press interview in January 2000 because of widespread concerns that the NHS was grossly under-funded. The vicious bullying I experienced from Alastair Campbell at the end of a telephone may seem quite trivial in the general scheme of things, but his venomous behaviour was a clue as to how far our politicians and their henchmen might be prepared to go to ensure that the press get the 'right message'.

How true is it that powerful press magnates are successful in telling people what to think in liberal democracies like ours? Do the public alter their views after reading about political issues in the newspapers? Generally speaking, until 1997 Britain's partisan press, with a few exceptions, had always favoured the Conservative Party, and when Labour has won general elections it has done so without the populist newspaper support that Labour leaders curried after the disastrous election of 1992. Newspapers undoubtedly have some influence on the way that people decide to vote, but probably rather less than our political leaders believe. And it should not be forgotten – here is an echo of what happened in Germany in the 1930s – that people tend to read and approve of a newspaper that chimes with their existing views.

But the belief politicians have in the power of the press, and their attempts to manipulate the media in consequence, have probably done serious damage to our democracy. Just six years after the 1997 election, there were attempts to manipulate press coverage with far graver repercussions. The attempts to put a spin on the case for an invasion of Iraq, the false claims that Iraq had a vast stockpile of weapons of mass destruction (as if Britain and America didn't) and Alastair Campbell's complaints that the BBC reporting was heavily biased have left a deep scar. Leaving aside an analysis of the effect of the invasion on the people of Iraq and on Middle East politics, within Britain itself trust in politicians and belief in their ability to make mature decisions in the national interest has been seriously undermined. And that world leader in public service broadcasting, the BBC, previously always a jewel of factual and international news coverage, seems to have become increasingly defensive following its battering by government – led to a large extent by Alastair Campbell. Many people consider it has been in decline since that episode, with a serious loss to the culture of Britain.

A few years ago, during Tony Blair's government, I travelled out of London from King's Cross with an acquaintance of my mine, who was then a junior minister. It was early in the morning, and for a sizeable part of the journey she pored over photocopied press cuttings from every single newspaper published in London, and one or two local ones as well. Every so often she marked a cuttings with a pencil for reference later. On my asking, she looked rather embarrassed. She told me this was a daily ministerial chore and that one or more of her civil servants prepared these cuttings for her. Apparently this was pretty well routine in every department of government. What was particularly depressing was to see how entirely trivial were many of the comments which my ministerial friend had marked or underlined. Such slavish attention to the press can warp good judgement and may lead to the corruption of good government.

As it happens, conventional newspapers seem to have had their day. Circulations are dwindling in both the United States and Britain as more people turn to the internet and to satellite and cable television reporting. More members of the public, particularly younger people, are keeping in touch with events using 'unofficial' electronic sources such as webcasts and blogs. All the main newspapers are finding it essential to increase their commitment to the World Wide Web. Is the dangerous activity of writing now more threatening in its electronic form?

Chapter 6
Digital Communication

THE HISTORY OF writing was changed by a nineteenth-century convict, John Tawell, though he would have had good reason to regret his contribution afterwards. Until the late 1830s written information was delivered to the reader by hand, sold as a book or journal, or sent by post. It was to be read at leisure, or occasionally rather more speedily if it was contained in an urgent document. As a young man Tawell, who had periodic religious urges throughout his life, was admitted to the Society of Friends in Suffolk and then moved to the East End of London where he worked in a pharmacy, giving him professional experience that turned out, unfortunately, to be useful in an amateur capacity in later life. Perhaps he did not live up to the high moral standards to which the Quakers generally aspire, for when he was twenty-two years old he seduced a servant girl in Whitechapel. Under pressure to support a family and strapped for cash, he was convicted of forging a £10 note and sent to Australia as a convict. After a spell working in New South Wales as a clerk in the Sydney Academy – apparently an exemplary employee – he was granted an official pardon by the Governor which allowed him to return to London, where he married. In 1838 his wife died (the precise circumstances are not recorded) and three years later he married a Mrs Cutforth, a Quaker widow and school-teacher. Meanwhile, Tawell had been conducting an ardent liaison with one Sarah Lawrence, whom he had apparently promised to marry – possibly while he was still married to his first wife. He set up a house for Sarah in Salt Hill, near Slough, and she changed her name to Hart.

One day in January 1845 Sarah Hart was found dead in the house; she had drunk a quantity of brown stout, some of which contained a liberal admixture of Scheele's prussic acid – a poison of which, it was later established, Tawell had purchased two bottles before the last of his regular visits. A neighbour had noticed a man leaving the house in Salt Hill earlier that day 'dressed like a Quaker'. The police were called and a prompt piece of detective work led them to the newly built Slough railway station, where a man answering this description had been seen to buy a first-class ticket and board the 7.42 p.m. train to Paddington. With considerable excitement the stationmaster announced that his staff could assist the police, and the recently installed electric telegraph was proudly called into action. Unfortunately, this new device could signal only twenty letters of the English alphabet; among the six letters for which there was no electric switch were 'J' and 'Q'. So the signal-clerk at Slough improvised the following message:

A MURDER HAS GUST BEEN COMMITTED AT SALT HILL AND THE SUSPECTED MURDERER WAS SEEN TO TAKE A FIRST CLASS TICKET TO LONDON BY THE TRAIN WHICH LEFT SLOUGH AT 742 PM HE IS IN THE GARB OF A KWA . . . KER WITH A GREAT COAT ON WHICH REACHES NEARLY DOWN TO HIS FEET HE IS IN THE LAST COMPARTMENT OF THE SECOND CLASS COMPARTMENT [sic].

His first few attempts were baulked: every time he got to the letters 'KWA . . .' the clerk at Paddington sent a message by return – 'repent'– until eventually a young lad in the Paddington office suggested that the signalman at Slough be allowed to complete his message. As it turned out, it was sent only just in time. Tawell was spotted alighting at Paddington and trailed by a detective to a local coffee tavern where he was subsequently arrested. Convicted for murder, he was hanged in public at Aylesbury.

Such notoriety attached to the circumstances of this case, centring on the novel use of the electric telegraph, that this newfangled apparatus, which until then had hardly been used, enjoyed extensive publicity. It had been invented by one William Fothergill Cooke, who teamed up with the eminent physicist Sir Charles Wheatstone, Professor at King's College and a Fellow of the Royal Society, to patent the device in 1837.[1] It consisted of five galvanometer needles which, when the current was switched one way or the other, could point to a group of four letters aligned on a lozenge-shaped panel. Because six letters could not be sent (the others being U, X and Z),[2] and operation of the telegraph required considerable skill on the part of its operator, within a very few years this device was superseded by Samuel Morse's telegraph, which received a patent in 1840. Morse died a very rich man in New York city; William Fothergill Cooke, although he amassed a great sum of money and was rewarded with a knighthood, died in penury in 1879.

It is extraordinary to think that we have been able to send instantaneous messages for over 160 years. Rather more remarkable is that the first patent to send, not merely letters, but actual images using electricity was registered in London by a Scot, Alexander Bain, in 1843. The unedifying story of the facsimile machine demonstrates an important theme which runs through this book – namely, that scientists can be just as unscrupulous as other people and may behave aggressively, particularly when in competition with one another. Competition arises most frequently when there is pressure to claim to be first with an idea. Sometimes there may also be conflicts of commercial interest, as there was in the case of Alexander Bain's revolutionary ideas.

Bain showed no academic prowess at school and, coming from a large and very poor family, was apprenticed to a modest clockmaker in Wick. Having learned the trade of clockmaking, he moved first to Edinburgh and then to London. By 1840 Bain was working with John Barwise, a maker of chronometers, at his premises in Wigmore Street. Here he came up with the novel idea of a clock that was driven by an electro-magnetic pendulum, kept in motion by an electric current instead of the usual spring or weights. Later, he had the ingenious notion of collecting the electricity by burying the positive and negative plates – made of zinc and copper – in the ground. Desperate for money, he showed his design, together with several other ideas, to Charles Wheatstone, from whom he hoped to get advice on how to exploit his inventions. Sir Charles is said to have dismissed Bain, saying, 'Oh, I shouldn't bother to develop these things any further! There's no future in them.' Just three months later, at a meeting of the Royal Society, Wheatstone demonstrated an electric clock that he claimed he had invented himself. But Bain had already filed a patent application. Wheatstone used his position and influence in an attempt to block Bain's patents, but failed. Later, when a bill to set up the Electric Telegraph Company was being considered in Parliament, the House of Lords asked Bain to give evidence. The bill was passed but the company was required to pay Bain £10,000, a vast sum by modern standards and enough to ensure Wheatstone's resignation as a director.

Bain's facsimile machine used his electric clock to synchronize the movement of two pendulums in line-by-line scanning of a message. To transmit the image, Bain used metal pins arranged on a cylinder made of insulating material. An electric probe that transmitted on–off pulses then scanned the pins. The message was reproduced at the receiving station on electrochemically sensitive paper impregnated with ammonium nitrate and potassium ferrocyanide, which left a blue mark when an electric current was passed through it. It was not easy to synchronize the clocks at the transmitting and receiving ends, and this early fax machine was not really widely used until 1906, when newspapers started to use the principle to send photographic images over long distances; Bain was a genius before his time. He made another contribution to instantaneous long-distance messaging: the 'ticker-tape machine', which employed a punched paper ribbon to send messages. The ticker-tape was still in use at the end of the twentieth century, by which time it had given birth to the teletype (or telex) machine, based on the same design.

Once we worked out how to send images and pictures instantaneously, we had a technology that changed the nature of human communication for ever. We now live in a connected world which would have been impossible to imagine at the time of Gutenberg, and was never remotely envisaged by Cooke, Bain, Morse or

Wheatstone. That connection has been hugely valuable, but holds a menace which grows as digital technology expands at an exponential rate.

ELECTRONIC LINEAR ABC

tmot atm afap &t+. L%k@w3 uwc np. Rumof?- lh62nte cos imb w/my ps? 831 x x x

Well, it isn't Linear B, but I wouldn't have had the foggiest about this either until recently. It's text language, meaning: 'Trust me on this at the moment as far as possible and think positively. Look at the World Wide Web and you will see no problem. Are you male or female? Let's have sex tonight because I am bored with my Playstation portable. I love you [*831* is 'eight letters, three words and one meaning'] many kisses.'

There are various estimates – all of them probably rather in-accurate – about how many text messages are sent each day. About 17 billion were sent worldwide in 2000 and just one year later the number was said to be up to 250 billion. One source suggests 500 billion SMS (Short Message Service) messages were sent in 2004, but this does not seem quite right: if the figure rose nearly fifteenfold in one year at the beginning of the decade, it seems surprising that it did no more than double over the following two years. Another source reports that 1.2 trillion messages were transmitted in 2005. In that year, the revenue generated was well in excess of $100 billion, and roughly 100 text messages are now sent for every person in the world. It may be slightly more accurate to record the position in the UK, where in 2008 roughly 5,000 text messages were sent every second. The figures go up during Christmastime and at the New Year, when something close to 300 million texts are sent. About 58 million picture messages are sent annually in the UK, and this is increasing as broadband becomes more widely available.

Text messaging can become pathological. Dr Howes, a psychiatrist colleague at the institution where I work, Imperial College London, describes the sad case of a young woman who used text messaging to stalk the object of her love.[3] After he ended their relationship, she sent her ex-boyfriend text messages repeatedly asking him to meet her, chastising him for leaving her, and expressing love for him. He told her to stop contacting him, but to no avail. Eventually he took legal action, but still the texts kept coming, increasing until she was spending four hours every day sending up to thirty or forty messages. Later she said that only when she had sent him a message were the tensions in her relieved. She ran up phone bills of over £100 per month and almost lost her job through poor work performance. Finally she sought psychiatric help, and after extensive behavioural therapy and a long course of drugs, the intervals between texting eventually increased until her compulsive behaviour finally ceased.

A recent report by researchers at the Queensland University of Technology has claimed that text messaging can be as addictive as cigarette smoking.[4] They reported problems in some individuals including depression, anxiety and changed sleep patterns. This fairly sober and carefully written report generated some rather grotesque and overblown headlines in the press suggesting that text messaging could become the most important addiction of the twenty-first century – typical of the lurid reporting of some scientific conferences, especially those on topics with a social context. Nevertheless, it is clear that in a small section of the population there is a growing trend of compulsive messaging. As Dr Mark Collins, the head of the Priory Clinic's addictions unit at Roehampton, points out, 'There has been a big rise in the number of behavioural addictions, and many involve texting.' One California newspaper reports how Greg Hardesty, a parent in California, did not LOL when he got his daughter Reina's phone bill.[5] In January 2009 this thirteen-year-old set what appeared to be a world record: a bill for 14,528 text messages in a single month, producing an online statement from the telecoms company AT&T that ran to 440 pages. This works out to approximately 484 text messages a day – one every two minutes of every waking hour. After her father complained, she sent text messages to her friends to boast about the number of text messages she had billed.

Text messaging has created an easier way to cheat during exams. In December 2002, at the University of Maryland, twelve students were caught cheating by this means during an accounting exam, and in the same month in Japan twenty-six students from Hitotsubashi University were failed for receiving e-mailed exam answers on their mobile phones. Many universities in the UK now ban mobile phones from examination halls.

Advertisers are now jumping into the text message business, and what is to many an irritating intrusion seems to be surprisingly effective. The broadcasting media, too, are using text messaging, and competitions such as the Eurovision Song Contest are now largely decided by viewers' votes cast in this way. Text messaging is particularly popular in Finland, where it first started, and the Finnish author Hannu Luntiala has recently published a novel, *The Last Message*, in which the narrative consists entirely of text messages. It tells the story of a fictitious IT executive, Teemu, who resigns from his job and travels throughout Europe and India, keeping in touch with his friends and relatives only through text messages. Step by step, it becomes clear that he perhaps is escaping from something, hiding something, and that he is not coming back at all. Each incident is limited to 160 text characters. Teemu's messages, and the replies – roughly 1,000 altogether – are listed in chronological order in the 332-page novel.[6] The texts are full of the abbreviations commonly used in regular SMS traffic – and of grammatical errors.

Text messaging in Finland has penetrated not only the most private levels of communication but the most elevated levels of society. The Prime Minister, Matti Vanhanen, regularly communicated by text with his girlfriend Susan Kuronen, whom he had apparently met on an internet dating site. A typical loving late-night message sometimes read 'Good night from the Chief Executive'. One wonders whether Ms Kuronen suffered from insomnia. Finally the Prime Minister broke off their nine-month relationship with a text. Sadly, exactly what he said in the message does not seem to have been recorded – apparently this is an invasion of privacy just a little too far by Finnish standards – but one rumour is that it read simply: 'That's it'. His ex-lover tried to get her own back by a more conventional method, publishing a kiss-and-tell book designed to embarrass Vanhanen; but his popularity soared as the affair became public, with 49 per cent of Finns wanting him to continue leading

the country, compared to only 11 per cent in favour of his nearest
political rival.

The taciturn Finns seem to accept text messages as a tool to
communicate even in the most private matters – though some have
reservations: the Finnish foreign minister Ilkka Kanerva was sacked
in January 2008 over a scandal concerning suggestive text messages
he had sent to an erotic dancer he met on a plane. The minister, aged
sixty, had sent around 200 messages to the 29-year-old leader of the
Scandinavian Dolls dance troupe. Jyrki Kateinen, the leader of
the conservative National Coalition Party, is reported to have
commented that 'Kanerva has shown a great lack of judgment . . .
he doesn't enjoy the full trust which a minister needs.'

The dangers of texting can be far worse than embarrassment or
even the loss of a career. On 12 September 2008, at around 3.40
p.m., a Metrolink commuter train left Union Station in Los Angeles,
the engine pulling three coaches on its way to Ventura County. After
leaving Chatsworth further down the line about forty minutes later,
it was carrying 222 passengers. The Metrolink engine is calculated
to have weighed approximately 113,000 kilograms and was
travelling at around 60 kph when it hit a Union Pacific freight train
pulled by two engines, weighing 227,000 kilograms. Travelling at
roughly the same speed in the opposite direction and emerging from
a 200-metre tunnel, the driver of the freight train applied his brakes
vigorously two seconds before impact. But it seems the driver of the
passenger train did not brake at all. The trains collided head-on,
resulting in carnage. Although legal proceedings continue, it is now
established that the driver of the passenger train, Engineer Sanchez,
had gone through a red light at the beginning of the single-track
section. Earlier in the day, just before leaving the station in Los
Angeles, Mr Sanchez used his mobile phone to order a roast beef
sandwich from a restaurant in Moor Park, where he was due to
terminate his last journey. On an earlier run that morning it is
established he had sent or received some forty-five text messages
while at the controls of his engine. Even after leaving Chatsworth
station on the fatal leg of the journey, Bob Sanchez is alleged to have
sent some five text messages, the last one – twenty-two seconds
before impact – to a railway enthusiast with whom he was
corresponding: 'yea. usually @ north camarillo'. Twenty-five

people were killed that afternoon and many more severely injured.

There is no doubt that mobile phones can be a lethal distraction. One growing problem is that more people – especially young people – are driving a car and using text messaging simultaneously. The UK's Road Transport Laboratory has pointed out the hazard in doing this and any use of a mobile phone while driving is now illegal in Britain, but the legislation hasn't stopped drivers who are otherwise perfectly responsible from responding to text messages at the wheel. There have been a few fatal accidents in the UK as a result, and a prison sentence is now inevitable for drivers who have been found guilty.

TEXT-SPEAK TO NEWSPEAK?

Both acknowledged and self-appointed experts on the use of language have waded into the argument about whether the use of 'text-speak' is damaging the mental competence of young people and their ability to express themselves clearly, and the wider, related controversy about the extent to which the English language is becoming impoverished. In the UK, the debate was opened by the Emeritus Lord Northcliffe Professor of Modern English Literature at University College London, the very eminent Professor John Sutherland. He is on record as arguing that texting is 'bleak, bald, sad shorthand. Drab shrinktalk ... Linguistically it's all pig's ear ... it masks dyslexia, poor spelling and mental laziness. Texting is penmanship for illiterates.'[7] Perhaps unsurprisingly, that wonderfully curmudgeonly journalist, my friend John Humphrys, was at his predictably provocative best when offering his support to Professor Sutherland. In an article in the *Daily Mail* headlined 'I h8 txt msgs: How texting is wrecking our language',[8] he wrote that texters are 'vandals who are doing to our language what Genghis Khan did to his neighbours 800 years ago. They are destroying it: pillaging our punctuation; savaging our sentences; raping our vocabulary. And they must be stopped.' Since then various other literary authorities on the English language (well, at least they write it pretty well – except perhaps when they are railing against texting) have come out in support.

Notable champions of the purest English are Will Self and Lynne Truss.[9] In an article criticizing text messaging in the *Guardian*, Will Self starts by promoting his own book, the slightly forgettable novel *The Book of Dave*. He says:

> I employ some of the orthographic conventions – if they can be so dignified – of soi-disant text speak. I did it not because I believe in the permanence of this script, or even because I think of it as having lasting significance – what could such a thing imply? But merely in order to counter what I term the *Star Trek* convention.

While I am sure some literary scholars might regard this opening as a paragon of clear expression, I find myself as a mere scientist at a bit of a loss in understanding exactly what he is getting at. He goes on: 'I can't help finding the bowdlerisation of texting quite insufferable. I'd rather fiddle with my phone for precious seconds than neglect an apostrophe; I'd rather insert a word laboriously keyed out than resort to predictive texting for a – acceptable to some – synonym.' His lovely colleague Lynne Truss, who admits to being a text addict, is also somewhat scathing about its effects: 'Texting is a fundamentally sneaky form of communication.' She at least doesn't take herself quite as seriously as the Good Will. She admits that etiquette forbids composing quick replies to text messages while in company, as it is quite rude to fiddle with a phone. Apparently, when the call comes, she excuses herself as soon as she can and texts feverishly from the nearest lavatory. This seems less socially offensive, even if she has to leave the room for at least twenty minutes (what with all the spelling out of long words, her most meticulous punctuation and so on).

But is there genuine evidence that texting is reducing our ability to communicate properly? David Crystal, an honorary professor and lecturer in linguistics at the University of Bangor, argues that much of the criticism is misconceived, and doubts that texting is harmful or causing a deterioration in the English language. There is, he admits, some evidence that we are less obsessed than we used to be with correct spelling, but does not believe this is a texting phenomenon. He cites the Bullock Report of 1975 as a good example. More than thirty years ago, grave concerns were expressed that children were no longer being taught to spell and write properly.[10] He points out that the reality is that people have always had a tremendous fear about the impact of new technology on language. When the printing press was first invented, he asserts, 'people thought it was an instrument of the devil that would spawn unauthorised versions of the bible ... Text messaging is just the most recent focus of people's anxiety; what people are really worried about is a new generation gaining control of what they see as their language.'

Surely the jury is still out. Fairly obviously, texting is too new a phenomenon for enough good research to have been done on its long-tem effects. Languages evolve over considerable time, and the mobile phone is a new piece of technology. But this is a genuinely important issue. In *Nineteen Eighty-Four*, the gripping dystopian classic by George Orwell, the totalitarian government falsifies records and propagandizes.[11] One of its chief tools is Newspeak, the government's official language. A corrupt form of English, Newspeak has a reduced vocabulary which gives the proletariat fewer intellectual symbols with which to think. Such terms as 'honour', 'justice', 'morality', 'democracy', 'science' and 'religion' have ceased to exist. If human inventiveness is down to language, as I suggest in chapter 1, then surely this is an excellent way for the state to repress its citizens in the future. The rebel in this novel, Goldstein, says of Newspeak:

> The keyword here is blackwhite. Like so many Newspeak words, this word has two mutually contradictory meanings. Applied to an opponent, it means the habit of impudently claiming that black is white, in contradiction of the plain facts. Applied to a Party member, it means a loyal willingness to say that black is white when Party discipline demands this. But it means also the ability to believe that black is white, and more, to know that black is white, and to forget that one has ever believed the contrary. This demands a continuous alteration of the past, made possible by the system of thought which really embraces all the rest, and which is known in Newspeak as doublethink. Doublethink is basically the power of holding two contradictory beliefs in one's mind simultaneously, and accepting both of them.

Newspeak has some similarities to texting, notably in its reduction of English to a small set of terse expressions. Time will tell whether or not we are entering a dystopian period in our civilization.

SOCIAL UNREST, SEXTING AND SADISM

One unsettling issue is that texting can be used to encourage civil unrest and promote illegal crowd activity. Cronulla Beach near Sydney was the scene of race riots and violence in December 2005 when a mob of some 5,000 people were whipped into a frenzy and, chanting racist slogans, assaulted anybody in the vicinity who was thought to be of Lebanese origin. Sunshine and alcohol were certainly partly to blame but the crowd, who turned on a number of innocent bystanders as well as police and ambulance officers, had been incited by a series of aggressive and racist text messages. Apparently the riot was provoked by allegations of attacks on life-guards a few days earlier; but essentially the various text messages encouraged any person looking for a fight to go to this beach area. Unfortunately, the violence was also promoted by Sydney's popular talk-back radio station 2GB, whose breakfast show host, Alan Jones, fanned the flames by repeating the text messages on air and advising local residents to take vigilante action to protect them-selves. In the ensuing chaos, police horses and special operations officers were bombarded with beer bottles, and ambulance men and their patients were injured by various missiles thrown by the crowd.

Erich Fromm has written that 'the core of sadism, common to all its manifestations, is the passion to have absolute and unrestricted control over a living being'.[12] Cruelty has traditionally taken a physical manifestation – perhaps exertion to beat or chase a victim. And the sense of control can be reinforced if this is a group activity. Humans are naturally cruel, and the obvious distress of the victim, for example in crying for help, can be a signal which encourages the group to increase its sadistic efforts. In her excellent book on cruelty, Kathleen Taylor, researcher into genetics at the University of

Oxford, points out that cruel behaviour, particularly when conducted within a group, may bring social rewards.[13] The urge to please one's comrades and to be part of the band is a strong one. As Taylor points out, a gang of children torturing a puppy is a powerful reinforcement to the whole group, and threatens punishment to those who don't join in. And the perpetrators of cruelty unquestionably get a psychic reward, a feeling of pleasure which is brought about in the brain. It seems unbelievable now to think of the cheering and clapping inhabitants of Kovno in Lithuania as they watched the German *Einsatzgruppe* detachments massacring thousands of their Jewish fellow citizens in 1941. It is no exaggeration to recognize that many of the same instincts are responsible for all bullying.

The electronic age is fostering a particular aspect of human cruelty that is not physical but psychological and emotional. Cyberbullying, a relatively new phenomenon, often takes the form of spreading rumours, gossip or sexually suggestive material by text message. Sometimes the psychic reward is increased by humour, a powerful social bond and, when shared in the group, another signal of power expressed by making fun of the victim. While cyberbullying is common at universities and in the workplace, it is increasingly a weapon used by schoolchildren. In schools, electronic bullying, which is somewhat less visible to the authorities and may be more difficult to police, though less common is increasingly replacing physical violence. 'Happy slapping', where the object of derision is subjected to violence or humiliation by one person and filmed by another, perhaps on a mobile phone, with the resulting image then texted to a complicit group of friends, is one mode of bullying used by teenagers in particular. More frequent is sending malicious messages by text or email, messages that often have a sexual content designed to harass or humiliate children of the opposite sex – a particularly harrowing experience to be on the receiving end of at the time of puberty or adolescence. Text messaging is the most frequent form of psychological aggression because it can be anonymous – the victim will not necessarily know who the perpetrator is. And now, with the advent of freely available accounts through a number of service providers, it is very easy for a person sending an email also to be anonymous. In a study of London schoolchildren, Peter Smith from Goldsmiths, London University, reported that girls were more likely to be victims than boys, that children who had previously used physical bullying were often likely to convert to cyberbullying, and that schoolchildren were more likely to experience such harassment outside school than in it, partly because most schools are increasingly restricting the use of mobile phones.[14]

It is clear that cyberbullying is more common in many other countries. One Turkish study suggested that 36 per cent of schoolchildren had at one time displayed cyberbullying behaviour, and that most of this bullying was directed at only a few selected victims. In Turkey, too, girls are more likely than boys to be victims. A review by June Chisholm of Pace University, New York, points out that children between the ages of eight and eighteen spend an average of eight hours daily using computer or cellphone (mobile phone) technologies. She argues that vulnerable or dysfunctional children are more likely to be victims of electronic abuse and more likely to suffer deeply; and that young women too are very vulnerable and can suffer humiliation over a long period of time.[15]

Sending sexually explicit or suggestive content between mobile devices may turn your partner on, but it may also be dangerous. Apart from the fact that communications of this kind may not be welcome, and can constitute or result in harassment, the combination of electronic monitoring systems and dishonest telephone company employees means that messages can be intercepted.

The National Campaign to Prevent Teen and Unplanned Pregnancy, an American organization, is worried that teenage pregnancy is too common in the United States, where about one-third of American girls are pregnant by the age of twenty. Together with the magazine *CosmoGirl*, it commissioned a survey.[16] This revealed that 'sexting' and other seductive online content is being used increasingly by teens. One in five of all the teenaged girls surveyed, and over 10 per cent of girls aged between thirteen and sixteen, have electronically sent, or posted online, nude or semi-nude images of themselves. One-third of teenaged boys and 25 per cent of teenaged girls say they have been shown private nude or semi-nude images. According to this survey, sexually suggestive messages (text, email and instant messaging) were more common than images, having been sent or posted by 39 per cent of teenagers; and around 11 per cent of girls said they felt pressurized into sending such messages. Well over one-third of recipient teens have shown such photos to friends, whether or not they have permission from the sender. Apart from the risk of attack, loss of privacy or blackmail, this is a legal issue. Children under eighteen are minors, and, in many jurisdictions, any nude photos they may send of themselves place the recipients in possession of child pornography.

In our society, sexual harassment – though there is legal redress – is very common. It is expressed increasingly in electronic form, often as a weapon against adolescent girls or vulnerable women. Adolescent girls, still exploring their bodies, their sexuality and their relationships, are particularly at risk. June Chisholm tells of one of her students who had attended a small Catholic school, and how her innocence was shattered when she moved to college in New York. This young woman, having discovered the website MySpace, joined it and gave personal details about herself, mostly so that friends with whom she had lost touch could get back into contact. Initially, she says, she enjoyed this hugely, but gradually she became aware that the site had a dark side. So many pages she came across recorded drug abuse and under-age drinking, and many girls' pages were filled with personal tales of their own sexual exploits. She found this deeply shocking, but when she read a bulletin (a mass message sent to all of a member's MySpace friends) calling one particular girl a slut and listing her real phone number 'for a good time', she realized her own privacy was at risk. Then she learned about some students who started graphic sexual rumours about one of her teachers. These rumours were inevitably passed around the college and the teacher was constantly harassed; but when the college made attempts to ban MySpace usage, this was found to be impossible because 'it was an infringement of students' human rights'. Eventually MySpace was banned from all college computers – but the rumour machine still flourishes because students can freely use online facilities at home.

Chisholm points out that there is now an online mechanism for reporting electronic bullying, though how effective the administrating organization is remains unclear. She sets out a number of rules to help young people to avoid this kind of harassment:

1   Never give out your password.
2   Never share any embarrassing secrets or photos online.
3   Don't send messages when angry.
4   Never pass on emails written by friends without permission.
5   Know that there are serious consequences for cyberbullies.

Her last point seems well-meaning but a bit weak. Do cyber-bullies really face serious consequences? Can effective suppression of such activity be assured? It is not clear, I think, whether requests to an internet service provider to block access by known bullies really are effective, nor what the police or a school can do about some cyberbullying. But clearly lax use of the internet, for example post-ing one's personal details on social networking sites such as Facebook or MySpace, can lead to unexpected consequences which are very hard to control.

THE WRITING ON THE WALL

The '9/11 Truth Movement' is a term covering a loosely connected band of individuals, groups, websites, film-makers and journalists. Characterized by considerable internal disagreement, they stand united behind one claim: that the events that took place in New York and Washington DC on 11 September 2001 were stage-managed by the US government in order to justify embarking on a war in the Middle East. Jewish workers at the World Trade Center were accused of having taken the day off, after being warned in advance of some event to be avoided. It is also claimed that US air defences were deliberately disabled, that the pattern of damage to the World Trade Center towers is consistent with a controlled explosion, not the impact of an aircraft, and that one of the hijacked planes, United Airlines Flight 93, was deliberately shot down.

Two of those claims, by themselves, show how easy it seems to be for some people to believe mutually opposing things at the same time. If United Airlines Flight 93 was genuinely hijacked, and had to be shot down, why would the government have deliberately disabled its own air defences at the same time? We might understand how such wild claims come to be accepted by vast numbers of people if we look in more depth at one of them.

Conspiracy theorists, broadcasting their views on websites, through newsgroups, in books and films, and on television, claim that debris from United Airlines Flight 93 was found up to 6.9 miles away from the crash site – a distance consistent with the idea that it had been attacked while it was in the air. In fact, the distance as the crow flies is only about a mile, exactly as far away as crash investigators would expect to find debris. The figure of 6.9 miles comes from typing the two locations – the crash site and the site where debris was found – into an internet route finder, which measures the distance by road. And therein, to my thinking, lies a key aspect of the problem.

In the middle of the nineteenth century the satirical publication *Punch* commented on the first telegraphic communication, sent between Camden Town and Euston Railway Stations.

> What horrid fibs by that electric wire are flashed about!
> What falsehoods are its shocks!
> Oh! Rather let us have the fact that creeps
> Comparatively by the post so slow.

The verse expresses a fundamental mistrust for the new technology, a feeling that intensified with development of the telephone, and the subsequent coining of the term 'phoney' for anything that was false. The *Punch* epigrammist, rather charmingly, looks back with nostalgic fondness to the days when the trustworthy written word was the preferred means of communication. Perhaps it's just as well he didn't know what was round the corner.

Today, people turn to the internet as an increasingly available, ostensibly free form of information on all topics, from choosing a mortgage to the alleged faking of the moon landings. And while this has undoubtedly led to the empowerment of individuals, it has also greatly increased the ability of some people to make others believe utter nonsense. As an NHS physician, I have been appalled to see how many patients have been misled by poorly researched, unsubstantiated material about infertility published on the internet. What has concerned me most is not so much the presence of this material on the Web as the readiness of people to believe it. The printed word, even electronically, carries a mystique, an authority, which stretches back, surely, to the time when writing and reading were the possessions of a privileged élite.

The sheer volume of printed material becoming available in the fifteenth and sixteenth centuries was a blow to the authority of the Church. Its Bibles and prayer books, often kept chained to pulpits, retained a sacred quality by virtue of being so rare. But people's readiness to accept the word of a book as unchallengeable truth has persisted. *The Protocols of the Elders of Zion*, first published in Russia in 1903, promulgated the idea that a secret committee of Jews controlled the world's media and financial institutions. Elements of it seem to have come from a novel written in 1868, in which a committee of Jewish elders meets in the old cemetery of Prague, summons up the devil and plots the downfall of humanity. By 1871, from France to Russia, sections of this story were being repeated in pamphlets presenting them as fact. The Okhrana, the secret police of Russia's tsarist authorities, disseminated the *Protocols* as an easy means of turning public resentment away from the monarchy and back towards the Jews. By 1920, English translations of the text were so popular that five editions sold out. Hitler referred to the book in *Mein Kampf*, and in recent years politicians and television broadcasts in Egypt, Iran and Saudi Arabia have presented the book – in parts and as a whole – as fact.

Fundamentalism – in its Christian, Islamic and Jewish varieties – is not much more than the belief that certain books, assembled and edited by humans over centuries, were in fact created and revealed by God. If anything testifies to the awe and reverence – sometimes misplaced – with which humans have treated writing, it's the fact that three of the world's largest faiths define themselves as 'People of the Book'. For our shared culture, this has resulted in a commendable respect for books and literacy. It has also, on occasions, resulted in a rigid, regrettable zealotry.

We must also ask whether the spread of literacy has truly promoted freedom. Although terrorist groups have been able to use the printed word and image to publicize their aims, they have been dependent on an industry with very different goals. When publishing became a business concern, subject to market forces, it fell under the sway of money-makers with an interest in maintaining the status quo. In 1855 Edward Lloyd, publisher of the hugely successful *Lloyd's Illustrated London* newspaper, invested so much of his wealth in the setting-up of the first steam-powered rotary press that it was inconceivable for his publications to do anything other than support the government and stifle dissent.

If anything, the availability of mass-produced literature removed people's revolutionary zeal, or at least their capacity to do much with it. The period from the fifteenth century to the nineteenth saw an increasing privatization and individualization of social life. The earliest books were too expensive to be widely owned, reading a skill too rare to be done except in large gatherings of people. As reading matter became cheaper, though, and the market more fragmented and specialized, it became increasingly possible for people to descend into separate, private worlds. So within the family, for example, Father could settle down with accounts of heroic derring-do, Mother with romances, and the children with fairy stories and morality tales – a trend which was deepened and enhanced in the late twentieth century with devices like the personal computer and the MP3 player. People's minds expanded, but not necessarily together. And when the older generation ceased to be regarded as the repository of knowledge and memory, our society became increasingly atomized.

That trend may, of course, be reversed by writing's latest incarnation, the internet. The image of the lonely nerd seeking solace through his computer has been replaced by the net-savvy communicator plugged into vast, global networks of friends, roaming endless corridors of information, shaping and exploring the world without ever leaving his or her chair. Like the history of writing as a whole, though, this new chapter has both light and dark sides to its pages.

THE WILD WEST

The internet, a gigantic lattice in which millions of computer networks link together hundreds of millions of machines, is a remarkable mesh that has been used by ordinary people for less than twenty years. Yet at the last estimate, in March 2009, it was calculated that about 1.6 billion people use the World Wide Web, just a little less than one-quarter of all the world's inhabitants.[17] About 75 per cent of North Americans and 50 per cent of Europeans are connected; in Britain about two-thirds of people have a connection. Poorer countries lag behind, but even in Africa about 5.6 per cent of people use the internet and the number is rising fast.

The internet transfers writing (or, in computer parlance, data) between computers in such a way that the machine (and any person operating it) at each end of a connection is usually quite unaware of the physical location of the machine at the other end of the connection. Moreover, the person using the internet normally has little or no knowledge of the route through the web that his or her data is taking. It can therefore be intercepted without the user even knowing.

Perhaps the most important aspect of the internet is its extensive global economic use, fundamentally changing the way banks, businesses and individuals make financial transactions. But it also radically changes how we get news, information and entertainment, and how we meet our friends and lovers. Unquestionably, this revolution in writing has been extraordinarily valuable; but, as the use of the web grows rapidly, so the threats it poses become increasingly apparent. One problem is that very few of us understand exactly how it works. Even if we use the internet very frequently, most of us do not have the slightest inkling about the technicalities – how we get access to information, how we pass it on, how we communicate with others. We may have a vague idea about the web browser we use, but we do not see the layers of technology that lie beneath the surface. The mechanism is effectively hidden. And, just as the technology is largely invisible to the internet's users, so are the risks.

These risks are profound. They threaten our security because we have only a very limited ability to control or restrict how the information that we have entered on our PCs or mobile telephones is used. For example, we lose control over data about ourselves if our details are fed into a machine operated by a business, a public service or the government. Government can be remarkably careless about the information it purports to store on our behalf. It may seem unbelievable, but in 2008 the UK government lost the National Insurance numbers of 17,000 of its citizens. In retrospect it seems farcical that this digitally stored data was sent by routine post. If that were not enough, the Ministry of Justice lost information affecting more than 45,000 people, in some cases revealing their criminal records and credit histories. The list goes on. Details of 25 million child benefit claimants also vanished last year, and the Home Office lost the personal details of 3,000 seasonal agricultural workers – including their passport numbers – when two CDs went missing in an envelope. In five separate cases, the Foreign Office lost information affecting about 190 people. And the Department for Transport misplaced personal data on six separate occasions, including 3 million records of driving-test candidates in May 2007. One might reasonably expect that where the defence of the nation is involved, government officials would recognize the need for particular care – yet the Ministry of Defence lost an unencrypted laptop computer which contained 620,000 personal records, including bank account and National Insurance numbers and information on 450,000 people named as referees or next-of-kin by would-be servicemen and women. The Shadow Minister for the Home Office, Francis Maude, drew attention to the government's own damaging admission that there 'has been a torrent of data breaches in Whitehall', asking pointedly in the House of Commons: 'How can we trust the Government to protect the privacy of law-abiding citizens when they systematically ignore their own requirements?'[18]

Admittedly, the breaches of security listed above – mostly involving the loss of CD-ROMs or computer hard drives – do not necessarily mean that the data will be used by any unauthorized person finding it. But the potential for damage to the individuals concerned is varied, substantial and long-lasting, ranging from financial losses to loss of reputation, now or at some unknown time in the future.

Moreover, if, using our personal computer, we have our formal identity stolen, we may even find ourselves being accused of crimes we did not commit and of which we have no knowledge. Other risks undoubtedly include threats to our personal safety – including the potential for physical or psychological harm. One high-profile threat is that posed to children by predatory paedophiles, who conceal their true identities while using the internet to 'groom' potential victims.

Every computer connected to the internet is identifiable with a unique number – its so-called identity or 'IP address'. Messages or information are sent to the appropriate machines by the web machinery 'reading' the destination IP address placed by the sender. Blocks of addresses (of varying sizes, from hundreds to millions) are allocated to internet service providers (ISPs) that then make allocations from these blocks to their individual customers. Thus a router only ascertains the address block and relays the packet of information to the appropriate ISP. It is the ISP that delivers the information to the correct machine.

If a message or data is abusive – for example, so-called 'spam' email advertising penis enlargement or Viagra – the source can be traced by determining which block of addresses it comes from, and therefore which ISP allocated the address. That ISP can then – if it wishes to – identify the customer to whom the IP address was issued. Since many ISPs allocate the same address to different customers at different times, the exact time of the connection will often be needed in order to identify correctly the particular customer who was using any of these 'dynamic addresses'.

The ability to trace an IP address does mean that the source ISP can be identified. If the ISP is prepared to cooperate, it can act to prevent further abuse or identify the customer account. In practice an ISP will not divulge this kind of information unless a formal legal application has been made appropriately in the jurisdiction in question. But unfortunately even this recourse is very limited because the source machine may be being used without the permission of the registered owner – for example, in an internet café, an airport lounge or a hotel. Equally, an unprotected wireless connection may be the source – for example, in a person's private house where a computer is in wireless range, perhaps from the street. Commonly, the source will be an identifiable consumer's machine but one which is insecurely configured or is inadvertently running a 'malicious' piggy-back program. Under these circumstances the source may simply be relaying information in all innocence. Tracing the true source of abusive content under these circumstances is a complex, and usually unsuccessful, task.

For me, the double-edged nature of the net is summed up best in the story of Mohammed Irfan Raja, a quiet, studious Ilford school-boy who used internet chat-rooms to make friends. Bespectacled and close to his family, he used the best aspects of the new technology to move beyond his background and his limitations, corresponding with people across the world. But during that process he was exposed to jihadist propaganda material and put in touch with persuasive radicals, who urged him to travel to Pakistan and train for a holy war. If he'd wanted to, he could have also used the internet to discover simple, effective and deadly recipes for explosives and poisons. He changed his mind about jihad, confessed all to his parents and, as I write, is serving a two-year jail sentence, having been found guilty of possessing articles useful for terrorism. It is fortunate that this young man did not meet with some other fate; as it is, he provides a powerful example of how the written word can be used, as it has always been used, to alienate as well as to unite, to distort and mislead as well as to enlighten.

Chapter 7

The Right Promethean Fire

THE NAME OF Karl Marx is associated with revolution, social upheaval, the forging of new ideologies and world orders. The incident marking his entry on to the stage of world history is less well known. In 1842 the 25-year-old Marx accepted a position as editor of the *Rheinische Zeitung* – a modest regional newspaper, popular with the area's burgeoning middle class and mildly critical of its Prussian rulers. He had published a series of articles under a pseudonym the year before, and their stance on government censorship had been so well received that the paper's circulation began to increase. It was a natural step to offer the young journalist a leading role on the paper, and it was in this capacity that Marx first met Friedrich Engels, a disaffected young man on his way from Germany to take over the UK branch of his father's textile firm. As well as commissioning a number of articles from Engels after that meeting, Marx struck up a lifelong friendship that was to steer the course of world history. Swapping ideas, observations and complaints, editing one another's works, Marx and Engels would weave the ideology that informed the Russian Revolution of 1917 and, subsequently, the half-century-long division of the world into opposing communist and capitalist blocks.

Marx's tenure as editor of the *Rheinische Zeitung* did not last. We might expect him to have lost his job over some inflammatory call for revolution. But Marx incurred the ire of the Prussian authorities over a more parochial matter – the right of local people to gather dead wood in the Rhine area's abundant forests. This had long been considered an ancient privilege, but a recent edict had declared the wood to be the property of the landowners. The landowners were not acting out of mere perversity: this was 1843, with industrialization at full pelt in Germany, and although, as we shall see, coal was the fuel which drove the pistons and drop-hammers of this era, all combustible materials suddenly acquired new value. The quick-burning chemicals in wood could be leached out, leaving slow-burning, efficient charcoal, a valuable fuel source for new factories. To protect this resource, fines were suddenly imposed on ordinary people caught gathering a hitherto free item to heat their homes. In a clever article, foreshadowing his twin interests in economics and social justice, the 25-year-old Marx hit out at the absurdity of this: how, he demanded to know, could the rising scale of penalties – so many marks' fine for so much firewood – reflect the 'worth' of something that had always been free? Circulated, as much literature of the emergent social-democratic movement tended to be, right across Europe, Marx's tiny local news piece reached the attention of Tsar Nicholas I. Alarmed at its seditious implications, Nicholas put pressure on his Prussian allies, who had Marx dismissed from his position and the *Rheinische Zeitung* closed down.

Marx, of course, was not silenced. Just five years later, the *Neue Rheinische Zeitung* started printing under Marx's editorial control – though that too was closed down within a year or so of its first edition. But as well as providing the spur for the great man's subsequent wanderings through the slum-lands of Europe, the story about fuel proved symbolic: because wood and coal – the sources of fire itself – stood in a palpable sense behind the upheavals of the industrial revolution, the political unrest of the late nineteenth century, and the subsequent ideological carving-up of the world during the Cold War. Man's exploitation of fire stands, too, behind many of the crises facing us today – from the rising sea levels to the obesity of our children and the enduring poverty of the Third World.

FRIENDLY FIRE

In her outstanding book *The Forest Dwellers*, conservationist Stella Brewer described her work with chimpanzee colonies in Senegal and the Gambia.[1] While training orphaned chimps in the various 'life skills' necessary to survive in the wild, she noticed something interesting. Far from being afraid of fires, chimps would wait cautiously until the embers cooled, and then pick through them for charred afzelia beans and other delicacies. There was a clear aware-ness that this potentially terrifying phenomenon – fire – conferred advantages, if treated in the right way.

This charming example of chimp-cookery gives us an idea of how our ancestors may have come to appreciate the value of fire – prob-ably long before subsequent generations discovered how to create and maintain it. The archaeological record provides ample evidence of fires occurring spontaneously – as a result of volcanic action, sun-light, build-ups of gases – from 350 million years ago. So our ancestors would have had plenty of opportunities to observe fires, develop strategies for coping with them, and even benefit from them.

Black kites have sometimes been observed 'starting' bush fires by carrying bits of flaming wood and vegetation and dropping them on to dry, combustible scrub. While this may be a rare means by which wildfires spread, it is unlikely these birds have the intelligence to make fire deliberately. It is more probable that they swoop on patches of flaming bush to scoop up any escaping rodents and insects, and occasionally include the odd scrap of burning matter in their haul. But they clearly see the useful aspects of a fire, too. Fire is a bonus for many species, destroying parasites and competitors, providing warmth and light, clearing land and improving visibility, and attracting and scattering nutritious prey.

It is not clear when humans began to use fire. Some debated find-ings by archaeologist Raymond Dart suggest that the australopithecines, distant ancestors of *Homo sapiens*, could have been using fire at Makapansgat, in South Africa, 1.5 million years ago. Evidence from Zhou Kou Dien cave, in China, puts the figure at only 500,000 years BP, which would make fire a new tool in the repertoire of *Homo erectus*. But some Israeli scientists made a novel find in 2004.[2] It seems that *Homo erectus* there may well have used fire some 790,000 years ago, and also that these early ancestors used surprisingly sophisticated tools. The researchers found numerous flint implements belonging to the so-called Acheulean tradition of tool manufacture.[3] Some of these were burned, while others were not. The researchers think the clusters of burned artefacts indicate the sites of ancient camp-fires or hearths. These experts consider that the control of fire encouraged social interaction, enabled dramatic changes in the diet of these proto-humans and gave them the ability to defend themselves against wild animals.

The use of fire should be an easy innovation to identify. Unlike language, or social organization, it leaves an obvious blackened trail in the archaeological record. But, as in the example of the avian 'arsonist' kites, when and what intentions came to be attached to the occurrence of fire are harder to read. It is likely there would have been an intermediate stage in our ancestors' careers when they began to control, manipulate and prolong those conflagrations they came across, but were not yet deliberately starting them.

Fire would certainly have offered early humans huge advantages in respect of survival and reproduction; so there would have been a strong incentive to learn how to create as well as control it. Sterkfontein, the South African cave excavated over a thirty-year period by the eminent archaeologist C. K. Brain, provides a classic illustration of how fire helped the balance of power to shift towards intelligent humans,[4] and away from brawny beasts. Early layers of the cave reveal humans to be the prey of big cats; in later ones, contemporaneous with evidence of human-made fire, the predators are being consumed by us.

Fire would have provided several advantages. By burning scrubland, fires enabled human hunters to see their prey more clearly. Cooked food was easier to chew and digest, and could also be preserved for longer, leaving more time for activities not related to hunting or gathering. Fire may also have become a useful element in the hunt itself. Evidence at Torralba in Spain suggests humans might have used fires to drive herds of large mammals – including elephants – off a precipice; a lazy way of butchering in volume.

The evidence from Torralba is debatable: the distinguished

American archaeologist Lewis Binford, among others, has pointed out that the charcoal deposits around this site could have come from natural fires, and the huge stash of animal bones from millennia of human fly-tipping.[5] But still the general picture cannot be ignored: fire lent us a massive advantage. The burning of scrubland also encouraged the growth of edible grasses and legumes – exactly the plants that humans would later come to domesticate. In addition to being edible by us, these first crops would have attracted hosts of small game to the site, which could then be picked off at will.

How tempting it is to think of the wide open spaces of the American prairies as a primordial landscape: but in fact they were created by the indigenous inhabitants, who set light to the woodland twice a year, creating lush grassland for the buffalo they hunted. This so-called 'firestick farming' – noted and admired by European explorers everywhere they came into contact with indigenous peoples – was probably a precursor to more sophisticated forms of land management, of which deliberate cultivation was the apogee. And it is impossible to imagine farming without fire. For a start, the cereal crops first domesticated were only truly edible as a result of fire – either boiled into a pottage (like the meal for which hunter-gatherer Esau sells his birthright in Genesis 25) or baked into a crude bread. Fire would have attracted small pack animals to the fringes of human settlements, where humans would have captured them and domesticated them. Most importantly of all, it cleared the land and replenished its resources. In numerous tribal groups land is still claimed by means of setting light to it: man establishes his per-ceived dominion over nature with fire, as he almost certainly did 10,000–12,000 years ago, when global warming coincided with a population bulge. At this point the need for new territory might well have necessitated mass torching of the land. As areas became settled, the occupants re-enacted the original 'claiming' fire every two years or so, aware that the ashes would revivify and enrich the soil. This classic 'slash and burn' technique continues across the world to this day, although the invention of the plough around 6000 BCE provided us with a different means of renewal.

As agriculture enabled humans to acquire food surpluses, some individuals would have been released from farming to undertake other technologies enabled by our control of fire. Using primitive kilns, potters endowed their communities with food storage vessels – insurance against a lean harvest – and smiths melted ores to produce metal. It is not clear exactly how humankind arrived at metallurgy. Perhaps people observed what happened to certain ore-containing rocks in the heat of a volcano. Or perhaps, when firing earthenware, they were intrigued to see what happened when they heated clay rich in metal. At any rate, working metal by fire provided vital tools and luxury goods – and deadly weapons.

A MAGIC MATERIAL

Gold was almost certainly the first metal that humans identified, as it can be found in river beds or occasionally on the ground as a pure metal. But it was really only useful as an ornament, the very malleability that lent it to decorative work making it inadequate for other applications. The first metal that humans really worked was copper, which was recognized by about 7000 BCE; it existed naturally in its pure form, and could be bashed into useful shapes when heated. It is harder than gold, a valuable property if you want to make a knife, sickle or arrowhead. Gold and copper have fairly similar melting points, around 1,060–85 degrees Celsius, but in those first days of metallurgy it would have been difficult to get a high enough temperature to melt copper completely. The limit of an ancient furnace was about 1,150 degrees Celsius, though ancient pottery kilns may occasionally have exceeded this temperature.

Eventually fire would have been used to heat copper-containing rocks like azurite or malachite. Such rocks, though they usually do not contain much elemental copper, would have been easily recognized because they generally have a distinctive green or bluish hue. When the ore is subjected to a high enough temperature, the liquid metal will ooze out. Thus fire became important for smelting ores to extract pure metal and for casting this molten liquid in prepared moulds. Around 3500 BCE smelting was used in Persia; we know this because worked copper objects have been found there. Regular supplies of metal would have depended on another technology, and the first mining may have been undertaken about 4000 BCE. Like the pure metals gold and copper, some mineral ores were found on the earth's surface in rocky outcrops. But in chipping away at them; men would have recognized that veins of ore ran below the surface.

One of the earliest mines is in Serbia at Rudna Glava, where Dr Borislav Jovanović, from the Institute of Archaeology in Belgrade, uncovered some twenty prehistoric mineshafts in the limestone massif. Here there were once plentiful supplies of copper carbonate. Prehistoric miners excavated shafts as much as 15–20 metres down from the surface – extraordinarily dangerous work – even using discarded stones and rock to shore up the walls and roof as they dug. They used alternate heating and cooling to break up the ore, first lighting wood fires with which they lined the wall face, then throwing water on to the hot rock, causing cracks which facilitated quarrying. Similar copper mines are found in the Negev in Israel and the neighbouring Sinai peninsula. Israeli archaeologists have found earthenware crucibles at these sites, evidence that excavated ore would have been smelted to extract the pure molten metal.

The next technological advance was the accidental discovery of metal alloys, heralding the bronze age. Ores of copper and tin are often found close together; tin is easy to extract, with a melting point of 232 degrees Celsius. Contamination of molten copper with tin gives the much harder material, bronze, when the mixture cools. Most bronze is about 90 per cent copper; brass is a similar alloy, but contains roughly 15 per cent zinc. As a weapon for killing, a bronze blade had a significant advance as it had a much sharper edge if worked properly. Bronze implements were also much more durable.

Bronze seems to have been first worked in Sumer, at Ur, in around 2800 BCE; but it may have been developed simultaneously in Anatolia (modern Turkey). The remains of bronze instruments dating from about 2500 BCE have also been found in the Indus valley; 500 years later bronze implements were being used in Europe. Given the speed of modern human development – dirigible balloons, powered aircraft and rocket ships all being manufactured in turn within a single century – it is interesting to think it took at least 500–1,000 years for this basic technology to be picked up, or reinvented, in Europe. Around the same time that the first bronze artefacts are found in Europe, crude instruments are seen in China.

Later, during the Shang dynasty around 1500 BCE, decorated bronze objects of remarkable workmanship were being produced. But in all these cultures bronze was a luxury and items fashioned from it used almost exclusively by the rich and powerful. Most people still had to rely on flint stones.

Iron has a melting point around 1,530 degrees Celsius – an unachievable temperature in the primitive furnaces of the bronze age. So although iron and its ores were always common on the surface of the ground in many parts of the world, iron tools and weapons are very rarely found in archaelogical sites dating from before 1500 BCE. About then, the Hittites started to work iron in Anatolia. It was still not possible to melt it completely, so repeated hammering of the hot metal was the best they could do. Pure iron is less hard than bronze and therefore less useful for weapons; it was not until the eleventh century BCE that the discovery was made that when iron was reheated with charcoal from the fire it absorbed some of the carbon, producing a harder metal. Quenching hot carbonized iron in water, cooling it rapidly, furthers tempering and makes steel. Around 500 BCE the Chinese fashioned much hotter furnaces capable of melting iron completely – and made the first cast objects. I repeatedly suggest in this book that similar technology is often developed simultaneously in different and unconnected parts of the globe; this appears, however, not to have been the case with metallurgy, as the first iron foundry in England was not established until well after the Norman Conquest.

Behind all these advances lay fire. From the bronze age onwards, fire enabled more privileged people to acquire vital tools, desirable *objets d'art* and more sophisticated weaponry. The institution of metallurgy marks a time when we became less interested in the conquest of nature and more concerned with the conquest of our fellow men.

Did all humans have fire? Early European explorers in Australia noted the tendency of some aboriginal groups to 'borrow' fire from each other, swapping a slow-burning stick or bundle of fibres for some food. Seemingly unfamiliar with these claims from Australia, in his 1922 account of life on the Andaman Islands anthropologist Alfred Radcliffe-Brown concluded:

the Andamanese are perhaps the only people in the world who have no method of their own for making fire. At the present time they obtain matches from the settlement of Port Blair, and a few of them have learnt, either from Burmese or Nicobarese, a method of making fire by the friction of pieces of split bamboo. Formerly, however, they had no knowledge of any method by which fire could be produced.[6]

Some anthropologists think Radcliffe-Brown was misguided in these claims. In his defence we might mention the Pirahã (see chapter 1), who have challenged many of the assumptions made about the attributes and technologies allegedly possessed by 'all' human groups. Stranger things have been known. But, tellingly, the Pirahã *do* know how to make fire. What the early observers in Australia saw was the *exchange* of fire, neglecting, perhaps, to see its importance in a network of social relationships. It may have been more important to give fire than to receive it, and erroneous to assume from that transaction that one party did not know how to create it. And what Radcliffe-Brown saw occurred after generations of unsettling contact with white men: at that point, few people in Sydney or Bombay would have known how to make a fire except with matches either.

So mysterious, powerful and dangerous was fire that traces of our initial fascination remain peppered throughout human culture. Zoroastrians keep a fire burning in their temple at Yazd in Iran which has not been extinguished for over 2,000 years. The Israelites burned the *Olah Tamid*, the burnt offering on the Temple Altar, over a fire which was continuously kept lit by the priests. In many Christian churches, a flame burns constantly inside a glass lamp: a symbol which goes back to the Jewish Temple where a lamp, a *Neir Tamid*, was kept lit with holy oil – until the Temple was destroyed by the Romans. The Herero people of Botswana keep a holy ritual fire alight in perpetuity – a symbol of life, prosperity and fertility. Secular festivals, such as Britain's Fifth of November, testify to an enduring affection for building huge fires and simply watching them burn. So, too, does the popular phenomenon of the barbecue – an unnecessary, inconvenient, inefficient and unhealthy method of grilling meat, usually conducted at the very time of year when we least want to be standing around a vigorous source of heat, or in my case when I am forced by family to go outside to cook miserably in pouring rain. But an undercooked bit of chicken is preferable to the midsummer feast of St John's Day which, until the eighteenth century, was celebrated in Paris by throwing live cats on to a vast bonfire. Afterwards, people collected the embers and took them home, believing them to confer good luck.

DEADLY FIRE

As pupils at St Paul's School in London, we were required to read and attempt to translate Thucydides' account of the Peloponnesian War. I regret that not all of us were entirely assured in rendering ancient Greek, and many of us used as a crib the translation composed by a much more learned past pupil of the school, Benjamin Jowett.[7] Here is a section from Book 4, written in 411 BCE:

> The Boeotians . . . now marched against Delium and attacked the rampart, employing among other military devices an engine, with which they succeeded in taking the place; it was of the following description. They sawed in two and hollowed out a great beam, which they joined together again very exactly, like a flute, and suspended a vessel by chains at the end of the beam; the iron mouth of a bellows directed downwards into the vessel was attached to the beam, of which a great part was itself overlaid with iron. This machine they brought up from a distance on carts to various points of the rampart where vine stems and wood had been most extensively used, and when it was quite near the wall they applied a large bellows to their own end of the beam, and blew through it. The blast, prevented from escaping, passed into the vessel which contained burning coals and sulphur and pitch; these made a huge flame, and set fire to the rampart, so that no one could remain upon it. The garrison took flight, and the fort was taken. Some were slain; two hundred were captured; but the greater number got on board their ships and so reached home.

Some people question the truth of Thucydides' description, but I see every reason to accept that this is an accurate contemporary account of a flame-thrower used during the Spartans' siege of Delium. Fire has been used repeatedly in warfare, often with exceptional cruelty.

Like the Spartans, but before them, the ancient Israelites seem also to have possessed a very strange fiery weapon. In a curious account in the Apocrypha, the Second Book of Maccabees – written around 445 BCE – relates how the exiled Jew Nehemiah, cupbearer to Ataxerxes I and subsequently governor of Judah, received permission from his Persian captors to return to the ruins of Jerusalem and retrieve what remained of the Great Temple fire. Allegedly, this fire had been preserved in the form of a viscous liquid which, when sprinkled on to wood and left in the sun, spontaneously burst into a roaring flame again. And it seems from the Hebrew, *nphth*, that this is where we get the word 'naphtha':

> When this matter became known, and it was reported to the king of the Persians that, in the place where the exiled priests had hidden the fire, the liquid had appeared with which Nehemiah and his associates had burned the materials of the sacrifice, the king investigated the matter, and enclosed the place and made it sacred. And with those persons whom the king favored he exchanged many excellent gifts. Nehemiah and his associates called this 'nephthar', which means purification, but by most people it is called naphtha.

Centuries later, historians would write in awe of the equally strange 'Greek fire'. A secret weapon of the Byzantine navy, this burning liquid could be shot out of tubes on to the ships of their enemies, and it continued to burn even when it landed on water. In the tenth century CE, Emperor Constantine Porphyrogenitus urged his son never to reveal the recipe behind Greek fire, claiming that 'it should be manufactured among the Christians only, and in the city ruled by them, and nowhere else at all'. So well-kept did the secret remain that when a quantity of Greek fire fell into the hands of the Bulgarians, in 812 CE, they didn't know what to do with it. So ghastly were the effects of this stuff – usually described as a compound of sulphur, saltpetre, naphtha and tar – that Pope Innocent II condemned its use as a deadly sin in 1139. This didn't stop Christian armies trying to make their own, frequently igniting themselves when they failed to factor in the direction of the wind. Fortunately, or unfortunately, use of Greek fire dwindled after the twelfth century, largely because of its limited possibilities and the increasing sophistication of other weapons – most notably the cannon loaded with gunpowder. Initially developed by Chinese alchemists seeking an elixir for eternal life, gunpowder – effectively a portable fire – was swiftly adopted by humans wishing to kill one another. As with so many great human innovations, something first designed to enhance life soon became a tool for mass destruction.

Fire could turn on its masters, too. The city of Rome grew up organically, with little planning or direction from architects and city fathers, and consequently its cramped lanes and teetering houses were a fire hazard. The earliest fire brigades began their work around 450 BCE, and one enterprising individual, Marcus Crassus, built up a small fortune by putting out flaming buildings and then purchasing their smouldering shells from the distraught owners. Rome's harshest fire, which occurred in 64 CE, was said to have been started by the Emperor Nero himself, in order to clear room for his grandiose parks and monuments – but it could just as easily have been started by some careless Neapolitan kebab vendor. It was concern over such fires that led Pliny the Younger, an official in the employ of Emperor Trajan, to appeal, in the second century CE, for funds to set up a local imperial fire brigade. Trajan's response was almost comical: 'remember that the province of Bithynia [part of central Turkey, where Pliny was governor], and especially city-states like Nicomedia, are the prey of factions. Give them the name we may, and however good be the reasons for organization, such associations will soon degenerate into dangerous secret societies.'

Repeated conflagrations led to a rapacious demand for new wood for rebuilding. Some historians have suggested that this set in motion the demise of the Roman empire. The deforestation of surrounding areas led to floods and soil erosion, making it harder to produce sufficient food to feed the population. This drove the empire into a phase of constant, costly territorial expansion, to the point where it could no longer defend itself effectively.

In the First World War, the German army used fire partly as an effective way of clearing trenches and partly to cause utter terror.

Like virtually all modern weapons, the *Flammenwerfer* was invented by a scientist, one Richard Fiedler. It used pressurized air (or, later, carbon dioxide or nitrogen – both of which it was hoped would cause less risk to the operator as they are inert gases) in a cylinder which belched a jet of burning petroleum or oil over a distance of 15 metres or more. Fiedler first designed what he thought of as a relatively lightweight machine, weighing just 38 kilograms, but amended it subsequently to produce a larger weapon with a greater range which required two operators to carry it. Having tested flamethrowers in 1900, the German army created three specialist battalions to use them during the Great War. A surprise attack on British troops was launched on 30 July 1915 at Hooge, 2 miles east of Ypres in Flanders. Just before any real light, at 3.15 a.m., the Germans made devastating use of these machines with their gas cylinders strapped to the back of the operators. The effect was terrifying to the British and their line retreated with the loss of 31 officers and 751 infantry. After the battle in the Ypres salient, the Germans launched more than 650 flamethrower attacks, but eventually the allied forces dealt with them more effectively. The man operating a *Flammenwerfer* could only stumble relatively slowly over the uneven ground between trenches with the massive weight on his back, and he was a sitting target. The device was nasty to handle and there was always the risk of the operator incinerating himself: sometimes the fuel cylinder exploded unexpectedly. Consequently, the average life expectancy of a flamethrower operator was a few weeks, and if taken prisoner these operators could expect no mercy.

Inevitably, the British and the French condemned such brutal tactics – but they were not to be outdone. At the Somme, the British army had static flamethrowers each weighing about 2,000 kilograms, with a range of 80–90 metres. They were placed in no man's land, just 60 metres from the German front line. However, it was impossible to aim them properly and they were an easy target for shellfire, so they turned out to be virtually useless; eventually they were abandoned.

Fire has continued to be a weapon of war right up to the present. We have only to think of the horrors of the incendiary attacks on the cities of Coventry and Dresden in the Second World War to remember the sheer terror that intense fire causes. And during the first Gulf War of 1990–1, the retreating Iraqi forces engaged in an activity their Babylonian and Sumerian ancestors would have fully understood – setting light to the land and its oil wells, causing fires which in some cases took months to extinguish, and caused massive contamination for hundreds of miles around.

SUBTLE ALCHEMY

Earth is the lowest Element of All
Which Black, is exalted into Water,
Then no more Earth but Water wee it call;
Although it seeme a black Earthy matter,
And in black dust all about will scatter,
Yet when soe high as to Water it hath clym'd
Then is it truly said to be Sublym'd
When this blacke Masse againe is become White
Both in and out like snow and shining faire
Then this Child, this Wife, this Heaven so bright
This Water Earth sublimed into Aire
When there it is it further will prepare
It selfe into the Element of Fire
Then give God thankes for granting thy desire.
(*Kelle's Worke* by Edward Kelley)[8]

The history of Sir Edward Kelley is convoluted and bizarre, and stories about him are contradictory. This enigmatic alchemist was born in 1555 and in his time he was regarded as a visionary; he was also a failed lawyer and an apothecary. But he was an alchemist with a difference. Most alchemy was pursued in attempts to turn base metals into gold, and this, of course, generally required fire. In classical times, and into the Middle Ages, philosophers identified four 'elements': fire, earth, air and water. Fire was the alchemists' dominant element. But Kelley claimed to have discovered how to transmute metals without the need for a large furnace.

By one account at least, Kelley is said to have been Irish; other biographers record him as having been born in Worcester. Possibly Kelley was an assumed name, because in Worcester and at Oxford University (where there is no record of his taking a degree) he was known under the name Edward Talbot. From Oxford he travelled to Lancaster, where he appears to have fallen foul of the local authorities. Then he moved on to Wales, apparently in search of a volume entitled *The Book of St Dunstan*. How he eventually managed to acquire this tome, the only copy known to be in existence of a much sought-after learned treatise, is puzzling. The book was said to be a notable alchemical textbook and valuable especially for its secret description of the 'philosopher's stone'. This stone, the object of much research by alchemists, was the agent with which an experimenter might be able to turn base metals into gold. More precisely, St Dunstan's book was believed to have a crucial section on how to derive the formula for a recipe: 'Red and White Powders for the Transmutation of Metals'.

In 1582, having left Wales, Edward Kelley approached John Dee, again giving his name as 'Talbot'. Dee was much esteemed and well connected – a respected Elizabethan philosopher, mathematician and founder Fellow of Trinity College, Cambridge. John Dee was fascinated by Kelley because Kelley claimed he had made contact with angels, something that Dee had been trying to do for some time. Dee was a very significant scholar, equally at home in Hebrew, Latin and Greek. He possessed one of the largest libraries in England, with around 4,000 books, and was noted for having pre-pared the horoscopes of both Mary Tudor and Queen Elizabeth. He was also a cabbalist, extremely interested in the occult, and some contemporaries later claimed that he was such a fine magician that he had conjured up the bad weather in the English Channel which resulted in the defeat of the Spanish Armada. His occult expertise may have been the inspiration for Shakespeare's Prospero in *The Tempest*. But Dee certainly wasn't clairvoyant. Although he soon discovered that 'Talbot' was an assumed name, he did not seem to know why Kelley always wore a hat which reached down to his neck, and which he never removed in public. This, according to some accounts, was because Kelley had had both ears cut off in Lancaster after his conviction for forgery.

Kelley entered Dee's service and remained with him for the next five years, acting as medium and interpreter in their regular spiritualist conferences. About a year after meeting Dee for the first time, Kelley mentioned that he had acquired *The Book of St Dunstan*, together with a bag of red powder; he promised that with this highly prized manual and the sample of powder he could make a tincture which would produce gold. In March 1583 Dee was invited to a reception at Greenwich Palace, where Robert Dudley, the earl of Leicester, was entertaining a Polish noble, Prince Albert Laski. Prince Laski had also experimented extensively with the occult, and it seems the Pole and Dee hit it off immediately, Laski even going so far as to offer to fund some of Dee's experiments.

John Dee and Prince Laski decided on a trip to Poland later that year, with Kelley in attendance, apparently in the expectation of finding the philosopher's stone that would enable them to make gold. They attended the court of King Stefan in Krakow, and then that of the Holy Roman Emperor, Rudolf II, in Prague, the European centre for alchemical studies at that time. Rudolf II, however, who had a deep personal interest in magic, was unimpressed by Dee. So Kelley and Dee left Prague together and eventually settled in Trebon, about 160 kilometres away, to continue their research. At some point Kelley had married a woman called Jane Cooper who had been an English courtier in Rudolf's polyglot, cosmopolitan court − some accounts say that they had already met and married while living in England.[9] But perhaps he tired of her soon after their journey to Trebon, because there he confided in Dee about an angel who had visited him. This angel, he said, had ordered them both to share everything they had, including their spouses. John Dee was, to say the least, somewhat concerned about this and briefly stopped his spiritual conferences with Kelley; he did, however, share his wife before he eventually returned to England. It is not recorded whether Kelley ever took his hat off in bed.

Kelley soon convinced a number of influential Poles that he had a method for transmuting base metals to make gold. His recipe, though secret, required very little fire − just gentle heating of the ingredients; the work was done mostly by mixing his two powders. He was now living rather well, having been given various estates together with a knighthood by the Emperor Rudolf, who was increasingly eager to manufacture gold on a large scale. But on returning to Prague, Kelley seemed to the Emperor to be procrastinating and Rudolf, frustrated, had him imprisoned in Krivoklat Castle. It never seemed to enter Rudolf's head that he might be the victim of a monstrous confidence trick – he apparently had no doubt that Kelley could make the gold he wanted – but he seems to have thought a period of imprisonment in the tower of Krivoklat, a rather attractive hunting lodge some 40 kilometres from Prague, might give Kelley a slightly greater sense of urgency.

Kelley now played cat and mouse. He agreed to produce some gold and was released; failing to come up with the goods, he was incarcerated once more, this time in Hnevin Castle – a rather more forbidding place that has some slight resemblance to the South Wing of Wormwood Scrubs. Imprisonment did not seem to stop Kelley from writing alchemical treatises, but none of them gave specific instructions for how to manufacture gold. At some point Kelley must have thought the game was up; he tried to escape down a rope from a window in the castle turret, but the rope he chose was not long enough to reach the ground – a rather basic error for a scientist. He fell, broke one or possibly both legs, and later died from his injuries.

Alchemy is the father of modern chemistry, and the practical alchemists of the sixteenth and seventeenth centuries were professional experimenters. The key obsession of most alchemists was the idea of making gold, and fire, their primal 'element', wreathed in magic and mystery, was the key tool which needed to be mastered. As the history of Kelley illustrates, many of those who considered themselves élite possessors of occult, privileged knowledge hoped to gull the foolish, blinding them with science.

In *The Alchemist*, Ben Jonson's play written in 1610, which satirizes both Dee and Kelley, Subtle the alchemist fools Sir Epicure Mammon by promising to make him some gold more pure than can be found even in nature:

> But these two
> Make the rest ductile, malleable, extensive.
> And even in gold they are; for we do find
> Seeds of them, in our fire, and gold in them;
> And can produce the species of each metal
> More perfect thence, than nature doth in earth.
>
> (Act II, scene 1)

Ben Jonson's response to the obfuscations of the alchemists is still relevant today. Even though there is now some recognition of the need for scientists to communicate their knowledge, many scientists still consider that they are above any requirement to explain and share the results of their work – which, incidentally, in most modern societies has been paid for by the taxpayer. Sometimes they go so far as to defend being obscure – even to the point of criticizing those of their colleagues who try to explain the principles and practice of the experiments they are conducting, and who dare to publicize the potential risks of the technology they are exploring.

NANOTECHNOLOGY: TINY PARTICLES, GREAT BENEFITS – AND GREAT RISKS?

There is a modern version of 'alchemy' which is a perfect example of the danger of keeping potential risks hidden. Those metals that fascinated the alchemists so much do indeed turn out to have many hidden and unexpected properties. Gold is a largely inert element: but when it exists as very tiny particles on the nano scale – with dimensions, say, around one-thousandth of the thickness of a human hair – it can be highly reactive. Not only can it be soluble, but its physical properties mean that it can be used to catalyse, or speed up, many chemical reactions. An opaque metal like copper in similarly tiny particles can become transparent, and a stable element like aluminium capable of burning furiously. It so happens that when many substances are produced as very small particles, their physical properties change in ways that are remarkable and mysterious – and also, sometimes, unpredictable.

One nanometre is one billionth (10^{-9}) of a metre. To give you an idea of just how tiny this is, the smallest life forms known (e.g. the bacterium *Mycoplasma* – see page 369) are around 2,000 nanometres in length. The DNA helix has a diameter of 2 nanometres, and an average human hair is approximately 25,000 nanometres thick. To put it another way, a nanometre is roughly the same size in relation to a metre as the diameter of a marble is to the diameter of the earth.

A number of changes occur in a material when its particle size is reduced to 100 nanometres or less. Its electronic properties change as the so-called 'quantum size effect' kicks in. At this size, the atoms or molecules that make up a nanoparticle behave differently from atoms or molecules in a bulkier piece of material. This is because chemical forces and hydrogen bonding have a much larger influence on the physical properties of nanoparticles than do physical forces such as gravitational strength. Also, when a substance is of nanoscale size, the ratio of its surface area to its volume changes, altering its mechanical, thermal and chemical attributes.

It turns out the idea of nanotechnology is not new. In a remarkable lecture given in 1959, the Nobel Prize-winning physicist Richard Feynman, of the California Institute of Technology, predicted the advent of this area of science and pointed out that the manipulation of very tiny particles might be highly useful in the future.[10] While he did not coin the terms 'nanoparticle' or 'nanotechnology', he argued that, with the use of particles at the atomic level, we might be able to make minuscule robots to carry out medical treatments inside a person's body, miniature computers and cameras, and much more powerful microscopes. He also recognized that many elements, in particles reduced to the nanolevel in size, would have many curious and unpredicted properties. But it is only relatively recently, as a result of advances in modern chemical processes, that scientists have been able to manufacture very small particles of substances and then assemble them to make a structure of any desired size, just as Feynman postulated.

How may nanotechnology be useful? It is already used in the clothing industry. Nanoparticles embedded in a garment – for example, a khaki shirt – can make it so stain-resistant that dry cleaning or detergents are unnecessary. In other areas, nanocomposites are being manufactured which result in scratch-proof, rust-free plastic materials for car body manufacture. These materials are unusually strong and extremely light, and their use should mean safer automobiles requiring less fuel. Metal bearings made of metal nanocrystals are likely to be longer-lasting and produce less friction, leading to greater engine efficiency. Research is now in progress to see whether nanocrystals have a value in the manufacture of photovoltaic devices which produce energy from sunlight, and thus reduce CO_2 emissions.

Previously insoluble substances sometimes become soluble at the nanolevel, so some nanoparticles can be dissolved even in cold water. Consequently nanoparticles can increase the solubility and permeability of other substances to which they are attached. These properties may eventually make them very useful for drug delivery. Nanoparticles are already in use in some zinc oxide sun creams as they have the valuable property of being effective in absorbing ultraviolet light. The small size of the particles means that these creams may be transparent rather than white, spread more easily, and cover wider areas so that less cream is needed. Some commercially available lemonades and fruit juices already contain nanoadditives, which give attractive colour to the drinks.

There are an increasing number of industrial applications of this new science. Aluminium nanoparticles are now used in very combustible rocket fuels which burn with high efficiency. Copper nanoparticles are being incorporated into motor oils because of their excellent lubricant properties, which help reduce engine wear. A tungsten-carbide–cobalt composite nanopowder is now being employed to make alloys as hard as diamond that can be used in drill bits, cutting tools, armour plating and jet engine parts. Nearly every industry which requires components that are hard and durable is exploring the potential of nanoparticles. They are also increasingly being studied for use in car stereos, mobile phones, power-efficient light-emitting diodes, cameras, laptop computers and televisions.

Great advances seem likely in the manufacture of filters. Disposable aluminium nanofibres can be made so fine as to enable virtually all viruses, bacteria and other noxious organic compounds to be filtered from water even at very fast flow rates. These filters are likely to be much more efficient than those made from conventional materials and probably will be increasingly useful for sterilization of biological or pharmaceutical fluids (such as plasma), for the separation of different proteins used in the drug industry, and for the detection of dangerous substances in the air – for example, bacteria used in biological warfare. Just one passage of water through an appropriate nanofilter is likely to be sufficient to provide large amounts of clean, sterile drinking water for immediate use, and may be very valuable in poorer countries with inadequate or contaminated water supplies.

Provided that their lack of toxicity can be guaranteed, nanoparticles may be of immense value in human medicine. Apart from facilitating the delivery of pharmacological agents or drugs into tissues (for example, without the need for injections), it seems that nanotechnology may be used in future to build repair machines which could enter and leave cells, destroy intruders in blood vessels, and check the DNA in a cell for any errors or mutations. Nanotechnology holds particular promise for cancer treatment: nanoparticles attached to appropriate antibodies or peptides might be used to target tumours and the blood vessels which feed them, enabling various anti-cancer drugs to be released inside the tumours or their blood vessels which would kill them. Nanoparticles attached to suitable dyes or indicators may be useful to detect diseases, particularly in cancer screening.

Nevertheless, it is fair to say that nanotechnology has not yet delivered massive benefits which could not have been achieved using existing processes. In general, we already have fairly safe technologies that do many of the jobs proposed for nanotechnology quite efficiently. However, the nanotechnology field is extremely promising and many scientists believe that it will prove to be one of the most important and exciting areas of scientific progress in the twenty-first century. But, like all progress described in this book, the use of nanotechnology carries significant risks, and possible dangers.

Jessica Ponti and colleagues from the European Commission's Nanobiosciences Unit and In-vitro Methods Unit recently reported that nanoparticles of cobalt are highly toxic to cells in the laboratory cell culture. While what happens to cultured cells by no means always reflects what happens in a living organism, the damage to DNA that she and her co-workers report is worrying because what they observed could just possibly increase the risk of cancerous cells developing.[11] This alarming finding is not an isolated result. Paresh Chandra Ray from Jackson State University in Mississippi and Peter Fu from the Division of Biochemical Toxicology, National Center for Toxicological Research, Jefferson, Arkansas, recently reviewed a number of research studies (including their own) on the toxicity of nanoparticles.[12] It seems that a number of elements which are normally not at all poisonous are toxic when reduced to particles with dimensions below 100 nanometres. Their review also implies that these dangerous effects are hard to predict – for example, some forms (for example, spheres) of nano-gold do not seem to be toxic, while others (perhaps rod shapes) may be biologically hazardous. Sometimes the toxicity that has been observed is much greater if the cells exposed to nanoparticles are already abnormal. So cancer cells may be much more likely than healthy cells to be poisoned by them – a finding which could be very valuable. But we still do not fully understand what makes a nanoparticle dangerous. It seems that when particles are reduced to the nanoscale their degree of toxicity may depend on their precise size and their shape – whether they are spheres, rods, tubes, or cone-shaped, for example. But the type of material is important, too: for example, nanosilver and nanocopper were found to be toxic to zebrafish and to their embryos, and to *Daphnia pulex*, the little water flea found in ponds. But under similar conditions, nano-titanium seemed harmless.[13]

It was established a few years ago that some nanoparticles disperse on the surface of the cells lining the lung after they have been inhaled. Once on the lining cells, they may be less likely to be cleared by the body's defence systems in people with chronic lung diseases. This phenomenon might be useful for treating some pulmonary conditions with certain drugs; but, as some nano-particles seem capable of causing extensive inflammation in the lungs of rats under experimental conditions, this is not a technology to be used lightly. Very fine carbon particles cross from the lung into the bloodstream very easily, and can also enter the central nervous system, where they may have an effect on the brain. One particular concern is that the structure of carbon nanotubes very much resembles the structure of asbestos. The inhalation of asbestos particles is known to cause lung cancer, and cancer of the lining of the chest wall and abdomen, so it is certainly possible that carbon nanotubes may carry similar risks.

One serious anxiety is that nanoparticles released into the environment may have adverse effects on ecosystems. We do not know how nanoparticles will behave, but it is clear that with the rising incidence of nanotechnology – for example, in sunblock creams, in factory emissions, or in discarded fabrics, paints and healthcare products – it is only a matter of time before water and soils are contaminated. Will these materials retain their original nanoscale size and structure and be as reactive as they are in laboratory systems where their effects can be carefully controlled? At present we have little idea whether the effect of nanoparticles on organisms in the natural environment will differ substantially from that of larger particles of the same material.

Nanomaterials may have strange effects on plants. Titanium oxide nanoparticles appear to promote photosynthesis and nitrogen metabolism and improve the growth of spinach plants, while aluminium oxide particles inhibited root formation in corn, cucumber, soya beans and carrots.[14] Nanoparticles also seem to alter the growth of seaweeds and algae in some circumstances. As yet we do not know what effect the ingestion of plants which have been exposed to nanoparticles growing in the sea, or on land, has on animals feeding on this vegetation. These particles may be highly toxic to many mammalian life forms, including humans.

All these examples are very good reasons why extensive research needs to be done before nanotechnology is implemented on anything except the most limited scale. And the spectre of the use of nanotechnology by unscrupulous governments to manufacture untraceable weapons of mass destruction is a genuine threat that we need to keep under careful review.

Nanotechnology provides a perfect example of why openness about science and an appreciation of what it can do, an understanding of the exploitation of ideas and the dark side of technology, are needed today as never before. So it is good that many scientists who are starting to take a lead in public engagement are focusing on the issues raised by the science of nanotechnology. In the UK, the Engineering and Physical Sciences Research Council (which is the largest body funding scientific research in Britain), the Royal Academy of Engineering and the Royal Society have shown commendable openness and willingness to communicate with the public and to hear and respond to their concerns about the complex issues which nanotechnology raises.

All of us must learn more about the advantages and disadvantages of technologies from which we hope benefit will result. Without our ability to harness fire and exploit its properties fully, humanity could not exist within the remarkably favourable environment we enjoy today. Almost everything we have around us

– our buildings, our comforts, our transport, our medicines – have needed that technology. In consequence of our ability to use it, we live more fulfilled and longer lives, and we are better protected from things which threatened humans in the past. Our new-found ability to manipulate metals and create minute particles may turn out to be yet another huge step forward in making our lives better. But the possible risks involved in this technology need careful evaluation and responsible control if we are to be safe. And ironically, as we shall see, our ancient ability to harness fire threatens our planet and could lead to our own destruction, if we are not vigilant.

Chapter 8

Sulphurous and Thought-executing Fires

WE HAVE USED fire for millennia and have a long tradition of manipulating it for our own ends, but it wasn't until the seventeenth century that we began truly to investigate it to find out what it is. Crucial to this journey of discovery was the work of the remarkable scientist Robert Boyle – a towering figure in the history of scientific investigation. The historian Professor Michael Hunter at Birkbeck College, London, has devoted much of his academic career to analysis of Boyle's life and work, and his eloquent writing about him is worthy of study.[1] Boyle, above all, believed implicitly in the importance of conducting experiments. The youngest son of the first earl of Cork, he was born at Lismore Castle in 1627, just seventeen years after Ben Jonson had written his highly satirical *The Alchemist*. The duke of Devonshire's family now own Lismore, and the Robert Boyle Science Room in the Heritage Centre nearby commemorates Boyle's origins and his commitment to experimentation.

FROM ALCHEMY TO EXPERIMENTATION

Boyle's father was Lord High Treasurer of Ireland, a position in which he amassed a huge personal fortune. So Robert grew up in a wealthy, aristocratic milieu, educated partly at home and partly at Eton College, and later on the continent of Europe. According to Professor Hunter, it was during his travels overseas as a young man that Boyle had a conversion experience, occasioned by an awe-inspiring thunderstorm. It seems this was highly formative and he remained a profoundly religious person throughout his life.

In 1649 he successfully set up a laboratory at his home in Stalbridge in Dorset (sadly, his house there is no longer in existence), and his experiments there transformed his career. His early work was clearly influenced by the alchemical tradition, but he was always intent on doing serious and reliable experiments based on accurate measurement. Later he moved to Oxford, where, assisted by the ample revenues from his father's Irish estates and his family connections, he settled comfortably, enjoying the company of many like-minded colleagues intent on gaining a better understanding of the natural world. During these years, Boyle developed a wide range of interests and published many papers. With the help of Robert Hooke he researched the nature of air, devising his most famous piece of experimental equipment, the vacuum chamber and air-pump. Using this, he discovered that a cat in a vacuum would suffocate, and he came uncannily close to discovering oxygen when he noted that both breathing and fire caused some of the air inside the chamber to be used up.

A contemporary of Boyle's, the German scientist George Stahl, had noted something similar about combustion, arriving at what was known as the 'phlogiston model' as a means of explaining what happened to substances that had been burned. In Stahl's view, phlogiston was a substance lost during burning, leaving any item burned lighter than it had been before. The reason, as Boyle had observed, that nothing burned without air was because air was necessary to absorb the escaping phlogiston.

It took a century for chemists to scrap the phlogiston model and realize that some substances actually became heavier after combustion. This wasn't down to any failings on the part of chemists or their discipline, but more because, as historian of science John Gribben points out, they lacked the necessary instrumentation. Physics could be carried out with balls on pieces of string – or even, if tales are to be trusted, little more than apples and trees. Chemistry, on the other hand, required reliable sources of heat, and accurate instruments for measuring temperature, as well as scales to record the weight of what was about to be burned and the weight of the ash that was produced. Although the potential existed for other thinkers to take Boyle's and Stahl's findings and develop them, they were hamstrung by the fact that the first really accurate mercury thermometer was not invented until 1714.

When it was, it did not take the French-born Scotsman Joseph Black long to exploit it. Black's initial interest was in the various dangerously caustic remedies used to dissolve kidney stones – or, rather, remedies ingested in the belief that they would. Investigating the properties of white magnesia in the hope that it might provide a viable and less harmful cure for stone, Black noticed that the substance lost weight whenever it was heated. He also noticed that it effervesced when combined with water, whereas other, more caustic alkalis did not. This led him to the conclusion that mild alkalis like white magnesia contain a quantity of 'fixed air' that can be liberated by burning or by combination with water. Using increasingly accurate weighing scales, Black examined and measured this 'fixed air' given off, concluding that it was different from ordinary air, but shared certain properties with it. Later, after publication had won him fame and fortune, he returned to these musings, concluding that 'fixed air' was also produced by the respiration of animals, the fermentation of liquids and the burning of charcoal. He had, of course, described carbon dioxide.

Black, who became Professor of Medicine at Glasgow University and a lifelong friend and supporter of James Watt, pioneer of the steam engine, was also interested in the properties of heat. During a decade of experimentation between 1756 and 1766, he developed two concepts which would be vital to those creating the technology of the industrial revolution. One was 'heat capacity' – Black having established that, to achieve the same rise in temperature, different substances required different quantities of heat. The second was 'latent heat' – the principle by which substances could change state, that is, liquefy, solidify, evaporate or condense, without a thermometer indicating any rise in temperature. In a style typical of the Scottish universities of the time, Black shared these ideas with all his students instead of publishing them in élite journals. They were to prove highly influential for the work of another scientist, Joseph Priestley.

or fall

In the late eighteenth century the environment was changing rapidly. People were not threatened by climate change and there was no obvious global warming as yet. But Britain was a country in rapid transition from a mostly agricultural economy into a mighty industrial nation. Corn and wool were being usurped by iron, cotton and coal. The provinces, particularly the Midlands, Lancashire and Yorkshire, were no longer poor neighbours of London but power-houses of the nation's wealth.[2] Birmingham was a natural centre for much activity, and around 1770 a group of 'ingenious philosophers', men of science, industry and letters, started to meet regularly. This club of influential figures called itself the Lunar Society, meeting for a dinner once a month on the Sunday evening (or, later during the club's existence, the Monday) closest to the full moon. The full moon was seemingly chosen because it gave them the most light to get home safely after dinner, when they would clamber on to their horses or into their carriages slightly the worse for wear after an evening of wine, food and conversation about the latest scientific and engineering ideas. The group, whose members often referred to themselves jokingly as 'Lunatics', included James Watt and Joseph Priestley, the potter Josiah Wedgwood, the inventor and political writer Richard Edgeworth, the industrialist Matthew Boulton, the doctor and botanist Erasmus Darwin (grandfather of Charles), and the soap and glass magnate James Keir. No women were allowed, and the men who provided technical help in their laboratories were seldom admitted to this rather exclusive group. (Does all this sound a bit like the Garrick Club today?) Their activities included logging the weather, four times a day for nine years; trying to extract alkali from sea water; and developing presses for stamping coins and time-punch clocks for monitoring factory workers. They even discussed the idea of an engine that used charcoal gas to power a piston by means of combustion, but dismissed it as too fanciful. The breadth of the Lunar Society's ideas reflected that of the interests and intellect of its members, epitomized by the figure of Joseph Priestley.

Priestley was an extraordinarily eclectic individual. By the age of thirty-four he had written an English grammar, a history of electricity about 250,000 words long (nearly twice as long as this book), and various pamphlets criticizing the government's handling of the American colonies. He studied the Bible in detail and came to the view that the idea of the Holy Trinity was absurd. 'I bless God that I was born a dissenter,' he once said, 'not manacled by the chains of so debasing a system as the Church of England, and that I was not educated at Oxford or Cambridge.'[3] He became political adviser to Lord Shelburne, a senior minister under George III, but was so outspoken on political issues that he was forced to move on. Being committed to the cause of the French Revolution he, together with various Dissenter friends from the Lunar Society, organized a dinner in Birmingham on 14 July 1791 to celebrate the second anniversary of the storming of the Bastille. The Lunar Men had become increasingly contemptuous of their own government, which they felt was oppressive. But Birmingham trade was suffering a temporarily severe economic downturn, parish rates were rising and people also thought they were being asked to pay a new tax to finance the Police Act. Conditions were ripe for a riot, and – religious Dissenters always being suspect – an organized mob, largely ignored by the authorities, went to the hotel where they hoped to disrupt the exclusive dinner party. But the diners, sensing trouble, had already left; so, having broken all the windows of the hotel, the frustrated mob attacked Priestley's house and pelted his family with stones as they abandoned it. After Priestley had escaped to a friend's home, they destroyed his library, scientific papers and various pieces of apparatus, and soon the house was ablaze. By the end of these riots, twenty-seven houses had been attacked, four meeting houses had been destroyed, and a policemen and eight rioters were dead.[4] Priestley himself fled to London, and a couple of years later, at the relatively advanced age of sixty-one, emigrated with his wife and children to the more tolerant climes of Pennsylvania, where he remained until his death in 1804.

Priestley identified a number of the gases involved in both breath-ing and combustion, and also was the first to identify oxygen as the component used by the body when air is breathed in. Obtaining pure oxygen by heating up oxides, he conducted experiments on mice to see how they performed both in ordinary conditions and in a rarefied atmosphere. Assuming that a mouse would run on a wheel for longer when the air it breathed contained more oxygen, he concluded that pure oxygen supported respiration four or five times better than ordinary air. Unwittingly, he had grasped that oxygen constitutes only one-fifth of the air we breathe.

Among the paintings on display in the National Gallery in London is the wonderful *An Experiment on a Bird in the Air Pump* by Joseph Wright of Derby, painted in 1768, which conjures up the fascination and terror of such exploration into science. A white cockatoo flutters helplessly in a large glass globe, from which the air is being sucked out by a scientist who is demonstrating his great expertise. The scientist pays no attention to the plight of the bird, nor does he engage with us, the viewers, at whom he stares out of the painting with a glazed look in his eyes. Around him, in the gloom of the laboratory, the fire from the experiment lights the faces of his audience dramatically. Two young sisters are horrified, the elder one averting her eyes; a young couple, apparently impervious to the cruelty of what is going on, are seemingly more interested in each other, while an older man, who refuses to look at the bird, seems to be pondering on the implications of what is happening.

Priestley was so much under the influence of Stahl's faulty 'phlogiston' model that he did not fully understand what happened to oxygen when things burned. That advance was made by the Frenchman Antoine Lavoisier, who observed – before the Revolutionary authorities, so beloved of Priestley, chopped his head off – that substances gain weight when burned. Using a vast magnifying lens 4 feet wide and 6 inches thick to concentrate the heat, he realized that during combustion oxygen from the air combined with the material being burned. This explained the gain in weight. Dismissing the phlogiston model for good, Lavoisier also examined the relationship between bodily heat and fire – concluding that both involved the consumption of oxygen and its conversion into 'fixed air', or carbon dioxide. This particular avenue of enquiry, advanced by packing an unfortunate guinea pig into a container surrounded by ice, was crucial in building the eighteenth-century view of the human body as a system no different from the rest of the universe, governed by the same forces and laws as those which controlled the falling of stones or the burning of candles.

The furthest-reaching implications of all these various studies of combustion were to make their mark not in medicine, but in industry. From the time men first observed the explosive actions of volcanoes, or the devouring properties of a bush fire, they could appreciate the energy locked up inside flame. In the eighteenth century, this observation became more practical and profit-oriented, as scientists pondered not merely the properties of fire but its potential for pulling, pushing, pounding and pumping.

A significant figure in this trend was the Glasgow University instrument-maker and member of the Lunar circle James Watt. Employed to assist those at the cutting edge of the new science, this shipwright's son was ideally placed to take the latest discoveries into the world of commerce. One day in 1759 – the exact date is not recorded – a university student named John Robison invited Watt to share his interest in some work being conducted on combustion. He would later become a professor of philosophy, and attract more fame for his off-the-wall theory about the French Revolution being caused by a Masonic conspiracy,[5] but at that time, Robison was principally interested in the properties of steam, and how they might be used to drive a small carriage.

This was hardly revolutionary. As long as humans have been boiling water, they have been observing the way a plume of steam can lift a heavy lid, or buffet the thatch of the roof above. In the first century CE the Greek mathematician Hero of Alexandria had developed a canny method for opening and shutting the doors of a temple by harnessing the steam from the altar fire. Of more practical use in later times was the invention in 1679 by the Frenchman Denis Papin of the pressure cooker, a 'steam digester, or engine for softening bones', described in a tract printed in Paris on *La Manière d'amollir les os et de faire couire toutes sortes de viandes en fort peu de temps et à peu de frais, avec une description de la marmite, ses propriétés et ses usages.*[6] In 1690 Papin published his first work on the steam engine, *De novis quibusdam machinis*, describing a mechanism built to raise water to a canal between Kassel and

Karlshaven. Soon afterwards he built another to pump water to a tank on the roof of a palace belonging to his patron, the Landgrave of Hesse-Kassel. Its object was to supply a head of pressure for the fountain in the castle grounds. And in 1707 Papin published *The New Art of Pumping Water by using Steam*. But Papin's steam engines leaked badly and the piston chamber was only cooled by air, which was quite inadequate for the purpose. So the apparatus hardly worked at all; although he was good on ideas, Papin was a poor engineer. It was fortunate that he also designed a safety valve for his machinery to prevent the pressure of steam reaching lethal levels.

Other projects of Papin's included the construction of a submarine – which seems to have sunk at its first outing, to the chagrin of his patron, the Landgrave, who had already paid handsomely for the invention – and an air gun and a grenade launcher for use during the War of the Spanish Succession. Several of Papin's papers seem to have been put before the Royal Society between 1707 and 1712 without his being properly acknowledged; one of them included a description of his 1690 steam engine, very similar to that which was finally built by Thomas Newcomen in 1712. Papin, who came to London as a Huguenot refugee, got little credit for his ideas, suffering from the political and religious intrigue that plagued science at the time, as well as personal rivalries (Papin was a friend of Leibniz, and therefore possibly at odds with Isaac Newton, President of the Royal Society). Such conflicts still beset scientists and their work today. Papin died, destitute, in 1712 and was buried in an unmarked pauper's pit.

In 1698 the inventor Thomas Savery patented an 'engine for raising water by fire', which he dubbed 'the Miner's Friend', on account of its usefulness as a pump for the mining industry. So, in various guises, the 'steam engine' had been around for a long time. But all this only meant that Watt, when he came to develop his famous invention, was working in a milieu with which he and others were very familiar. Steam was almost mundane. Its consequences would not be.

INDUSTRY ON THE MOVE

Sent in 1842 to a branch of his father's thriving textile business in Manchester, the Rhinelander Friedrich Engels was a keen observer of social life. Bored by business, contemptuous of the affluent bourgeoisie, he set about compiling a record of his impressions that would form the basis of a seminal work. *The Condition of the Working Classes in England* was to seal Engels' reputation as a chronicler and thinker, and provide Karl Marx with material for his own assault on capitalism. While Marx's *Capital* seems to me, for the most part, an extremely dull book, Engels' prose has considerable descriptive power:

> Passing along a rough bank, among stakes and washing-lines, one penetrates this chaos of small, one-storied, one-roomed huts, in most of which there is no artificial floor; kitchen, living and sleeping-room all in one . . . Everywhere before the doors residue and offal; that any sort of pavement lay underneath could not be seen but only felt here and there with the feet. This collection of cattle-sheds for human beings was surrounded on two sides by houses and a factory . . . Everything which here arouses horror and indignation is of recent origin, belongs to the industrial epoch.

Later, Engels described the condition of child labourers in the glass industry: 'many of the children are pale, have red eyes, often blind for weeks at a time, suffer from violent nausea, vomiting, coughs, colds and rheumatism. The glass-blowers usually die young of debility or chest infections.'

Engels was not alone in being moved by the appalling poverty of England's industrial poor. In 1883 the writer Andrew Mearns compared the condition of urban slum-dwellers to the 'middle passage of a slave ship' and called for pity upon 'these thousands of beings who belong, as much as you, to the race for whom Christ died'.

Writers like Engels and Mearns were not slow in pinpointing the cause of this human misery. In England, a shift in economic patterns had brought about polluted, overcrowded cities, filled with workers who were utterly dependent on factory-based labour for their own subsistence. It could not have been possible without the harnessing of fire.

Many writers have devoted time and thought to considering why the industrial revolution should have occurred in England first. In the first place, being an island meant that it had limited supplies of the timber being so greedily snapped up by landowners and industrialists in Marx's Rhineland. This compelled the English to look elsewhere for sources of power. And criss-crossing our land underground were abundant, accessible seams of coal – a portable and super-efficient fuel. Also, while much of Europe was still hamstrung by ancient feudal institutions, which invested power in the hands of a few nobles, thwarting the ambitions of the middle classes, in Britain many of these privileges had been swept away in the Civil War, and there were fewer obstacles to the energy of those who lacked inherited wealth but had received sufficient education to go out to create their own. The after-effects of the Reformation also helped: Protestantism, to which the bulk of the population had converted, imbued believers with the conviction that God could be worshipped in and through daily life, even by busily making money. The mildly ascetic leanings of Protestantism, meanwhile, encouraged people to plough whatever money they made back into their businesses, rather than squandering it on extravagant displays of wealth. This, in turn, predisposed Protestant-owned businesses to a process of continual expansion.

But this revolution was not begun with coal. First, in the 1760s, came a weaving machine which could accomplish the work of several men, drawing its power from the hill streams of the Derbyshire peak district. But hydropower had severe limitations. Factories had to be sited close to the power sources – most often in remote river valleys, which posed a problem both for the transportation of the raw materials and the onward shipping of finished goods. Water, like wind, could be harnessed, but not summoned or even particularly well controlled: a harsh winter followed by a sudden thaw could bring a flood; an overlong summer and the power source would dwindle, along with the factory-owner's profits.

Coal was a different matter. It was portable – so the places that needed it could be situated wherever was most convenient and economic. It provided a regular source of power that could be shut down or switched on at the will of man. Transformed into steam, and harnessed to drive machines, coal marked a point where man was freed from his reliance upon nature, and became its true master. Or so he might believe.

In Greek mythology, Hephaestus – Vulcan to the Romans – was the god of the metalworkers. Depictions on vases show him crippled, sometimes with his feet facing backwards, or walking with a stick. His features, too, were deformed, perhaps demonstrating classic signs of arsenic poisoning. These features of the lame god may have had some reference to classical society. Metalworkers – such as smiths and miners – worked in appalling conditions, exposed to danger, noxious fumes and fierce heat. In addition, their skills were so valued that it wasn't uncommon for their masters to lame them deliberately to prevent them from running away.

The bulk of the earth's valuable resources are – as the German word for mines, *Bergwerke*, suggests – buried underneath mountains. Mining is perilous, because it is necessary to travel some way from the earth's surface in order to find the precious ores. As well as predisposing the mine to flooding, this means that there needs to be some form of power to transport miners and the stuff they mine to and from the surface, and to despatch it onwards.

As early as the fifteenth century, German miners – for it was within the territories of the Holy Roman Empire that the bulk of Europe's valuable ores were to be found – had developed a partial solution to this problem. A wheeled cart – known as a *Hund*, or dog – was placed on wooden tracks and guided along them by means of a metal nail that slotted into a gap between them. Power was still provided by men pushing or pulling the cart – but the task was much easier because it remained on a fixed path.

These proto-railways had been in use for a good hundred years before England's forests became so depleted (mostly for ship-building) in the reign of Queen Elizabeth I that coal began to replace wood as a fuel. The lack of wood also brought increasing demand for bricks as the new material for construction – and bricks required intense fires for their manufacture. Uneven growth of population, meanwhile, meant that the bulk of the demand for coal was in the south, while the coal was mostly concentrated in the mines of the north-east.

Prospector Huntington Beaumont was the first to think of a solution, leasing land in 1603 from Sir Philip Strelley in order to transport coal from his Nottinghamshire mines to the River Trent 'along a [2-mile] passage now laide with Railes, and with suche or the lyke Carriages as are now in use for that purpose'. These fixed wagon-ways meant that coal could be transported more swiftly and in greater quantities than by packhorses. Geography and gravity also lent a hand: most pitheads were much higher up than rivers and seaports, so the carriages of coal needed little force to move them.

Strelley, realizing what tidy profits Beaumont was making from his lands, appealed to Parliament to let him revise the terms of his rental agreement. Tired of the wrangling, Beaumont simply transferred his interests to Northumberland, where he built three fixed wagon-ways to take coal to the River Blyth, whence it could be taken by boat to the North Sea and then on to London. He lost almost all of his money in the process – but a precedent had been set, and others were quick to follow over the next two centuries. Of all the mine-to-sea wagon-ways built in those years the most significant was Killingworth, which first came into use in 1806. A young man called George Stephenson was employed to operate the brakes on the trains of loaded coal carriages.

But before Stephenson, another name would prove to be important – that of Thomas Newcomen, who in 1712 took Papin's steam-powered piston apparatus and developed it into a working engine. Whether the Devonshire ironmonger and preacher had any knowledge of the Frenchman's work is unclear, but his device worked along quite similar lines.

I am not sure whether climatic conditions were especially harsh in England in the early eighteenth century, but certainly many mines were abandoned in these years because of flooding. This was a particularly acute problem in low-lying Staffordshire, where Newcomen's steam engine was first used in 1712 to pump out water at the Dudley Castle mine. This machine, housed in a 40-foot-high brick building, required considerable capital investment, but its benefits were rapidly obvious to mine owners across the soggy land. Within three years, similar engines were being constructed throughout England – and by the time Newcomen died in 1729, his engines were to be found in Hungary, Belgium, Slovakia and France.

With the coal-mining industry now itself dependent on coal and fewer mines out of action through flooding, Newcomen's pumps resulted in a rapid acceleration of traffic down the fixed wagon-ways. In a beautiful feedback loop, the increased availability of coal enabled Abraham Darby to perfect the process of iron-smelting at Coalbrookdale in Shropshire, which in turn meant that Newcomen engines could be built using cheaper materials than the copper and brass of the earliest models.

Newcomen's engine was amazing, but it had shortcomings. It fitted the task for which it was created – pumping out water – but attempts to transfer this system to other contexts fell flat. The Newcomen engine, with its reciprocal up-and-down action, could not be used to make a wheel rotate efficiently. At that time, coal wagons were being pulled to the surface by means of a winding drum, the power being supplied by horses. Attempts to harness the rather erratic force of the Newcomen engine for the purpose proved unsuccessful.

Flooding was not, of course, only or even the most serious danger in the coal mine. While all kinds of injury, usually from falling rock, were common, by far the most terrifying threat came from fire, in particular explosion. At 11.30 a.m. on 25 May 1812, a massive explosion caused by the ignition of 'firedamp' – methane gas – erupted through the dimly lit pit at Felling, County Durham. The coal face was more than 600 feet underground and two shafts were in use: the William on a small hill and, about 550 yards away, the John shaft.

Two eruptions of fire burst from the mouth of the John shaft, immediately followed by one from the William shaft. Profuse quantities of dust, bits of coal and rock rose high into the air; eye witnesses described how the sky grew (and stayed) dark from the coal dust. The earth was felt to shake about half a mile away, and the noise of the explosion was heard at distances of 3 or 4 miles. The winding gear of both shafts was destroyed, the pulleys at the William disintegrated and the wooden frames were set on fire. Luckily the winding pulleys of the John pit – which were slung on a crane outside the blast area – remained intact, and half an hour later, thirty-two dazed men and boys were winched out through this shaft. But 128 men had been down the pit when the explosion occurred.

Distressed relatives and friends ran to the pithead, the miners' wives knowing almost certainly that many of them were already widows and possibly childless, too.[7]

Forty-five minutes after the explosions, with no more survivors emerging from the pit, a number of remarkable men courageously volunteered to go down to see if they could find anyone else alive. Nine miners descended the John shaft and reached the bottom, 660 feet below ground level, but could make very little progress through the low corridors where the roof had collapsed in places. The workings smelt burnt and gas was clearly still about, so reluctantly the rescue party retreated to the shaft bottom, ready to ascend the rope. Just as five of these brave rescuers had reached the top safely, with two still in the shaft and two at the bottom, gas in the mine exploded a second time. The men on the rope felt the heat wave blast past them, but, though singed badly, managed to cling on. Their two comrades below 'threw themselves on their faces and kept firm hold of a strong prop' and escaped without serious injury. Two further attempts at rescue were made, but it was clear that further descents were not only risky but utterly hopeless. Both shafts were sealed to block the oxygen in order to extinguish the fire below.

The ninety-two miners killed by the explosions, eventually interred at St Mary's Cemetery, Heworth, were not buried until July because it was unsafe to retrieve the corpses from the still smouldering mine earlier. Isaac Greener, aged sixty-five, a hewer, was the oldest miner who was killed; also lost were two of his sons, aged twenty-four and twenty-one. The youngest children who perished were Joseph and Thomas Gordon, aged ten and eight years old respectively. Their father, Robert, also died. Joseph and Thomas are in the mine's records as 'trappers'. Trappers stood at doors in various parts of the pit to open them to allow the trams of coal to pass. They then had to close them again immediately, to allow the air to flow down side corridors. Trappers started work at 2.00 a.m. and usually stayed in the pit for eighteen hours, for a daily wage of five pence. These little boys would be alone and in total darkness the whole time they were in the mine, except when a tram was passing.

The Felling disaster acted as a stimulus for attempts to improve safety in coal mines, in particular to find ways of avoiding methane explosions. The eminent Humphry Davy, knighted for his services to science that same year, was approached by a group of clergymen who wanted to avert more mining catastrophes. The result was the Davy lamp, where the flame was isolated inside a tube of wire gauze. This would prevent heat from the flame propagating, and therefore any methane in the atmosphere would not ignite. The lamp had another useful feature: the flame changed in size when methane was about, enabling miners to detect its presence. While Davy sought no profit from his lamp, it was still a matter of dispute about who had the idea first, as both George Stephenson and one Dr William Clanny claimed prior invention.

But while certainly an improvement, the lamp was hardly ideal. The light it gave out was very dim and many men, working long hours in semi-darkness, had their sight severely affected. Nor was it foolproof, for if the gauze rusted, or was not kept tightly sealed at the edges, a leak might allow methane to come into contact with the flame and an explosion could still occur. In any case, just the sparks from the pickaxes the men used were occasionally enough to ignite the firedamp. Even when 'safe' lighting was used, fire was a terrible risk. At Senghenydd Colliery in Wales, a pit lit by electricity, a massive explosion on 14 October 1913 took the lives of 439 miners.

Apart from the risks associated with fire and flood, and injuries to head and limbs, miners were also exposed to a long list of serious diseases – and still are, in parts of the world where coal is mined. Various lung diseases, including pneumoconiosis, emphysema, tuberculosis and cancer, affect miners very commonly, and while only a few thousand men have been killed in explosions or rock falls in the last hundred years in Wales, tens of thousands have suffered severe damage to their lungs. While the extent to which these lung diseases have caused premature death is debatable,[8] there is no dispute about the chronic ill health that many coal miners face. The health of mining communities is noticeably worse than that of non-mining communities with similar levels of income. For example, in Aberdare in South Wales in 1911, infant mortality was more than four times that in the local general population and, because so many of these families lived in very unhealthy conditions, those surviving infancy often had chronic disease, of which tuberculosis was the biggest problem.[9]

Mining remains a very dangerous job. In the United States, there is an average of 315 mining deaths each year, giving an annual fatality rate of 30.1 per 100,000 workers. This compares with the national average of 9 deaths per 100,000 workers. In China, the chance of a miner dying is thirty-seven times higher than in the United States, and nearly 6,000 coal-workers died there in the year 2000. This is many more than all the deaths resulting from accidents in nuclear power stations worldwide.

THE PERFECT ENGINE

In 1791 – almost another epoch, given the pace of technological change – the Albion Flour Mill in London's Bermondsey went up in flames just five years after opening. Flour grains and dust in the air are very inflammable and flour mills are prone to fire, but in this case the evidence points to arson. Powered by one of James Watt's new rotary steam engines, the Albion was London's first great factory. It could churn out an unprecedented ten bushels of flour per hour, putting many rival millers out of business. As Thomas Crump notes,[10] a great many small London businessmen had cause to rejoice at what happened on the night of 3 March, a conflagration captured in a famous painting by Charles Augustus Pugin.[11] The blackened shell of the mill remained standing for a long time, and was a regular haunt of Lambeth-based poet and artist William Blake. Some speculate that this sight may have fed into his often quoted (and often misquoted) polemic against the industrial revolution, 'Jerusalem', in particular influencing the line: 'The banks of the Thames are clouded! The ancient porches of Albion are Darken'd.'

The culprits – if there were any – were never found, but it was the desire to improve upon Newcomen's engine that had given the Albion Mill such a massive, jealousy-inducing advantage. In partnership with the entrepreneur Matthew Boulton, James Watt had moved away from the steam car and devoted his attentions to the needs of industry. He kept a patent covering possible applications of his improved steam engine for transport, but Watt's first major achievement was the development of his rotary machine. Once again, the benefits this offered to industry resulted in an increased demand for coal – putting further pressure on England's congested waterways.

Watt's contribution to the industrial revolution was significant, but must be kept in perspective. Most scientific discovery, and most of the technological developments stemming from it, are made in small, incremental jumps, but so often just one individual gets singled out and given credit. Watt was not, as we have seen, the first to ponder subjects like heat, fire and pressure, nor even the first to apply steam power to the needs of industry. But he did make that power truly attractive in terms of cost and efficiency. In keeping with his staunchly Calvinist upbringing, Watt abhorred waste. For long periods of his life he kept detailed personal accounts, recording every minute transaction from the extraction of teeth to the sale of 'an old big overcoat'. He transferred this personal concern with loss and waste into his work on the Newcomen engine – a miniature working model of which had been rusting away in the Glasgow University stores.

This model did not work very well, but even so, if fired up it required a considerable amount of fuel – a complaint echoed by those who used full-scale functional versions in the mining industry. Although he had a commendable obsession with creating what he called 'the perfect engine', with minimal losses of steam and heat, and maximum power for the least possible fuel, Watt's interest was always partly financial. His father had lost most of his money, so he had no inheritance, his university work providing only a modest income. Watt perceived, astutely, that the 'perfect engine' would sell, because its physical advantages were replicated economically: maximum gain for minimum costs.

A Newcomen engine was not a complex piece of kit. Inside a vertical iron cylinder with an open top was a piston, connected by chains to a quadrant, shaped like a quarter-circle. This quadrant was attached to one end of a wooden beam, pivoted in the centre, the other end of which was connected to a pump. When the furnace was fired up, steam from a boiler was allowed into the bottom of the cylinder, and then mixed with water. This would then condense, leaving a partial vacuum underneath the piston. Atmospheric pressure would push the piston downwards, causing the attached beam to work the pump.

Simple as its principles were, the average Newcomen engine came with a lot of complications in practice. It needed a water supply, a tank to contain the water that was mixed with the steam, a pump to get that water into the cylinder, a valve to regulate its injection, a boiler to provide the steam, packing material to make the cylinder airtight. The first engine was 50 feet high, and came with its own brick house – along with a stoker and an engineer to run it.

Watt realized that the Newcomen engine was inherently inefficient when he attempted to run the university's small-scale model, which needed more steam than could be produced within its boiler to keep it going long enough for a useful classroom demonstration. None of his initial attempts to solve the problem worked. He tried to make more steam, by boiling the water more quickly. He also put the cistern higher up, so that the water would enter at a higher pressure.

Long after he had attempted to modify the model engine, Watt continued to mull over the problem, studying various full-scale Newcomen engines to identify where heat and pressure were lost. He concluded that the issue was how many cylinders of steam were needed to make the piston rise and fall, and whether the vacuum was secure. Consulting his friend Joseph Black – although Watt, Black and Watt's backer Boulton would later quarrel over whose idea it was – Watt worked out how much steam was needed by a Newcomen engine for each stroke. He then calculated how this could be reduced, which would correspondingly reduce the amount of fuel necessary to make the engine run.

Some potential sources of loss were easy to spot. As Black's work on latent heat had shown, boilers and cylinders made of different metals acquired and passed on heat at quite differing rates. To create the perfect vacuum, a cylinder which was cooled quickly was needed. Watt had also to measure how much cold water was needed to cool the cylinder sufficiently to condense the steam. This gave rise to a puzzle. In theory, it was possible to assess how much cold water was necessary by calculating the amount of heat required to raise the cylinder's temperature (its 'heat capacity') and then multiplying that figure by its weight. But when Watt applied that figure in practice, he was surprised to find that yet more cold water was required. And after the cold water had done its job, it was far hotter than he would have expected.

He also encountered another problem. In working to obtain the perfect vacuum, Watt had assumed that the cylinder and the steam didn't need to be cooled much below boiling point, since that was the temperature at which steam condensed into water. But when the pressure in the cylinder was low, hot condensation water turned back into steam because water at low pressure boils at lower temperatures. This destroyed the vacuum.

Discussions with Black led Watt to understand something of the concept of latent heat, and how it applied to steam. It occurred to him that steam contained more heat than water which was on the brink of boiling, which is why it took more fuel to turn water into steam than it did just to make it boil. This, in turn, was why a small amount of steam raised the temperature of the cylinder, and required more cold water than he had expected to cool it.

The problem, in a nutshell, was this. At least one full cylinder of steam was needed to move the piston. But a cylinder full of steam was hotter than boiling water. Cold water then had to be injected, not just to condense the steam, but also to bring the cylinder down to such conditions as to produce a vacuum. When steam was let back into the cylinder as the piston rose, most of this steam was used in reheating the cylinder rather than in filling it. The constant cooling and heating in Newcomen's engine was the cause of continual losses in terms of steam, fuel and vacuum.

Watt was striding across Glasgow Green one Sunday afternoon in 1765 when the solution hit him. His idea was this – the cylinder would not need constant heating and cooling cycles, and steam need not be wasted, if the steam were condensed in a separate vessel connected to the cylinder by a pipe with a valve. This separate 'condenser' was the first of a series of improvements. Heat loss, another source of waste, was prevented by wrapping the cylinder inside another one, the space between them filled with steam and the outer surface covered in an insulating 'jacket' of wood.

For a while, Watt struggled to develop the idea into a working pump-engine, with cautious backing from a local mine owner, John Roebuck. But using his skills as a civil engineer proved more lucrative, and work on steam engines halted until 1767, when Watt had a propitious meeting with Birmingham buckle-maker Matthew Boulton. Boulton's sizeable works depended for power upon a water-wheel, which, as we have seen, is an unreliable source of energy. He had already considered using a Newcomen engine to pump water up and over the water wheel, but abandoned the idea. Talking with Watt convinced him of the need to return to steam as a source of power.

Roebuck and Boulton were as keen to back Watt as they were cautious of one another. Nevertheless, in 1769, when the first working engine was in place at his mine, Roebuck offered Boulton a one-third share in Watt's patent. Boulton quietly declined, which turned out to be a wise decision because by 1770 Roebuck was in serious financial trouble. Boulton eventually stepped in as sole backer, persuading Watt, who was grieving for the loss of his wife and frustrated over teething problems with the working engines, to leave Scotland for Birmingham.

In 1775 the two men formalized their partnership, and began to supply Watt engines to mines as far afield as Cornwall and Scotland. In Cornwall the new technology was truly put to the test. Tin mines were deep; also, coal had to be shipped in at considerable cost. About sixty Cornish mine managers had purchased Watt engines by 1775, and they kept records of how well their machinery performed, often comparing notes. A peculiar system of payment was in place, whereby mine owners calculated how much money they saved by using a Watt engine and paid the Watt–Boulton company a third of that sum. This led to constant disputes and Watt found himself so often in Cornwall that, despite loathing the county's scenery, its inhabitants and their methods of business, he purchased a house there.

A desire to escape being a sales rep, debt collector and troubleshooter impelled Watt back to the drawing board. He knew from his direct association with Boulton that a simple, reciprocal-action engine was of little use to the factory owner. A circular motion was required – but not the jerky, inefficient one provided by the treadle, which for centuries had converted the vertical action of a foot-driven pedal into power for a circular wheel.

By 1781 – after several false starts, and one spat with a rival engineer, Matthew Wasborough – Watt had a solution, which he called his 'sun-and-planet' device. In time, it would revolutionize British industry and society. It consisted of two cog-wheels of the

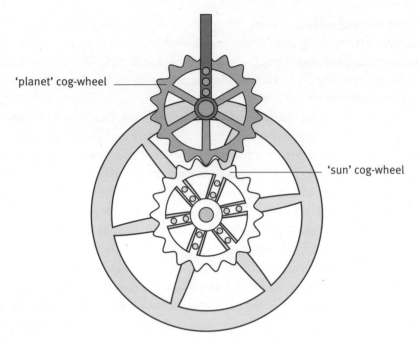

'planet' cog-wheel

'sun' cog-wheel

Watt's 'sun-and-planet' mechanism

same size. One of these – the sun, as Watt called it – was attached to the end of an axis which supported a flywheel. The other – the planet – was fixed to a rod, whose other end was joined to a pivoted wooden beam. Unlike the sun, the planet wheel could not rotate on its own axis, but was fixed into an 'orbit'. With the engine fired up, the beam see-sawed and the top end of the connecting rod moved with it, pulling the planet in a circle around the rotating sun. And since the two wheels had interlocking teeth, the sun made two complete revolutions for every single orbit of the planet.

Ever concerned with waste, Watt was troubled that his engine produced power only on the upward stroke of its piston. By doubling the apparatus, so that steam was let into both the bottom and the top of the cylinder, he obtained a piston that provided power on both strokes. The idea was immediately attractive to high-output industry, although not, as the Albion Mills fire suggested, to the competition.

At this stage, although engines were being used to pull, push and pump, the idea of using one to propel a wheeled vehicle was beyond consideration, because the sheer size and weight of the apparatus would require almost its own weight in fuel to move it. But in 1799 the Cornish mining engineer Robert Trevithick launched the high-pressure engine, in which steam was generated in a stronger chamber. Having dispensed with the heavy condenser, which made his engine cheaper, lighter and less bulky, Trevithick began to ponder its suitability for transportation. On Boxing Day 1801 crowds gathered at Beacon Hill, outside Trevithick's home town of Camborne, to see the 'Puffing Devil', the first steam carriage. A second run took place on 27 December, but due to over-liberal consumption of Christmas spirits the carriage was driven into a gully. The party and onlookers retired to an adjacent tavern to steady their nerves, enjoying roast goose and suitable liquid refreshment – at such length that the engine they had left outside overheated and the entire apparatus burned to a cinder. Nevertheless, Trevithick had established that steam-powered transport was possible, and in 1802 he was granted a patent.[12] Two years later, Trevithick's steam engine was used to pull loaded carriages 10 miles from a South Wales ironworks to the Glamorganshire Canal. Steam transportation was effectively already a decade old when, across the other side of the country, a self-educated mine-pump operator called George Stephenson was instructed by his employers at the Killingworth colliery to build a coal-fired device to take carriages down to the River Tyne.

Although it would exist for another 170 years, the north-east's mining industry was in something of a crisis when Stephenson set to his task. Yields from the pits with closest access to the sea and inland waterways were falling. Prospectors, meanwhile, were eager to exploit the rich seams of the central Auckland coalfields, but the costs of getting coal from there to the rest of the country would eat too much into their profits. To address this problem, the burghers of Stockton and Darlington commissioned Stephenson – now joined by his son Robert – to build a railway linking the coalfields to the waterways in 1821.

This was the most complex task so far required of any railway. The route was not only a long one, about 26 miles, it also involved steep inclines. So there was considerable wonder when the crowds gathered to see the line in action on 26 September 1825. Nor was the excitement confined to the various feats of civil engineering on display: Stephenson's engine, the *Locomotion*, was attached to twenty-one specially converted coal carriages he called the *Experiment* – we, today, would call it a passenger train – tickets for which had been allotted to 300 lucky citizens of the area. Such was the excess of enthusiasm that almost twice that figure surged on to the train. It wasn't exactly a quick journey – the first 19 miles took two hours to cover. Nonetheless, Stephenson's train completed a circuit, shuttling both coal and people between Darlington and Stockton, arriving back at Stockton Quay at 3.45 p.m. to a seven-gun salute and a night of unabashed celebration.

A BRIGHT NEW AGE?

On all sides, the fires of the foundries are shining, on all sides you hear the dull humming of the engines, the steam is whistling, the dark smoke billows in black plumes along the track, or the vapour rises up in a white thin column. The steam boats glide by you on the river; the locomotives thunder past you with their trains.

So wrote the Dutch socialist historian Henrik Quack, describing his first visit to the Belgian city of Liège around 1860. We can forgive him some hyperbole, I think, for the spread of the fire-fed industrial revolution was no less significant than its consequences. Aided by cheaper and swifter transportation, and drawn by the prospects of factory work, people flocked from the countryside to the towns and cities. Europe's major conurbations became melting-pots for people of different regions and nationalities – and amid the smoke and the fumes and the sparks of this new, optimistic age, ideas were swapped and fresh ideologies forged. Steam-powered presses produced, as we saw in chapter 5, a wealth of ever cheaper literature and newspapers, while the need for safety on the railways spawned the invention of the telegraph for fast, effective global communication. The world became a smaller place, not least because steam-powered ships could take goods, ideas and people themselves across its oceans in increasingly shorter times.

It was a giddy age, and that spirit is reflected in the ambitious, overreaching and slightly pompous tone of its culture. In men's fashion, towering top hats and lavish side-whiskers seemed to make their owners larger than life – outward expressions of their confidence and certainty. Buildings like Liverpool's St George's Hall and Manchester Town Hall recalled, without a blush of modesty, the grandeur of ancient Rome, not least because the city fathers who commissioned them saw themselves as founders of a new civilization.

Changes took place at other levels, too. The acrimony that disfigured Huntington Beaumont's business arrangement of 1603 with the baronet landowner Sir Philip Strelley foreshadowed later, more bitter conflicts between the aristocracy and the arrivistes. When George Stephenson embarked on his next major English railway project – the construction of a line between the port of Liverpool and the cotton mills of Manchester – his route passed over the ancestral lands of Lords Derby and Sefton. Both were so opposed to the railway – or, more accurately, to the prospect of mere commoners making money from their lands – that they encouraged harassment of, even attacks on, Stephenson's surveyors. But however much they struggled, the nobility could not stop wealth shifting from those who owned land to those who owned machinery. With the opportunities now open for global trading, it was a different kind of wealth, too, rapidly making the assets of the old landed classes look puny. In turn, the power and the influence passed to those with the most at stake. The nineteenth century was an era of unprecedented peace in Europe, because the industrialists had more to gain by cooperation than the old warrior classes had from fighting. War became thought of, for the first time, in terms of its economic costs, with banker James Mayer de Rothschild pointing out in 1830 that a war would make his rents drop by 30 per cent. (This was, incidentally, the same James Mayer de Rothschild who had recently bankrolled Louis-Philippe's takeover of the French throne.)

When Stephenson's Liverpool to Manchester line was eventually opened on 15 September 1830, it was not greeted with the same scenes of jubilation. Liverpool MP William Huskisson had invited the Prime Minister, the duke of Wellington, to attend the event as a public relations exercise. Wellington, the 'Iron Duke', took a notoriously high-handed attitude to the lower orders, and was steadfastly opposed to anything amounting to granting privileges to the growing industrial poor. Nevertheless, seeking to improve his public image in the northern cities, the duke accepted the invitation, which necessitated a string of complex preparations for the day's events. This resulted in the duke's train spending some time in a siding. The unfortunate Mr Huskisson, his legs still stiff from having just sat through the lengthy funeral of King George IV, took a stroll along the track, and was promptly killed by Stephenson's *Rocket* trundling up the other line.

Wellington considered abandoning his onward journey to Manchester, but was advised that this might be taken as a snub by the crowd waiting in the rain for him and decided to go ahead. He arrived late, mainly because hostile, jeering crowds had blocked the way for the final mile, and when the train came to a halt it was mobbed on the platform. The Iron Duke refused to get out, and police had to chase away the crowds before the train made an immediate return to Liverpool.

The angry scenes which greeted the Duke of Wellington at Manchester were due in no small part to the conditions created by this fire-fuelled industrial revolution. The use of coal may have made us less dependent on unpredictable forces of nature, like water and wind, but it made us increasingly dependent on each other. As we saw in chapter 3 in relation to the food industry, the link between the craftsman and his product was broken. The village carpenter who had previously produced chairs in his back room, and the baker who had sold his bread at the market, had become separated from what they made and from their customers. Factories atomized production so that men and women might spend a lifetime doing nothing more meaningful than drilling a tiny hole in an unrecognizable component of some object they would never see in its entirety. This was the root of the 'alienation' of which Karl Marx wrote so persuasively in *Das Kapital*.

In those days, capitalism was an inherently chaotic system, with long, complex chains of suppliers and agents and investors and distributors, all constantly scrabbling for bigger slices of the pie. This frequently resulted in overproduction, leading to surpluses, price drops, mass unemployment and intolerable suffering for those at the very bottom of the chain – the workers. They had merely swapped the insecurity of the harvest for the insecurity of the wage-labour system, and at the same time lost a considerable amount of freedom.

The worker's whole life – the way he organized his day, his week and his calendar – and the lives of his wife and children all became subject to the rhythms of machine-led production. Confined to filthy, insanitary, overcrowded cities, his old rural pastimes eroded and forgotten, the most fortunate working man of northern England could only look forward to 'Wakes Week', when he and his co-workers might go, en masse, to a seaside resort like Blackpool, while his factory underwent essential maintenance. Even that was a hard-won privilege.

Marx called this the *Verelendung* of the common man – pauperization or, literally, the process by which he was made wretched. And not all the well-to-do were unaware of its horrors. After a visit to Manchester in 1835, the French political thinker Alexis de Tocqueville noted: 'here is the slave, there the master; there is the wealth of some, here the poverty of most; there the efforts of thousands produce, to the profit of one man, what society has not yet learnt to give ... Here civilisation makes its miracles, and civilised man is turned back almost into a savage.'[13] When we read of the endemic suffering inherent in the factory system, we can understand the appeal of Marx's ideology, and may feel that those who overthrew the tsarist regime in Russia in 1917 were justified. But 3.5 million people died under Lenin, around 50 million under Stalin, and perhaps 40 million under Mao Tse Tung. And the Cold War cost $13.1 trillion.[14] If those figures are too great to comprehend, then consider the 125 mostly young, healthy people who were shot dead in Berlin between 1961 and 1989 for no crime, simply for wanting to pass from one part of their city to another. In non-revolutionary countries like Britain, the privations of the industrial revolution spawned great things: the labour movement, universal suffrage, greater concern for public health and education, the welfare state. But these benefits, in terms every capitalist would understand, must be weighed against the costs.

PROGRESS OR POISON?

It is easy to find the town of Dombivli, in India's north-western state of Maharashtra. It has some 100,000 residents, mainly employed in the local industry – which is producing intermediary chemicals for the dye industry. The town is visible for miles around, because it is perpetually swathed in choking smog. Residents claim that hazardous wastes are dumped in nearby fields, from where they are washed into people's houses during floods. Doctors are continually asked to treat respiratory, skin and eye disorders. According to one source, the fifty different chemical units operating in the area co-ordinate the release of poisonous gases, such as chlorine, so that no single operator can be identified as the culprit.

Maharashtra and the nearby state of Gujarat are jewels of the new India's crown – its two most industrialized states, together responsible for more than a third of the income created by factory-based manufacture. Governments in both states boast of the booming prosperity and opportunities enjoyed by the region, and do everything to attract new industries. But travel into the countryside around Ahmadabad and Mumbai, and progress is not the word that first comes to mind. Chimneys belch choking fumes into the sky. Industrial units dump toxic sludge by the roadside. In the villages, farmers speak not only of poor health, but of bright red water which kills their crops and destroys their livelihoods. Forced off their land, they have little alternative but to turn to the polluting factories for work – a transition which guarantees their continuing lack of opposition to what the industries are doing.

The Indian government has little incentive to tackle the pollution. Maharashtra generates the highest amount of tax revenue, and the highest GDP, of all the Indian states. Handcuffing industries with over-zealous legislation would result in financial loss. Accordingly, in this powerhouse of industrial progress, it is estimated that every major town has air deemed unsuitable for breathing. Since 1985, the state has had the highest number of reported chemical accidents; 80 per cent of its 83,000 industrial units are shown to pollute water, 15 per cent to pollute the air. As its urban areas become too cramped, prospective businesses simply move to the countryside, resulting in further environmental degradation.

A similar picture emerges of manufacturing in China, Africa and the former Soviet Union – so similar, in fact, that a combined corpus of environmental groups recently issued a 'top ten' of worst-polluted places. Top of the list was Sumgayit, in Azerbaijan, where cancer rates are estimated to be around 20–50 per cent higher than the national average. In third place was Tianying, in China – the city that produces a third of the country's lead. The ten sites, in seven countries, are estimated to affect the health of more than 12 million people – and they all have one thing in common. The damage is due to extensive mining, heavy industry and the commercial pressures which drive them.

Mere descriptions cannot do justice to the state of affairs in China, where respiratory and heart complaints are a leading cause of death. The Chinese are inveterate smokers of high-tar cigarettes, of course, but poor air quality alone causes up to three-quarters of a million fatalities every year. It is so bad that in Zhejiang province in 2005 some 30,000 people clashed with police in a pollution riot. It had begun peacefully, when a group of pensioners established a makeshift camp to protest at the number of factories that had sprung up in the area. They claimed the turnip crop had been affected by emissions from the factories to such an extent that, as in India, local farmers had been forced to turn to the polluting factories for work. When rumours spread that an elderly protester had been killed by police, the embittered local population grew violent.

There is a link, certainly, between the relative youth of an economy and its dirtiness. From the 1970s to the 1990s, Western governments tightened up on the environmental responsibilities of industry; but by investing in factories in the developing world, and championing capitalism in the communist one, they passed the baton of pollution. In many countries, lack of government regulation, and sometimes corruption, are significant factors. But the peaks and troughs of the global markets play a part. In new and relatively unstable countries, small economic ripples are felt as major waves. Buffeted financially, new industrialists seek to cut costs and secure profit – frequently avoiding safety and pollution controls. And the problem is also more complex than that. In Maharashtra, during periodic recessions, bigger factories have offered cash to smaller and worse-hit businesses to dispose of their waste. Chemical plants in remote areas, facing high costs, have falsely labelled their waste as processed goods and 'sold' it to plants closer to the sea, where it is then dumped.

It is not only in the developing world that industry is choking the

environment. Acid rain, that demon of the eighties ecology movement, has been dramatically reduced in Europe, but it continues in
the United States, where coal-fired plants still provide 50 per cent
of the country's electricity. The rain – occurring when pollutants
from coal-burning processes, such as sulphur dioxide and nitrogen
oxide, mix with water in the air to form sulphuric and nitric acids –
can destroy whole ecosystems, from trees to fish and insects. And
humans suffer, too. One of the few creatures known to benefit from
acid rain is the blackfly, which thrives in toxic conditions and, in the
absence of competitors and predators, takes over in unprecedented
volume. Leaving behind painful bites and ugly welts on human skin,
a blackfly infestation causes whole areas of countryside to be hardly
inhabitable – with significant knock-on effects for tourism and other
local industries.

Acid rain also affects human life in other, less obvious ways. In
Pennsylvania, the Gettysburg Memorial, commemorating the 1863
battle – the turning point of the American Civil War – is among 300
historic monuments degraded by acid rain. In the Mexican state of
Veracruz, ancient stone carvings testify to the extraordinary
sophistication of the Totonac civilization, which erected the temple
city of El Tajin. Its many treasures include an unprecedented number
of depictions of an ancient ball-game, believed to have been an
initiation ritual for young warriors. It is unlikely we shall ever find
out, not least because the carvings are disappearing at an unnerving
rate. The finger of blame is pointed principally at heavy industries
near Mexico City, just to the west, but it is hard to be certain.

Mexico City is a particularly tough case. Here, in one of the
largest cities in the world with a population of close to 20 million
inhabitants, human activity combines with topology to produce a
poisonous scenario. The city is 2,240 metres above sea level and sits
in the crater of an old volcano. At this height, oxygen levels are
reduced and those little fires in the internal combustion engines of
motor cars are not so efficient. So unburnt fuel, particulate compounds and carbon monoxide pour into the thin atmosphere. Ozone
levels are also extremely high. Because Mexico City lies in the
tropics, not very far from equatorial regions, there is intense sunlight
much of the time. The smog produced as a result prevents better
light penetration, and heavily contaminated air just hangs over the

city. It is not merely statues which are a cause for concern: the air that hangs over the Mexican capital is thought to do more damage to children's lungs than cigarette smoke, and it possibly increases the risk of lung diseases when they are adults. Dr Lilian Calderon-Garciduenas and her colleagues from the Instituto Nacional de Pediatría in Mexico City paint quite a grim picture.[15] Apart from their concern about lung damage in children, they have found evidence of inflammation in the brains of some dogs living in Mexico City which closely resembles the effects of adult human neurodegenerative disease. They argue that there is a theoretical risk that some children exposed to this environment may carry an increased risk of developing Alzheimer's disease when adults. I know of no evidence yet suggesting that adult dementia is more common than average in this part of the world; but the increase in pollution is relatively recent, so it may be some years before its full effects can be adequately assessed.

Undoubtedly, acid rain has increased in Mexico, as it has in other places around the world. The pollutants that cause acid rain can travel hundreds of miles – one reason why national or state-wide legislation is useless and global control is needed. Whatever the origins of the problem, it is a poignant example of how industry, that hallmark of civilization itself, is rubbing out our own history and blighting our present.

In Europe, at least, the smoking chimneys and blast furnaces of the nineteenth century are relegated to museums: in the modern power station, combustion takes place in one room, fumes go into another, heat comes out of a third. But fire, and the products of fire, are still present, albeit subliminally, in all our surroundings – in our brick walls, in our tiled bathrooms, in our light bulbs and cookers. Fire still plays a huge part in our society: most of the energy we use still comes from combustion of fossil fuels. The motor car is a classic example of the way hidden fire continues to dominate our lives – combustion and heat are involved in almost every stage of its manufacture and transport, while the engine propels it by hundreds of small explosions. We do not think of fire when we turn our ignition keys, but that is what we use and, in doing so, we echo the actions of our earliest ancestors.

As the industrial revolution gathered pace, factory-owners looked for ways to maximize production. An easy fix was to run factories all night, a move which prompted the next generation of engineers to look at cheap and safe ways of providing lighting. It would not prove to be the final solution, but along the way oil presented itself as an abundant and efficient commodity. It was not long before people noted its wider potential as a fuel source – and the consequences of this human innovation could even bring about the end of all the civilizations we have built.

Chapter 9

Oil that Maketh Man's Heart Glad

IN 1901 A PHARMACEUTICAL salesman called Sidney Reilly set out for Tehran with a case of patent medicines. Urbane and handsome, he swiftly became a regular face at the dances and cocktail parties of the eastern city's various international legations, as well as an intimate of the Grand Vizier to the Shah, Mozaffer ad-Deen. After staying for several months, he departed with substantial payments and a book full of orders – all of which were subsequently fulfilled. It was only much later, when Reilly's grateful clients tried to order medicines and medical appliances a second time, that people began to scratch their heads. The man couldn't be traced. His company didn't exist. And nor, for that matter, did Reilly.

Sidney Reilly was the *nom de jeu* of a Russian Jew named Georgi Rosenblum, who left Odessa in 1893. The man himself was a chameleon – at times he claimed to be the son of an Irish seaman, at others the son of a Catholic clergyman, and at yet others the son of a Russian aristocrat from Tsar Alexander III's court. He was arrested at the age of eighteen by the tsarist secret police, faked his death in Odessa harbour and stowed away on a British ship to Brazil. After spells as a roadmender, brothel doorman and docker in

Rio de Janeiro, he ended up working as a cook for a British
intelligence expedition in 1895. Most historians claim that he saved
the expedition and the life of its leader by seizing an officer's pistol
when natives attacked them, but Andrew Cook, his biographer, is
not so sure.[1] Whatever the truth about this enigmatic man, it is clear
he was paid as an informant by Scotland Yard's Special Branch and
in later life served as the clandestine head of the British Secret
Service Bureau in Russia. His real mission in Tehran was to look
into the activities of one William Knox D'Arcy, an Australian gold
miner who had obtained a £10,000 concession from the Shah of
Persia to exploit the land's putative oil resources.

   The British government, appreciating the potential of this new
fuel source, and keen to protect its own access to it, was nervous.
Oil had been discovered not far away around Baku, in the Russian-
owned province of Azerbaijan. Persia lay in the buffer zone between
Russia and British India, whose ongoing cat-and-mouse struggles
were dubbed 'the Great Game' by Rudyard Kipling. Despite the
name, the diplomatic tensions between Great Britain and Russia in
the nineteenth century were serious. For Britain, the Middle East
covered vital transit routes to the eastern colonies of India and
Malaya. For Russia, it was an ambivalent and troublesome territory
at the fringes of its own, mostly Islamic, colonies. Both sides had
pressing reasons to seek and keep the upper hand – and the presence
of oil intensified them. With his Russian background, Reilly was
considered the perfect man to investigate the situation – gauging not
only the likelihood of an oil strike in the region, but also the extent
of the Russians' interest in it. On returning to HQ, he confirmed
that the discovery of oil was imminent. Russia, he proposed, could
be bought off with territory in Persia's northern provinces. The two
powers could effectively carve up the land between them.

TURNING ON THE TAP

Nothing was done about Reilly's recommendations until 1905,
when the British government learned that the prospector D'Arcy
was bankrupt. He had struck oil in Persia, but not in the quantities

he had hoped. He was now in Europe, trying to find someone to whom he could sell his concession from the Shah. Britain was determined D'Arcy should sell to them. It was equally determined that its rivals, France and Russia, should get no wind of this until the deal was done. Reilly was despatched to Cannes, where D'Arcy was rumoured to be undertaking negotiations with the French branch of the Rothschild banking dynasty. Disguised as a Catholic priest seeking donations from wealthy holidaymakers for his orphanage, Reilly kept watch on the Rothschild yacht, which D'Arcy was visiting each day. The mission now tapped into many of the agent's own deep-rooted fantasies and neuroses. According to some accounts, he had been traumatized to discover as an adolescent that he was not the heir to a line of Russian landowners, but merely the illegitimate son of a Jewish doctor. Getting the better of Rothschild seemed, in some way, to offer compensation for the way his life had been turned upside down. In this version of his story, Reilly's exile from Russia had begun shortly after his discovery, and now he saw a potential route back – not by the back door, in disguise, but as a hero of the nation, the man who'd brokered a major deal between the tsar and the king.

Seizing his moment, Reilly stormed on board the Rothschild yacht in his priestly gear, begging his reluctant hosts to give alms to his orphanage. Taking advantage of the confusion his unsolicited entry caused, he pulled the astonished D'Arcy aside and told him that he had a message from the British government. They would pay double whatever the Rothschilds were offering. He had just enough time to convey this, and to arrange a meeting at a hotel that evening, before the stunned yacht party agreed to make a sizeable donation to the fake priest and his 'orphans'.

History does not record what Reilly did with the money. It does, however, record the deal that D'Arcy struck with the British government. This led to the formation of the Anglo-Persian Oil Company – later British Petroleum – in which, on the advice of a young minister named Winston Churchill, the British government acquired a 51 per cent share. Oil was struck in Persia again – at 4 a.m. on 26 May 1908 – and this time in such quantities as to ensure Britain decades' worth of access to a vital fuel resource. Its abundance influenced Churchill in his decision to convert the British naval fleet from coal to oil, giving it a significant advantage in the First World War – one matched only by the power of the oil-fuelled tank on land.

Reilly's covert missions had a palpable effect on global history, an effect that continues to this day – not least because they confirmed our devotion to carbon fuels. Continuing his own remarkable and varied career as a British agent, Reilly was promoted to the rank of lieutenant in the Royal Canadian Flying Corps and a staff case officer in His Majesty's Secret Intelligence Service (SIS). He was awarded the Military Cross in 1919. After many espionage operations, he was eventually arrested in the Soviet Union and interrogated by OGPU. After a short spell in prison in 1925, he was executed by Soviet intelligence in a forest near Moscow. Reilly may never have been rehabilitated in his homeland, but he lives on as the inspiration for Ian Fleming's creation of James Bond.[2]

BLACK GOLD

The use of oil goes back thousands of years, to the ancient empires of the Near East, which used it for waterproofing ships, for covering roads, and possibly as a military weapon – crude oil may have been a major ingredient of the 'Greek fire' referred to in chapter 7. However, at this time its use was confined to a few purposes, and production to a handful of locations where the substance happened to seep up to the surface. Widespread exploitation of the earth's oilfields, and the almost universal employment of oil in activities like transport, manufacturing, plastics and agriculture, is a very recent phenomenon. And it began in a rather surprising way.

The factories spawned by the industrial revolution needed to be manned, and hence illuminated, around the clock in order to justify the huge expenditure involved in setting them up. The initial fuel of choice was whale oil, but as a result of overfishing in the Atlantic this had become very costly by the middle of the nineteenth century. In addition, this marine-derived light source gave off a revolting smell and burned with a constant hissing sound – irksome for a single household, a massive drawback for a factory with hundreds of lanterns. Coal gas was also in use, but this was toxic and highly flammable. So there was a ready market for the refined version of oil patented in 1854 as 'keroselain' or kerosene by its inventor, Canadian Abraham Gesner.[3] This had the added attraction of being a cheaper, more effective machine lubricant than animal fat – so industrialists obtained two products for the price of one. Despite these advantages, the market for oil failed to take off because the available methods for extraction and refinement were costly and not suited to large-scale production.

This introduces an important concept in the story of oil and other energy sources. Some scientists call it ERoEI – 'Energy Returned over Energy Invested'. Simply put, if it costs me $1 to extract and refine a barrel of oil which I can sell for $10, I'm in business. If it costs me $9 50¢, I may still be in business, if I can sell enough barrels. If, on the other hand, it costs me $10 or $11 to extract a barrel of oil, then I'm going to be bankrupt. When scientists use the term, of course, they are referring to the amount of energy used to obtain a fuel source, not the monetary cost – but the two are often related. As we shall see, this simple axiom has had a powerful effect on the planet, its economies and its cultures, and may even have a bearing on its future survival.

It certainly had a powerful effect on an arthritic former railway conductor and steamboat night clerk called Edwin Drake, who in 1859 took a recuperative break in New Haven, Connecticut. There he encountered a banker named George Bissell, who had recently formed a consortium with the aim of extracting oil from the muddy hills around Titusville, Pennsylvania. Demand from the burgeoning textile industry had inflated the price from 75¢ a gallon to around $2. Bissell took a shine to Drake, appointing him coordinator of the extraction project, and dubbing him 'Colonel' so that future investors would be impressed.

Odd though Drake's career might have been to that point, he knew a thing or two about drilling. Copying a type of oil rig used in Azerbaijan in 1847, he erected a steam-driven wheel, which drove a steel bit in and out of the ground – a venture which produced around thirty-five barrels of oil. Along the way he proved responsible for one of the most enduring symbols of the oil industry, by choosing to store the slippery product in wooden whisky barrels. To this day, the price of oil is measured in barrels.

Drake's first precious cargo sold at $40 a barrel, and it sparked a rush of interest. The region was flooded by 'wildcatter' prospectors – so called because they drilled in remote areas, where only the calls of the local wildlife could be heard, and they hung up the pelts of the feral cats they killed from the tops of the derricks they built. Freebooters, long disillusioned with the California gold rush, were given new hope. This surge of interest drove the development of the oil industry so fast that by 1861 the world's first oil refinery was established and Britain's factories received their first shipment of lamp oil by sea. Four years later, on the eastern side of the United States, what was probably the world's first oil pipeline was completed at Titusville, Pennsylvania, running for 5 miles from Edwin Drake's oilfield near Titusville to the Union City & Titusville Railroad.

But the glut of oil pushed prices into a tailspin – to a point where, sometimes, the $2–3 wooden barrels cost more than their contents. There was no science to drilling in those days: when the black gold appeared, a rush of prospectors swarmed in, using it up in a matter of months, and often damaging the geological substructures of the area in the process. When a hot spot ran dry, the shortages meant that prices shot up again, and in such a wildly fluctuating market fortunes were made and lost and made again in the space of weeks. Drake sank all of his own reserves into a new oil well which burned to the ground, and he later died in penury.

The chaotic character of this new industry, and the destruction of lives and livelihoods, appalled an Ohio Baptist book-keeper called John D. Rockefeller. The boom–bust cycle could, he surmised, be avoided if one company controlled all the wells and all the supply lines. So he set about making that the case – acquiring, in one period known as the 'Cleveland Massacre', twenty-four out of Ohio's twenty-six oil companies in a single month. His methods were anything but violent. He offered handsome prices for his acquisitions, and compensated the former owners with well-paid management jobs and shares in his own company, which he named Standard Oil.

Some of Rockefeller's methods, however, would bring him into disrepute. These were largely down to his assistant Henry Flagler, upon whose desk a plaque read: 'Do Unto Others As They Would Do Unto You. And Do It First.'[4] Flagler entered into secret agreements with the railroad companies, which were vital, in the absence of pipelines, for the transportation of the oil, to pay them more for carrying Rockefeller oil. A network of covert Flagler agents was spread around the country, keeping a watch on new drilling sites, railroad depots and oilmen's saloons. If a prospector, flush with a new discovery, was able to undercut the Standard Oil price on a barrel, then Rockefeller knew about it, and could afford to go lower, simply by pushing up his prices in some other part of his growing empire.

It was Flagler's machinations, in the end, that loosened Rockefeller's grip on the oil industry. America was founded on principles of fairness. The discovery, at the end of the nineteenth century, of massive new oilfields in California, Oklahoma and Texas heightened people's desire to gain a slice of the pie. And when the public – some of whom were immigrants who had fled from auto-cratic, secretive dictatorships in Europe – learned how Standard Oil was manipulating the market, with its agents and its underhand agreements, there was an outcry. In the popular press, the name of Rockefeller became synonymous with unbridled greed and ruthless-ness – which was in fact a shame, because the man himself lived modestly, and was one of capitalism's most philanthropic employ-ers. Nonetheless, in 1911 President Theodore Roosevelt responded to the public mood by ordering the break-up of Standard Oil. Its constituent parts live on, in the shape of well-known firms like Exxon, Mobil and Texaco. And even more significantly, perhaps, the legacy of Rockefeller's hated monopoly lives on in the form of the alliance of oil-producing countries, OPEC.

A SEISMIC SHIFT: THE MOTOR CAR

It was more than just public feeling that undermined the Rockefeller empire. Financed by the Rothschild family, a man named Marcus Samuel had developed a railway to take oil from Baku to Batum on the Black Sea. He had also led the construction of a fleet of tanker ships, on to which oil could be piped directly, without the need for barrels, and which were narrow enough to pass through the Suez

Canal. Going it alone after 1897, Samuel named his new business in honour of his father, who had sold ornamental boxes made from dead marine life. Today, it is known worldwide as Shell. And just as Samuel opened up routes to and from the East, the Royal Dutch Company struck oil on the island of Sumatra. Marcus Samuel eventually became Viscount Bearsted and was a noted philanthropist – as a medical student, I used to sit in the rather uncomfortable Victorian lecture theatre he donated to the London Hospital.

It is ironic that Rockefeller's monopoly should have faltered at the very moment when it might have been on the brink of greater things. Oil's use as a form of illumination was limited, particularly after Thomas Edison unleashed his light bulb on the world in 1882. But as this major change to daily life occurred, a bevy of small inventors was at work on the internal combustion engine.

In choosing a starting-point for the stories of the combustion engine and the motor car – which are woven so tightly around that of oil as to be almost indistinguishable from it – we have a lot of options. The very first internal combustion engines were firearms – propulsion being achieved by a sudden change in air pressures inside the barrel of a gun. In 1680, inspired by advances in human anatomy, the Dutch physicist Christiaan Huygens trialled an engine powered by gunpowder, in which expelled air was powered along special 'arteries' made from tubes of leather. A number of explosion-driven pumps were to follow over the next century. Then, in 1791, the English inventor John Barber patented an engine based on the combustion of oil vapour.

Although Britain would, at a later stage, become world-renowned for its motor industry, advances in the technology were stalled for a long time by public opposition. Barber's invention took second place to the massive advances being made by British inventors in the field of steam engineering and so, when it came to road transport, people naturally thought in terms of that particular means of propulsion. But the heavy, inefficient steam carriages damaged the roads, the boilers were dangerous and quite likely to explode, and the coal fires they generally used produced a thick screen of smoke; these early vehicles were consequently charged stiff tolls by the nervous opera-tors of the 'turnpike trusts' who were, in addition, anxious not to lose the business of the stagecoach operators. The punitive 'Red Flag

Act' of 1865 restricted the speeds of road-based steam carriages, forbade them to cross bridges or carry passengers, and compelled each
to employ a crew of three, one of whose job it was to walk ahead,
waving a red flag to warn people of the carriage's ponderous
approach.

With such a negative attitude towards mechanized road transport
in Britain the advantage passed to Europe. In 1863 Etienne Lenoir
attached an internal combustion engine of his own design to a
carriage, and drove it 10 kilometres through Paris, at a speed of
8 kph. The four-stroke engine followed – a more efficient
mechanism, in which the piston performed four moves for every one
act of combustion. These four strokes are the basis of the modern
car engine. The first move of the piston downwards sucks the fuel
mixture and oxygen into the cylinder through an inlet valve. Then,
as the crankshaft rotates, the piston is forced back upwards, com-
pressing the fuel mixture into a small space. The third movement is
when the piston is forced downwards by the explosion of the
mixture after a spark has ignited it, and the momentum of this
power turns the crankshaft, forcing the piston back up again and
shunting waste products from the explosion out through an exhaust
valve and into the exhaust pipe.

Enthusiasm for internal combustion was not widespread. Trying
out his own proto-motor-car in the suburbs of Vienna, inventor
Siegfried Mercus was threatened with arrest if he didn't cease
troubling his neighbours with noise and exhaust fumes. And when
the German Karl Benz – appropriately enough, the illegitimate son
of a railway-engine driver – tested his own car, the first to become
commercially available, crowds gathered to mock the enterprise.

The tide could not be stemmed, however, and by the early
twentieth century motor cars were in production across Europe
and the United States. They remained toys of the rich, though, and
became a focus for class-based antagonism. In 1910, a petition sent
to Queen Mary from 'the village women of England' begged for

> relief from the motor cars. We are sure your Majesty cannot know
> how much we suffer from them. Our children are always in danger,
> our things are ruined by the dust, we cannot open our windows, our
> rest is spoiled by their noise at night. If they could be made to go slow

through the village, it would be a great thing, but we are only poor people, and the great majority of those who use motor cars take no account of us.

Regrettably, Queen Mary's husband, King George V, responded to this petition that same year by purchasing a Daimler, the preferred car of the British monarchy from the time the Prince of Wales, later King Edward VII, first rode in one in 1896 right through to the mid-1950s. This particular limousine was appreciated for its comfortably large size, high quality and dignified appearance. A staggering 17 feet long and 7 feet 6 inches tall, it was powered by a smooth and quiet 9.4-litre six-cylinder Knight sleeve-valve engine that produced 57 horsepower. Perhaps not surprisingly, it now graces a museum in California.

Death tolls from road accidents continued to be high – one person a day by 1914 – until more stringent regulations were introduced in 1930, and newspapers often made much of the vast disparities between wealthy, speed-crazy motorists and their humble, penniless victims. Licence plates and speed limits were enforced from an early date – but associations like the AA formed in opposition, providing bicycle patrols to alert motorists of police speed traps.

The great gear-shift in motoring history took place when Henry Ford introduced the 'Tin Lizzie', otherwise known as the 'Model T', in 1908. An active philanthropist who even won the admiration of Russian leader Lenin, Ford worked hard to produce models ever more in step with the earning power of the average worker. In the year it was launched, the car cost $890 and the average American's annual salary was $500–570. By 1924 it cost just $290.

In Britain, the petrol engine swiftly had its effect on the country, and on the daily lives of its inhabitants. As people could travel further to their places of work, towns spread into the countryside. Mechanized farming meant job losses in these rural areas, and a further swelling of the urban populations as a result – many of whom, particularly in the Midlands, found new work producing the motor cars themselves. As the production of food became limited to an ever smaller proportion of the land, people became increasingly dependent on mechanized transport to get hold of their daily bread. And yet, in spite of a few lone protests – like the heartfelt petition to

Queen Mary quoted earlier – nobody seems to have questioned the slow seepage of oil-based power into so many aspects of their society. The tale was one of steadily increasing dependence, heightened not just by an oil-fuelled victory in the First World War but by the subsequent demobbing of tens of thousands of men and women trained to drive the new vehicles. The statistics for vehicle use at the start and close of the conflict are striking. When the British army took off for France, it took only 827 motorcars and 15 motorcycles with it. By 1918 it had 56,000 trucks, 23,000 cars and 34,000 motorbikes. But by now, alongside this technological rush, there were warnings of doom. People began to speak of the possibility of the oil running out. In 1919 the head of the US Geological Survey announced there would be none left in his country by 1928.

It wasn't until the Buchanan report in 1963,[5] when there were fewer than 7 million vehicles on Britain's roads, that there was some public recognition of the harm being wrought by motor cars. The picture Buchanan and his committee painted was apocalyptic in tone: 'It is impossible to spend any time on the study of the future of traffic in towns', said the steering group, 'without at once being appalled by the magnitude of the emergency that is coming upon us. We are nourishing at immense cost a monster of great potential destructiveness [the motor vehicle], and yet we love him dearly. To refuse to accept the challenge it presents would be an act of defeatism.'

Buchanan, the Professor of Transport at Imperial College London, the university where I work, and both a keen cyclist and owner of a motorized camper-van, emphasized that tough decisions were needed. The report recommended that cars should be banned from parts of some towns and cities; existing buildings and roads should be demolished to allow traffic to drive at ground level with shops and pedestrians being placed on a level above. Controls should be placed on parking, and a 'congestion tax' should be levied on car owners. And, the report emphasized, in inner cities cheap public transport was going to be essential.

As it happened, Buchanan's predictions were unduly pessimistic: the report forecast that there would be 27 million motor vehicles on Britain's roads by 1980, but in fact the total did not rise this high until twenty years later. His 'saturation' point of 40 million vehicles

on Britain's roads, which he thought would be reached by 2010, is
now very unlikely to be reached.

A GUSH OF DISCOVERIES

The name of Fallujah, a town some 70 miles west of Baghdad, is
now synonymous with the sadness of war. When we hear of it, we
think of the 200 civilians who were killed when British bombs
landed upon a busy marketplace in the first Gulf War. We think of
its buildings, of which more than half have been reduced to rubble
by that war and the one that followed it. Before the second Gulf War
it was in a relatively unscathed part of the country, but it became a
hotbed of Sunni insurgency and the location for nightly battles with
American troops. We should also remember a happier, more tolerant
age, when Fallujah was called Pumbeditha, one of the world's fore-
most academies of Jewish learning from around the second century
CE. Just how important the town was as a centre of Jewish life after
the Babylonian captivity is clear from the episode related in the
Talmud in which the Gaon (the head of the academy) of Pumbeditha
was asked whether some stolen wine which had been recovered was
still kosher (in Jewish law, if handled by a non-Jew it might have
been used for some idolatrous purpose, making it unfit for use by a
Jew). The Gaon opined that there was no need for concern as he had
concluded that most of the thieves in Pumbeditha were Jewish.[6]

Oil is at the heart of the conflict that tore Fallujah apart; and it
began not in 2003 or in 1991 but just after the First World War.
Alarmed by American reports of dwindling oil supplies, the British
adopted a concerted policy towards the Middle East of encouraging
Arab opposition to their Ottoman rulers in a bid to secure loyalty
and hence access to whatever oil reserves might be under their lands.
In the pursuit of this duplicitous strategy, Britain – through the
ministrations of one T. E. Lawrence – funded an Arab Revolt, while
secretly plotting with France to carve up the former Ottoman
territories between themselves. When the Turks retreated and
sovereignty over Iraq passed to Britain, there were mass revolts in
1920. The British responded harshly in towns such as Nasiriya and

Fallujah, and some 10,000 Iraqi civilians were slaughtered. Winston Churchill, now Secretary for the Colonies, came up with a conciliatory policy of puppet government, establishing de facto British dominion behind the throne of King Faisal.

Something rather similar happened in Central America. A massive Mexican oilfield was found in 1901, and the rights to it were sold to Shell. After the deposition of the unpopular Mexican ruler Porfirio Diaz, things became tougher for the foreign-owned oil companies. So US interest spread to Venezuela, where Juan Vicente Gomez, a despot of equal turpitude, was happy to seize his people's mineral rights and sell them to foreign oil conglomerates. By the eve of the Second World War, the oil reserves of Persia, Mesopotamia, Venezuela and Mexico were owned by a tiny handful of European and American companies.

The picture was about to change, though, with the introduction of a crucial element of instability. It began in 1925, when Harry St John Bridger Philby – father of the more memorable Kim – stood down as head of the Secret Service in the British Mandate of Palestine, converted to Islam, and went to work as adviser for King Abdul Aziz ibn Saud.[7] Ibn Saud was in the process of unifying his Arabian desert kingdom, subjugating its fierce Bedouin tribes with the support of the Ikhwan – a religious militia, devoted to a puritanical form of Islam which they felt was incompatible with the wandering Bedouin lifestyle.[8] Unfortunately for King Saud, the Ikhwan also felt many of his own modernizing policies were incompatible with their faith. Their hatred for such 'foreign outrages' as mirrors, cars, telephones and telegraphs spilled into an all-out war which left the king victorious but broke, depending on the meagre income from pilgrims to Mecca and Medina to run his kingdom.

Under Philby Senior's guidance, King Saud permitted the oil company Chevron to enter his territory, granting it a sixty-year concession on the al-Hasa oilfield, which still holds some 20 per cent of the world's remaining oil reserves. The world now saw a huge shift in oil production. In part, this was due to the opening up of Saudi Arabia and nearby Gulf states such as Bahrain and Kuwait. But none of those territories could have been explored or exploited so well by the oil companies had it not been for a gear-change in the science on which they depended. The development of anticline theory explained how pools of oil, natural gas and water became trapped in subsurface porous rock, which formed upward bulges in the earth's crust. The ability to spot these anticlines led to a major oil discovery in Oklahoma in 1913, after which all the big companies brought geologists on to their staff. Around 70 per cent of oil discoveries have been within anticlines.

Another great oil discovery in Oklahoma – Seminole, in the 1920s – was achieved through the new science of seismic prospecting, whereby the echo pattern of explosive charges was used to build a map of underground structures. Advances in oil refining also helped – specifically, the technique of thermal cracking invented by a Russian engineer, Vladimir Shukhov, in 1891. This involves heating the oil to high temperatures and thereby breaking up large hydrocarbon molecules to make more suitable fuels, such as petroleum. Thus more usable fuel could be obtained from every barrel of the stuff extracted. And extraction itself improved after the introduction of gas flushing to eke out the secondary residues of an oil well.

The burgeoning oil market was swelled by the unplugging of the famous 'Black Giant' deposit in Rusk County, East Texas, a story commemorated in the museum in the county town, Henderson. The strike was made by the decrepit seventy-year-old 'wildcatter' Columbus Marion Joiner. He had moved from Oklahoma and, simply on a hunch, bought up several thousand East Texas oil leases. In 1926 he came to Rusk County to see what he had squandered his last dollars on. His drilling rig consisted of a few rusty pipes, a leaking boiler for power, and old tyres and cordwood for fuel. The following year he leased the 975-acre farm of widow Daisy Bradford and, rather taken with her, promised to drill his first well on her property. But he had to abandon the courtship and his drill bit when the whole thing jammed and could not be dislodged. He shifted his derrick about 30 metres, but drilling failed again. On the third try, he moved his derrick a further 100 metres and one of the ramshackle beams supporting the derrick snapped with a loud noise – but no loss of life.

On 5 September 1930 his little crew of farmhands and oddball drillers seemed to have reached some oil-saturated sand at 3,536 feet

Petroleum is crude oil. Cracking is employed to produce petrol (gasoline).

below the surface and they all got very excited. On 3 October the news broke that they might strike oil imminently, and 8,000 people from miles around flocked to the wellhead on Daisy Bradford's farm. Most of them had nothing much better to do and they were generally sceptical. But this was fun, snacks and drinks for sale made it all a bit of a jaunt, and Joiner was a character – a dreamer, a self-promoter, a sometime poet and a failed wildcatter. All experts had agreed there was no oil to be found in East Texas. But the crew of farmhands kept drilling while crowds watched, hoping for a gusher at any moment. Darkness approached and no oil came forth. So they all went home – at least they'd had a good outing and a few tots of illegal corn whiskey.

A few hours later, after the crowd had gone to bed, Joiner struck oil – a massive gusher. Within days, other drill holes were being made nearby. A fortune-teller told a certain J. Malcolm Crim, a merchant from Kilgore, that he would find oil 9 miles north of the Joiner well, and on 30 December he did. Twenty miles further north and twenty-seven days later, drillers at a third site struck oil. As the oil shot upwards, the spectacle was watched by an audience of 18,000 cheering farmers and townsfolk. Together, these wells immediately yielded over 20,000 barrels of oil a day. Thousands of feet below the pine-clad hills there lay a single black giant, a monster ocean* of oil, 45 miles long and up to 12 miles wide, totalling about 140,000 acres. Over the following decades it would yield 6 billion barrels of oil.

East Texas changed almost overnight. In the first month, Kilgore grew from a sleepy town of 700 to a bustling city with a population of 10,000 residents. The entire area became a mecca for fortune-seekers. Prices of everything except oil shot through the roof. A barrel of oil cost a mere 10¢ but a gallon of water cost $1. Eventually, 31,000 wells would be drilled – in an orgy of waste, because by August 1931 one million barrels of oil a day were being recovered: about twenty-six times as much oil as the world could use at that time. The state governor, Ross Sterling, ordered a shutdown, sending in 1,300 cavalry troops from the Texas National Guard to enforce it.

Freebooting oil operators fought regulation and sold 'hot oil', produced illegally. Sometimes unscrupulous prospectors secretly

* Oil does not exist in a liquid 'ocean' but in porous rock

tapped into other oilmen's pipes. Other ingenious oilmen installed a 'left-handed' valve, so that when the troopers came to enforce the law by turning the valve 'off' and padlocking it, they had actually sealed the pipe open. And one resourceful individual erected a fine brick building over his derrick and registered it as his house. Any time the authorities appeared, he simply pulled up the portable steps inside. Crime, violence and drunkenness became the norm and the local police were powerless. Eventually, the mayor of Kilgore called in the most famous of the Texas Rangers, 'Lone Wolf' Gonzaullas. Handsome (in a Daniel Craig kind of way), powerful, wearing two pearl-handled pistols and a large white Stetson, he rode into Kilgore on a black stallion to meet the mayor. Gonzaullas secretly recruited several other rangers and on his first night netted 300 'badmen'. He marched them down Main Street and into the Baptist church – Kilgore's jail was as yet unfinished – and shackled them all to a long, heavy chain. Over the next month, Gonzaullas averaged 100 arrests a day. Eventually nearly all of the freebooters and outlaws were expeditiously thrown out of town.

By the mid-1930s the Market Demand Law, the Confiscation Law, the Refinery Control and Tender Bill and other legislation had been enacted and big companies started to move in. The oil money enabled the local people to build one of the most up-to-date state schools in the whole of Texas, in East London, Rusk County. But just after 3.00 p.m. on 19 March 1937, as the children were getting ready to leave school to go home and the smaller ones waited on school buses for their older brothers and sisters to join them, an industrial crafts teacher turned on a sander in his classroom. A spark ignited the odourless natural gas that had leaked from pipes under the school and had been trapped in the basement. The school erupted; people living 4 miles away heard the explosion that literally blew apart 298 children and teachers. Some of the bodies were never recovered. The hazards of the oil industry were beginning to be understood.

The US government responded by requiring all natural gas to be given an identifying smell and the substance mercaptan, which smells like rotting cabbage, was added. It also responded with regulations that restricted oil producers to fixed quotas; a necessary move, but one with an extremely dangerous symbolic legacy. People have solid historical reasons to believe that oil supply depends only on the whims of governments and producers. They can turn the tap on and off, alter prices, shorten and lengthen the queues at the petrol stations, as they see fit. While that might once have been the case, the world's oil reserves are now under pressure for an entirely different reason.

PETROL POLITICS

The Second World War, as the American writer James Howard Kunstler puts it, 'was fought with oil and for oil'.[9] And it was won, we might add, by the simple expedient of denying it. The island empire of Japan sought access to the oilfields of Dutch-owned Indonesia, and it was scuppered more by the US navy's targeting of Japanese oil tankers – which starved the whole operation of fuel – than by the bombings of Hiroshima and Nagasaki. Likewise, Germany's attempts to get to the oilfields of Baku ended up in the stalemate of Stalingrad in 1943, which not only broke the Reich's morale but left it dependent on synthetic fuels, manufactured expensively from coal. This *pas de deux* of oil and history continued throughout the twentieth century, for the bulk of which the United States was the leading oil producer and exporter.

In the 1950s, conditions of peace combined with a glut of oil supplies to create a passing illusion of stability and prosperity for many. During this honeymoon period, many regrettable decisions were taken. With abundant cheap oil as the feedstock, much of the world shifted to a reliance on plastics, synthetic fibres and artificial fertilizers. With the motor car – sprayed with cheap oil-based paints, its interior fitted out with light, oil-based fittings – so widely accessible, patterns of living altered irrevocably. In the United States in particular, and to a lesser extent in Britain, city centres emptied of all but the poorest residents, with an outpouring of people to the growing suburbs. Many of their houses were built from oil-based materials, with the latest labour-saving devices – washing machines, refrigerators – equally dependent upon the oil economy. To get into work, or school, or even to buy a newspaper or a bag of groceries, they needed their cars, just as the newsagents and grocery stores needed lorries to ship the goods to them. Everyone needed oil. And nobody wondered what might happen if it disappeared.

Oil has been an excellent fuel source, for the most part relatively easily obtained. Though flammable, it is not particularly toxic. It burns efficiently and quietly, and as it is a chemically stable liquid it has been easy to transport and to store over long periods of time. It is no wonder we preferred it to wood and whale fat.

What exactly is this substance that has so transformed our pattern of living? Put simply, oil is the product of sunlight on plant life: one ancient source of energy converted – over millions of years, at high temperatures, tremendous pressures and great depths within the earth – into another. It is thought to come from algae, which bloomed in prehistoric lakes during a climactic warm spot between 300 million and 30 million years ago. Dragged under the earth by movements in the crust, these tiny plants were subjected to such temperatures at depths of between 2,000 and 4,500 metres underground that they were cooked into a hydrocarbon-saturated sediment.

Occasionally bits of this pool of primordial soup can seep up to the surface or close to it, owing to a mixture of soil erosion and the shifting of tectonic plates. But this happens rather rarely. Much of the oil extracted since 1859 has come from a finite, relatively narrow reservoir underneath the ground. Unlike our oceans, this reservoir does not run all the way around the planet; it is located in a handful of small areas, probably places where ancient rivers flowed into bays, or where rainfall so affected the salinity of the sea water that algae could flourish. This basic geology has some important implications that were overlooked by the early prospectors in their enthusiasm for the black gold. The world's oil wells will not be 'topped up' by some deeper source of the fuel elsewhere inside the planet. Nor is it likely that we shall discover any significant new pockets within the 'oil window' at 2,000–4,500 metres. It is likely that we have had, and have used up, the bulk of the world's oil, and that we shall not find much more.

A heady cocktail of complacency and panic over this issue has been the guiding force of global politics since the end of the Second World War, and particularly since the 1970s. It was during this latter period that the United States' own production of oil began to taper off. At the same time, its usage of oil was still increasing, calling for ever greater quantities to be imported. Previously, when Saudi Arabia had attempted to impose an oil embargo against supporters of Israel during the Six Day War of 1967, it had been an empty gesture. America had enough oil to bust the sanctions and make sure everyone, at home and in Europe, had enough petrol to get the kids to school and pick up the shopping. But now, in the 1970s, a vital shift occurred – because control of the world's surplus passed to the Middle East.

The consequences were obvious when, in 1973, Egyptian and Syrian forces – backed by the Soviet Union – attacked Israel on their most holy day, Yom Kippur. This was a very short-sighted strategy. On most days of the year, the traffic in Israel is chaotic; but on Yom Kippur people tend not to drive and nearly everybody attends synagogue, so it was exceptionally easy for Israel to mobilize its largely citizen army very rapidly. As the United States and its allies voiced increasing support for the Israelis, the Saudi-led Organization of Petroleum Exporting Countries (OPEC) put pressure on Western oil companies in the Middle East to raise their prices by 100 per cent. When they refused, the Arab members of OPEC announced they were breaking free of the cartel to set their own prices. As the Israelis – partly funded by America – pushed the Egyptian forces back, the Arab oil ministers announced a total embargo on oil sales to the United States and a 70 per cent price hike on sales to Western Europe. The price of a barrel of oil quadrupled. There were near-riots at petrol stations. A 'Giles' cartoon of that era in the *Daily Express* depicted a fat woman guiding an equally corpulent schoolboy by the hand as he took tentative steps on the pavement, with the family car rusting in the drive. 'That's it – one foot, and now the other,' she cooed. '*Nasty* Arabs!'

The shock waves were felt at all levels of the Western economies. The prices of food and manufactured goods – which used oil in their creation as well as their distribution – shot through the roof. The value of the American stock market dropped by 15 per cent in one month. The 'Big Three' of the American automobile trade – General Motors, Ford and Chrysler – suffered a knock from which they never recovered. Interest rates soared, and deep recessions set in.

By this stage, a number of voices were urging a rethink of our relationship with oil. President Jimmy Carter put money into research for alternative fuels, as well as giving a tax boost to hydroelectric power. As early as the 1950s, specialists such as the geologist M. King Hubbert had indicated that economic and political tensions would only increase as the world's oil supplies passed their peak – an event he expected to occur between 1990 and 2000.[10] It was a remarkably prescient view, but at the time very few took him seriously. Much more recently others, including Colin J. Campbell, retired research chief for Shell, and Kenneth Deffeyes of Princeton, have put the peak at somewhere between 2007 and 2010.[11] These theories suggested that, if we viewed the world's history as an hour, the period of our dependence on oil would correspond to less than a millisecond.

We ignored these messages. Discoveries of oil in Alaska and the North Sea provided an illusory fillip, bolstering the fiction that, whenever the oil started to dry up in one part of the world, the tap could just be turned on in another. Incredibly, this basic optimism was unshaken when, in 1979, a hard-line Islamic government took over Iran and ceased producing around 50 million barrels of oil a day available for export. That loss of 5 per cent of the world's oil reserves led to a repeat run of the scenes at petrol stations, and inflation soared again. In the meantime, the secular president of Iraq, Saddam Hussein, exacted revenge on Iran for stoking up Shi'ite insurgency within its borders, and attacked Iranian oil ports. Iran responded by shutting down all but one of Iraq's pipelines, removing another 8 per cent or so of the world's oil from the global markets. More price hikes, more recessions.

Someone might, in this second period of instability, have wondered if we should have been looking at alternatives, investigating ways for our society to endure without a slavish dependence upon this mercurial substance. But perhaps we cannot be blamed for what author Erik Davis calls the 'consensus trance'.[12] The world we have created rests on oil as a film of oil rests on water. To imagine a life without oil and the products of oil is . . . well, until recently, it has been unimaginable.

Through the 1980s the trance could endure because the Saudi oil-fields, combined with those in the North Sea, ramped up their production to make up for the losses from Iraq and Iran. The same happened when Saddam Hussein took over Kuwait in 1991 –

ostensibly in revenge for Kuwaiti oil operations having drilled into Iraqi territory. There was a brief bottleneck in the world's supply, a flurry of panic at the petrol pumps, and then the Saudis stepped in to make up the shortfall. In Europe and America, abundant oil fuelled the global village. With transport so cheap, we could afford to manufacture and grow what we needed in the developing world, where labour was cheap, too. It was as if, whenever we received a reminder of how inconstant a resource oil was, we increased our reliance upon it.

In the early years of the present century – against a background of overtly increasing hostility towards the West from parts of the Islamic world – it began to seem as if we were hearing warning bells. As demand for oil from China increased, the price kept rising. In 2004 came a scandal, as Shell Oil admitted to shareholders that it had overestimated its own reserves by some 20 per cent. From Saudi Arabia, meanwhile, there were reports that production was not rising, because it couldn't. The massive al-Ghawar oilfield seemed to be drying up, requiring increasing amounts of sea water to be flushed through it in order to extract ever-diminishing returns. In 2004 the price of oil hit $50 per barrel. When I started to write this chapter, in early 2008, it had just hit $100. Now, in the middle of 2009, it is around $40 per barrel as the global economy declines and reduced demand results in the latest recession.

To many people, these unwelcome changes have been intimately connected with the events following 9/11. It is no coincidence that the Middle East has been and remains a political flashpoint, for the precise reason that it is where the oil is. Most of the world needs something that is owned and controlled by very few people, and that essential imbalance contributes to the current mood of global anxiety. America's greatest fear is not that al-Qaeda will commit another atrocity on its soil, but that it will succeed in overthrowing the house of Saud, turning off the oil tap for good. The war in Afghanistan contains an oil-based sub-plot, because America hoped this country could be the site of a pipeline connecting oil-rich Central Asian reserves to the Indian Ocean, bypassing the troublesome Middle East for good. Iraq, similarly, was about establishing an American police station in the very region where its fuel had to come from.

The anxiety is felt across the Middle East. Arabia is a dry, hot region generally not conducive to economic development, and without this crucial 'cash crop' opulent mega-cities like Dubai and Bahrain would still be fly-blown trading-posts. The sudden rush of oil wealth into the area has created cultural confusion, along with the fear that it might all come to an end. These important psychological factors feed into and fan the various political tensions in the region: the existence of the State of Israel, the displacement of the Palestinians, the corruption of the house of Saud. Wahhabism, the austere form of Islam endorsed by Osama bin Laden, has a stronghold in Saudi Arabia, not least because people there are aware that the current conditions cannot endure. The oil cannot last, nor those bloated revenues with it. Nor can the air-conditioned shopping complexes, the architectural showpiece buildings, the hosts of foreign workers, the populations far in excess of what a desert region can naturally support. You can scarcely blame some people for wanting to return to a medieval theocracy. It must seem a lot safer than the future.

ALTERNATIVE FUELS?

In chapter 10 I discuss some of the remarkable advances in medicine, and point out how many of them were arrived at by chance. Many of the innovations and inventions described in this book have the same character – the product of hard work and human ingenuity, yes, but also of something else: a freak event, a synchronicity, a wind blowing in the right direction. This has been greatly to our advantage, but has also sown the seeds for a certain complacency. Everything will always be fine, so the thinking goes, because it always has been. Something will turn up. This attitude is especially marked when we consider our dwindling oil reserves.

On a recent plane journey, I got into discussion with some fellow passengers. Both intelligent, well-read individuals, their comments summed up the current mood. 'They're always finding more oil,' said one. 'They'll find some more.' The lady on my left took a different view. 'There are plenty of technologies that can step into

the breach,' she said. 'They just need to invest more money in them.'
These viewpoints, commonly voiced by a range of people who
should know better, are both woefully wrong.

It's unlikely that 'they' – the mythical body of beings who ensure
that the world will always be fine – will find more oil, because
geology suggests it is a limited resource. There may, of course, be oil
reserves at deeper levels. There exists the potential to flush out
'dead' wells with huge quantities of high-pressure water, or to
refine the tar sands of Alberta for their oil content. But there we
encounter the principle of ERoEI – Energy Returned on Energy
Invested. It will require more money and energy to get at this oil
than we will gain from its use.

There is a range of alternative energy sources on offer – but it is
unlikely that the answer will just be a matter of flicking a switch and
diverting revenue to them so that they become viable. Most of them
are just not that useful – and even if they are, they depend so heavily
on an oil-fuelled world that we will not have the reserves to make
use of them unless we start immediately.

Take gas. Like oil, it is created under the earth in specific conditions of heat and pressure. It burns efficiently and cleanly, it can
be transported at air temperatures in pipes, and it needs no energy
to get it out of the ground. But it is running out – at a rate of 5 per
cent per year in the United States. And for a country which has
invested heavily in gas-fired electricity plants, this is a worry. Unless
there is some major technological advance, gas can only be used on
the continent where it was extracted. In special pipelines, it can
move around at air temperature with very low energy costs. But to
get natural gas from overseas, it must be liquefied, shipped in
pressurized special low-temperature containers, decanted at specialized port facilities and then turned back into a gaseous state. The
process requires large amounts of energy – which could be supplied
by cheap, abundant oil, but may not be so readily available once
that cheap oil runs out. And, as we have recently seen, gas pipelines
are vulnerable – both to political forces, when the taps from another
country may be turned off, and to terrorist action. In January 2009,
just when Europe was experiencing bitterly cold weather, Hungary,
Slovakia, Bulgaria and Romania were hit very hard indeed, and
Germany, France and Italy were threatened by diminished fuel

supplies. The European Union gets about 25 per cent of its gas from
Russia, via the pipeline through Ukraine, so the decision by
Gazprom, the giant Russian gas company, to switch off supplies in
the course of a dispute with Ukraine was very serious. And the
Russian President Vladimir Putin was clearly in no hurry at all to see
the supplies reinstated because he claimed that the Ukrainian
government was stealing some of the gas sent down the line to
Western Europe. In a worsening scenario, the Ukrainians began
to insist that this yet again showed Russia as a bully, using gas as a
political weapon.

At the root of the dispute was partly the issue of profit. Russia
maintains that its gas is worth more than it is receiving in revenue,
while Ukraine argues that the transit fee it receives for piping the gas
is not sufficient. But this interruption of energy supplies was about
more than simply a commercial disagreement. Russia has accused
Ukraine of backing Georgia during recent uprisings there and
supplying arms to the insurgents, and the serious political instability
inside Ukraine is making the Russian government very nervous. For
now, an uneasy solution has been reached. But, essentially, until
Europe diversifies its sources of energy it will always remain a
potential victim of the disputes between Russia and what was once
the rest of the Soviet Union.

Hydroelectric power

How beautiful and arresting is the image of a hydroelectric station
– the power of nature, clean and serene, harnessed for our benefit
and the enduring health of the planet. Hydroelectric power ran the
first factories of the industrial revolution, and has been providing a
source of electricity since 1870. Once the massive dams and
installations have been built, using water flow to spin turbines
which power a generator produces no harmful pollutants or greenhouse gases. Currently, hydroelectric power provides 19 per cent of
the world's electricity – most of that generated in China, Canada
and Brazil.

But though the image is impressive, water power is never going to
be anything approaching a universal source of energy, and the
associated damage to the environment is usually quite profound.
Also, dams require vast amounts of concrete, and those materials

and the heavy transport needed to get them to the site of energy production have a significant carbon footprint. Dams can only be placed at a handful of locations, where water is found descending from a great height. In most countries, Britain among them, the majority of these sites have already been exploited, so there is no way we can ramp up hydroelectric power to provide much more than the current UK figure of less than a fifth of total energy consumption. Many of these sites have been in use for a long time, too, resulting in soil erosion and silting which will eventually leave them inoperable. If we could, in a best-case scenario, clean all the old plants up and build at all the remaining unexploited sites in the world, we still could not accomplish that without a huge supply of cheap oil – oil to manufacture the parts and ship them in, and to move the workforce around.

Wind power

Charles Francis Brush from Cleveland, Ohio, who died in 1929, was probably the first person to harness wind successfully to make usable electricity. A largely self-educated man, he invented the electric arc light and pioneered improved dynamo technology for producing electricity commercially – making a substantial fortune in the process. He also found a highly efficient method of making lead-acid batteries for storing the electricity he generated. In the winter of 1887–8 Brush built the first automatically operated wind turbine. His windmill, erected in his sizeable back garden, was 50 feet across and contained 144 rotor blades made of cedar wood. Amazing as it sounds, the turbine was reliable, running for twenty years. Brush used it to charge batteries in the cellar of his mansion where his electricity was stored. Because the windmill rotated slowly, it only generated around 12 kilowatts – but that was still enough power for his entire household.

The winds are weak where Brush lived – a bit of a problem. But in the last few years governments of many countries where there are higher or constant wind speeds have recognized wind as potentially highly valuable. Britain is a windy island – in Scotland the onshore average wind speed 50 metres above the ground is around 27 kph, and tends to be greater than 40 kph offshore – so in terms of potential energy production has one of the best resources in Europe.

The maximum output from a giant wind turbine scales up with the cube of the wind speed. So if the wind doubles in speed, the output is increased around eightfold. Moreover, the wind speed tends to be greater higher up: on average, raising the blades from 50 metres to 100 metres above ground level gives around 40 per cent extra energy.

All this sounds really good; but there are major problems. Big machines are difficult to build and face massive forces in higher winds. And if the wind is too fast – over around 70 kph – the turbine cannot cope and the generator will cut out. At the other end of the spectrum, even in the windiest parts of the UK there are days when there will be no wind at all. At present, the biggest wind turbines produce about 2–3 megawatts of energy. It may be possible to build larger turbines,* perhaps fixed in the sea bed offshore, which generate even more electricity, but securing them under water – particularly under corrosive sea water – presents considerable technical problems; indeed, a number of scientists are very dubious that this is even practicable.[13] So, for the time being the UK government has concentrated on promoting onshore installations, even though nearby residents do not always like the alteration of their local environment. It is calculated that our total onshore wind resource is theoretically around 110 gigawatts, and by 2010 we expect to be have a capacity of a total of 6.5 gigawatts from wind power, using about 3,200 wind turbines. But because not all the wind can be used – for example, because the speed is too high, or particular machines will not be in service all the time – it is calculated that only 30 per cent of the capacity can be harnessed. This amounts to around 2 gigawatts, which is about 5 per cent of the UK's total needs at present.

So harnessing this most abundant source of power presents serious problems. Wind turbines are extremely costly to make and difficult to install. Their manufacture produces large amounts of carbon; the cabling they need is complex and expensive; and storage of the electricity they generate would require massive batteries. Fixing wind turbines offshore may be impracticable; they could constitute a hazard to shipping, and their radar profile means that they might obscure hostile incoming aircraft over our shores. Both onshore and offshore, they present significant environmental problems. Most people, for example, do not want them in deserted beauty spots – where they might be most useful. But perhaps the

* The length of a turbine blade is limited by the need to avoid the blade tip approaching sonic speed.

biggest problem they pose is that the power they produce is discontinuous: high output when the wind is good, and no output during calm weather. Thus connecting them by cable to the national grid presents a very real challenge, because sudden surges of power risk 'tripping' it.

Wave and tidal power

Rising and falling tides can be used to generate considerable energy. This is not a new idea: mills for grinding corn were powered by tides along the coast of Brittany as early as the twelfth century. Tidal energy may be harnessed by allowing sea water to flow into a bay or lagoon as the tide rises; when the tide falls the water flow can be used to turn a paddle wheel. Various experiments were done in France in the 1920s, but the first plant generating electricity was completed at La Rance only in 1967. The maritime estuary there, where the River Rance meets the Channel, has exceptionally high tides, up to 13.5 metres. The French built a dam across the estuary, cutting off the sea. Water passes through sluice gates in the dam, filling the estuary and powering turbines. When the sluice gates are closed, 18 billion litres of water are trapped in the tidal basin. Once the tide is going out, the sluice gates can be opened again, allowing the water to spill back across the turbines to the sea. At La Rance, reversible turbines mean that electricity can be generated with both the rise and fall of the tide, producing around 240 megawatts.

Britain has a long coastline and big tides, so the potential for tidal power is big. Even so, there are not many places where a similar system might work. One is the Severn Estuary, where the tide rises around 13 metres: various proposals have been on the drawing board since 1950, but although a barrage could generate 6 per cent of the UK's energy needs, it would take more than ten years to build and would cause massive disruption to the marine and coastal environment.

Other schemes include floating pontoons moored at sea, which can generate electricity as the waves move them. One such is the Cockerell Raft, patented by Sir Christopher Cockerell. The raft is essentially two buoyant pontoons, hinged together by a hydraulic piston. Although the original raft did not work well, it has been substantially modified. Theoretically a raft 20 metres long by 8 metres wide floating in waves 2 metres high could yield around 35 kilowatts of electricity. Waves of 6 metres high could produce 1.6 megawatts. With more pontoons linked together a very large amount of electricity could be generated. But the ocean environment is highly corrosive, storms can generate enough energy to destroy the installations, and it is very expensive to bring this discontinuous source of power to land by any cabling method. And these installations present a major hazard to shipping – as shipping clearly presents a hazard to them.

Biomass fuel

Biomass can be used to generate heat which can then be converted into electrical energy using a turbine. Typically, biomass may be wood by-products from the agricultural or forestry industry, domestic waste, or processed fuels such as wood chips and sawdust. It is preferable to use biomass to generate energy rather than just burying it in the ground in a landfill site. It is a relatively efficient process, too, as the ash from biomass burning can be used to make fertilizers.

Biomass is underused. Perhaps one-third of the UK's straw production (8 million tonnes a year) could be used to generate enough energy to meet about 3 per cent of the UK's current electricity needs. Chicken litter, cow slurry and pig excrement – all resources that at present are largely wasted – could all be burned in this way (it is amazing to consider that the average cow produces 20 tonnes of manure annually). Once again, the ash could be used for fertilizer. At present, most cattle slurry is spread directly on fields as fertilizer, where it risks polluting rivers. Used as manure, slurry also produces methane gas, which contributes to global warming through the greenhouse effect.

The potential of rapidly growing energy-producing crops is even greater. For example, the perennial grass *miscanthus* could be a valuable energy source, as could willow branches from coppice production.[14] Five hundred hectares of farmland could produce 1 megawatt of continuous electricity, even when the biofuel is burning at only 25 per cent efficiency. It has been estimated that around 5–6 per cent of the UK's total energy requirement could be generated from just 350,000 hectares of land – and if the process is made more efficient by using improved farming and better chemistry in the power plant, the output could be even greater. One improved technology is 'gasification', where wood chips are heated in an air stream and carbon monoxide, methane and hydrogen are produced. When mixed with air these gases burn at a higher temperature than the wood chips, thus yielding still more energy. In the UK there are approximately 1 million hectares which could be used to grow biomass, but currently less than 2,000 hectares are under cultivation for energy-producing crops.

Biodiesel

The Bavarian inventor Rudolf Diesel considered biofuel as a potential source of energy as long ago as 1900. At the world exhibition in Paris that year, Diesel demonstrated that his engine could run on peanut oil; and later, Henry Ford had expected that the Model T would run on alcohol produced from corn. But mineral oils were so easily obtained that few people took vegetable oils seriously. Then, in the 1970s and 1980s, the idea of using biofuels was revisited in the United States, largely because of the Clean Air Act which stipulated more careful regulation of emissions such as sulphur dioxides, carbon monoxide, nitrogen oxides and ozone. The troubles in the Middle East and rising petroleum prices gave further impetus to its use, as we have seen.

Biodiesel is relatively environmentally friendly, being bio-degradable and producing fewer emissions than fossil fuels; for example, its use in vehicles reduces carbon dioxide production by nearly 80 per cent. It helps to lubricate the engine, thus decreasing engine wear, and can be used in almost any diesel with little engine modification. But biodiesel does increase nitrous oxide production and its solvent property can cause fuel filters to become jammed with the deposits. It is also associated with a decrease in fuel economy and power, and is more expensive to produce than conventional diesel.

Solar energy

Solar energy is obviously a limited resource in the UK because of our rather poor weather, but globally it offers significant opportunities. Solar photovoltaic (PV) panels, which can convert light into electrical energy, are expensive and their manufacture involves considerable carbon emissions. But the conversion of sunlight into electricity does not involve any moving parts, and once a PV panel is installed it is silent and pollution-free. The panels are usually rigid, rectangular and 1–2 metres across. Some are flexible and may measure 7 metres or even more. The supply of PV panels worldwide has increased by 20 to 30 per cent annually since 2004 to keep up with a new demand for this source of renewable energy. Currently, however, PV units are still rather inefficient, converting only 7–15 per cent of the light energy they receive into electricity. Imperial College in London is experimenting with transparent solar panels built into windows, which might be a good source of energy in office and commercial buildings, particularly if the efficiency of power generation can be improved.

Since the sun in North Africa is twice as strong as it is in southern Europe, just 0.3 per cent of that sunlight is enough, at least theoretically, to power the whole of Europe. According to Arnulf Jaeger-Waldau of the European Commission's Institute for Energy, a massive installation in the Sahara covering an area as large as Ireland could be a feasible solution to Europe's dependence on oil.[15] It has been claimed that 100 gigawatts could be generated by 2050, but this is likely to be too late to save the planet from the ravages of global warming. Needless to say, such an installation would be hugely costly, and there would be significant technical problems transporting the electricity generated to where it is needed. It is also true that such a large installation would probably be extremely vulnerable to sabotage or terrorism.

Solar energy can also be used to heat water in panels on roofs, or oil in pipes. In Granada, in Spain, on a plateau about 1,000 metres above sea level, is the Andasol solar power station, Europe's first commercial parabolic solar thermal plant. This is not a photovoltaic plant but uses a thermal storage system, which absorbs focused heat from the sun into a mixture of sodium and potassium nitrate. The tanks containing this salt are huge – 14 metres in height and 36 metres in diameter. When the sun's rays are concentrated with the parabolic mirrors very high temperatures can be maintained – around 400 degrees Celsius. The molten salt gives this heat to a turbine which operates at night-time or when the sky is overcast.

The installation is pretty large, covering a surface area of 51 hectares. The Andasol plant feeds electricity into the Spanish national grid and is said to have potential to supply power for up to 200,000 people, though at considerable cost.

Hydrogen

In November 2007 a short piece in the *New Scientist* reported that a lighthouse off Britain's North Sea cost was being illuminated with a clean, cheap and abundant power source. A consortium led by the engineering firm CPI had installed a hydrogen cell at the South Gare lighthouse, whose remote and inhospitable location made it prone to frequent power cuts. This article was one of many celebrating the benefits of hydrogen as a power source. Another, in 2006, plugged the work of chemist Don Gervasio, of Arizona State University, who had previously unveiled a prototype laptop battery, based on a novel way of storing and releasing hydrogen.[16]

Some of the optimism surrounding hydrogen is appropriate. Hydrogen is all around us and the only by-product of burning it is water vapour. The technology which promotes its conversion into an energy source could be understood by any schoolchild. Hydrogen, simply put, combines with oxygen to produce water vapour and electricity, which can then be harnessed to do work. Nevertheless, the tendency on the part of many people to believe that we have found the answer to our energy problems should be resisted. Hydrogen isn't really a fuel, so much as a means of making energy. You need another power source actually to manufacture the hydrogen itself – and, under present conditions, you need to put in more of that power than you get back. ERoEI, once again.

Hydrogen comprises about 73 per cent of all the matter in the universe. But here on earth, it's usually bound to other elements – as in water. Hydrogen can be 'cracked' away from water by running an electric current through it, but an energy source is needed to provide that electricity. It can be obtained from natural gas by treating it with superheated water – but that process rests on a fuel source for the heat, and an abundant reserve of gas.

In addition to this, hydrogen is a poor relation to oil when it comes to storage and transportation. It has a low density and takes up a lot of space, so if we were to try to run a car on hydrogen, the bulk of the vehicle would currently be given over to the fuel tanks. It is possible to reduce the amount of space the gas takes up by storing it at very high pressures, but this could result in a very dangerous vehicle. Hydrogen is also very hard to contain and highly corrosive – all of which makes storage and transportation more complex and costly. It could not be moved around the world in our existing oil and gas pipelines because they are too narrow, and they would corrode on contact with the gas.

The key advantage of hydrogen as an energy source, for example in the motor car, is that it does not involve carbon dioxide generation where the energy is used – on the street. The manufacture of hydrogen does produce significant amounts of carbon dioxide; however, the hydrogen gas could be produced in centralized plants where the carbon dioxide given off could be captured and sequestered, so that it would not be poured into the atmosphere like the exhaust from conventional motor car engines on the road.

Carbon capture

It is terrifying to consider that at present 25 per cent of our energy in the UK is still produced using coal, the most polluting source of power. Worse still, the United States generates around 50 per cent of its power from this source, India 70 per cent and China 80 per cent. While alternative fuel sources, like those described above, are urgently needed, one technology which ought to be very beneficial is carbon capture.

Carbon capture, also known as carbon sequestration, is a process which is actively being researched at present.[17] Among the options being explored are pumping carbon dioxide into caverns under the sea bed – for example, into old emptied oil deposits or coal mines – and storing it in porous rocks in the form of a carbonate. It has been calculated that there are enough geological spaces around the UK to store our carbon dioxide emissions for about forty years. Alternatively, it is also possible to sequester carbon dioxide as a carbonate by an industrial process combining the gas with a fine slurry of mineral particles; this method could be employed at manufacturing sites where carbon dioxide is currently released. Research into this process is still at a relatively early stage.

But whatever method is used, in order to capture carbon dioxide for sequestration energy is required – which obviously makes any power plant less efficient. If sequestered in its solid state, the carbon dioxide needs to be turned into a gas, which may then, for example, be captured by a solvent. Later, further energy is required to release the carbon dioxide so it can be transported to the place where it will be finally stored. One issue that has not yet been fully resolved regards the use of pipelines to carry carbon dioxide to the final storage site. For example, if a pipeline is on the sea bed (and storage is being considered in caverns under it), what problems may be caused by a gas leak? The local environment would become acidified, which could affect marine life. In the deep ocean the acid would be rapidly diluted, so this might not be such a problem. But the waters around the UK are relatively shallow (the sea around Viking oilfield off the coast of East Anglia, for example, is only 30 metres deep). However, the strong tides there might dissipate any spillage very quickly. Uncertainties like these are typical of the problems which dog this subject at the present time.

While on the subject of carbon capture, I must briefly mention the recent surge of interest in what is termed 'geoengineering'. Broadly, two strategies are proposed. One involves not merely carbon sequestration, but removal of CO_2 from the air around us – 'scrubbing' the atmosphere. Such proposals may seem fanciful but two in particular are now being actively researched. One mimics the ability of plants to absorb carbon dioxide through their leaves. The Institute of Mechanical Engineers (IME) has recently argued that motorways could be lined with artificial trees, which might absorb vehicle emissions most efficiently. Each of these 'trees' would have 'leaves' shaped like ribbons which would stretch between two vertical poles, supported on a single central post. They would resemble massive fly-swats in shape and would be about twice the height of a double-decker bus. As it is claimed that just one of these 'trees' could process around 3,650 tonnes of CO_2 each year (a real tree of equivalent leaf area processes about 0.03 tonnes each year), the IME calculates that about 100,000 of them would be sufficient to capture all emissions from transport, most industry and homes in the UK. But it still isn't clear how effectively the CO_2 could be sequestered and disposed of once it had been absorbed by these artificial leaves, nor how energy- and carbon-efficient the manufacture of such trees would be.

The other geoengineering strategy involves reducing the amount of solar radiation reaching the earth by reflecting some of our sun's rays back into space. The methods which have been proposed include using reflective materials on the earth's surface, enhancing cloud formation immediately above us, injecting aerosols into the lower atmosphere, or placing shields and mirrors in orbit around the earth to reduce the amount of sunlight reaching us. But methods of reducing solar radiation are a very good example of the unknown risks of implementing new technology, for it is widely agreed by scientists that the consequences are uncertain and the risks may be significant. So for the time being it is likely that we shall concentrate on various methods of carbon capture. But we should not forget that deforestation continues apace around the planet, contributing around 16 per cent of global emissions – about 1.5 gigatonnes of carbon a year. Nature's own methods of carbon capture are something we ignore at the earth's peril.

Geothermal energy

Temperatures even hotter than those on the sun's surface are continuously produced 6,400 kilometres below the surface of the earth. This is mainly the result of the slow decay of radioactive particles, a process that happens in all rocks. The earth has a number of different layers. In the centre is a solid iron inner core; the outer core is made of very hot melted rock, called magma. Around this is the mantle, which is about 3,000 kilometres thick and made up of magma and rock. The outermost layer of the earth is the crust, roughly 8 kilometres thick under the oceans in most places and up to 56 kilometres thick on land. Some of the immense heat generated deep underground is available for use where the earth's tectonic plates meet.

The earth's central heating has been used by humans since history began. During the third century the Chinese built a stone pool at Lisan containing warm water, which still has a constant temperature of 43 degrees Celsius. The Romans used geothermal energy for underfloor heating in Bath and charged an entrance fee for the privilege of combating the miserable weather in the West Country. Geothermal energy was used to heat the town of Boise in Idaho in 1892, and the Icelanders have been using similar sources for heating over the last eighty years. Although geothermal energy has hitherto been limited to volcanic regions, a number of people are now trying to exploit it by drilling holes in the earth's crust and pumping water into cavities around hot rocks below the surface, which is then retrieved as a source of energy. Geothermal energy is used in some places simply to circulate hot water through heating systems or to heat buildings sitting on top of a superficial heat source. Electricity generation uses very high-temperature steam or water (at around 150–350 degrees Celsius), pumped to the surface, usually from geothermal reservoirs a mile or so underground, to power turbines. Some remarkably hopeful neighbours of mine who bought a domestic property near Hampstead Heath have expensively drilled down under their newly reconstructed and refurbished house with the idea of tapping into this free source of heating. I have not felt any seismic disturbances in the area since they began drilling – yet.

About 10 gigawatts of geothermal electric capacity is installed around the world, which is perhaps enough to meet 0.3 per cent of global electricity demand. Iceland, Kenya and the Philippines generate more than 15 per cent of their total energy from geothermal sources. More geothermal electricity is generated in the United States than in any other country, but the amount is fairly insignificant – less than 0.5 per cent of the total electricity needed in the US. Four states have geothermal power plants: California (producing 90 per cent of the country's geothermal electricity), Nevada, Hawaii and Utah.

Is nuclear power an option?

Given that oil is running out, and that all the renewable sources of energy may not meet the world's needs in the next fifty years, we have to ask – are we doomed? Are we going to be forced to return to the way of life we had before the industrial revolution? Shall we work closer to our homes, live close to our food supplies, be reliant upon natural materials for our clothes, our buildings, our material culture, and depend on the tiny amounts of power we can generate from wind and sun and water?

Some of these changes might not be such a bad thing. As James Howard Kunstler puts it in his apocalyptic tome *The Long Emergency*, our social life has been eroded by the oil economy.[18] In

America and, to a lesser extent, Europe small businesses went to the
wall as major corporations used their purchasing power to fly or
ship in cheaper goods from the developing world. With those
businesses went the class of people who cared most about the lives
of their communities – the people who worked as councillors, school
governors, magistrates. In return for such glittering delights as the
ten-pound DVD player and year-round strawberries, we sat back
and watched as our countries turned into retail parks, sealing their
dependence on a fuel source already in decline.

There now seems to be a raw sense of longing for what we have
lost. Recently, two of our most popular BBC television series,
Cranford and *Lark Rise to Candleford*, have relied for their impact
on their depiction of small, close, simple communities, living in an
age without oil. If any set of changes, however unsettling, can induce
us to return to a simpler, less excessive, more harmonious and more
responsible way of life, driven by different values, then they ought
to be welcomed. The message is clear: we shall need to change our
behaviour.

There is also an answer, or at least part of an answer, for every
energy need short of the motor car. As I write, nuclear power looks
set for a revival in Britain, with a government White Paper making
provisions for a new generation of reactors. There are teething
troubles – in particular, disagreements over who will meet the waste
management costs and foot the bill for dismantling the reactors at
the end of their lives. Nonetheless, it is likely that construction – most
of it on existing nuclear power station sites – will begin in 2013.

Changes will be necessary. There may have to be a radical down-
scaling of society. But with nuclear power, that process can at least
take place without losing the benefits of heating and light. So much
power is obtainable, at such reasonable cost, by the method of
nuclear fission that the other 'alternative' energy sources barely look
like alternatives at all. This is such an obvious point that even leading environmentalists, such as the creator of the Gaia theory, James
Lovelock, and Patrick Moore, a co-founder of Greenpeace, have
given it their backing.

The astronomer Patrick Moore provides an interesting case,
because the organization he helped to set up remains opposed to
nuclear energy. Greenpeace has gone to the courts to fight the British

government's plans to revamp the country's nuclear power capacity. It argues that the public consultations were skewed with leading questions. It points out that the safety record for the nuclear industry remains poor, that power stations will be prime targets for terrorist attack and that there are no clear solutions for the disposal of waste.

Nuclear power does create waste. A reactor works by generating heat, which produces steam to drive turbines. There are no pollutant by-products like ozone or carbon dioxide. Building a reactor certainly produces them, but in quantities that, arguably, would be negated by the potential future benefits in reducing carbon and other emissions. Around every two years or so the fuel rods, containing pellets of uranium, need to be changed, and these rods remain highly radioactive for a long time. But there are solutions. Spent fuel rods can be 'recycled' to obtain more usable uranium. Since the first power plant came online in 1957, the total amount of nuclear waste generated is around 9,000 tonnes. The Finnish nuclear authority, Posiva, has come up with a disposal method – involving setting the rods in cast iron, encasing them in copper and then burying them in a borehole filled with bentonite – which metallurgists estimate will render them safe for a million years. Tragedies like Chernobyl hang like spectres over nuclear power, but that reactor was of a specific type, infamous for its lack of safety features. No new reactor would ever be constructed in such a way (although a threat remains from the handful of other models still working around the former Soviet Union). Just as public buildings have taken on board the need to provide for the risks of terrorist attack, so it will very soon be possible to construct reactors to be impervious to assault by bombs or hijacked jet planes.

It is also important to see the threats caused by nuclear waste in the context of other sources of energy. Many people have pointed out that the potential problems caused by nuclear waste do not approach the problems of fossil fuel waste. According to the WHO, around three million people die every year as a result of outdoor air pollution from vehicles and industrial emissions. Solid fuels, used indoors, account for another one and a half million deaths. How these statistics are obtained is, unfortunately, open to doubt; it is, of course, extraordinarily difficult to get accurate estimates, and many rather anecdotal figures abound. But the claim by various journalists, including Don Hopey of the *Pittsburgh Post-Gazette*, that in the US alone fossil fuel waste kills 20,000 people each year, rings true.[19] And it should not be forgotten that scientists have recognized for many years that burning coal in a power plant can release radioactive materials, mostly as uranium and thorium,[20] which may produce around 100 times as much radiation as a nuclear power plant of the same wattage.[21] Gordon Aubrecht of Ohio State University estimates that during 1982 coal burning in the US released 155 times more radioactivity into the atmosphere than the Three Mile Island accident.[22] We should also remember that coal burning releases other dangerous but non-radioactive elements into the atmosphere, including beryllium, arsenic, cadmium, chromium and nickel.

One of the problems about nuclear power is that the extremely negative press it has attracted means that comparatively little research has been undertaken on improving the technology we have used since the 1960s. This is in marked contrast to other technologies in constant use, such as the motor industry, where there is constant innovation because the industry is so high-profile and so widely accepted.

There is another source of nuclear power. The stars and our own sun are powered by nuclear fusion. In the sun's core, hydrogen atoms are transmuted into helium by fusion and this reaction is exothermic – that is to say, it generates heat, the heat of the sun that sustains all living creatures on our planet. Because it is more efficient to use isotopes of hydrogen, scientists are currently trying to fuse the two hydrogen isotopes deuterium and tritium. The fusion reaction, which requires immense quantities of energy to get it started, forces the protons of the hydrogen isotopes together. The fused hydrogen protons then break apart to form a nucleus of two protons with two neutrons – helium. Excess energy from the fusion reaction is released. But for fusion to occur, exceptionally high temperatures are required – greater than 100 million degrees Kelvin. At these temperatures, the deuterium–tritium gas mixture becomes a plasma (a hot, electrically charged gas). In a plasma, the constituent parts of atoms become separated, with electrons being stripped from atomic nuclei, which thus acquire a positive electrical charge and are called 'ions'. For the positively charged ions to fuse, their temperature (or energy) must be sufficient to overcome their natural mutual repulsion caused by the electrical charge.

One issue is how to control plasmas – how to heat a plasma to the extremely high temperatures required and how to confine it in such a state, while sustaining it so that the fusion reaction can become established. Any conventional vessel would just burn, of course. The two most promising methods of containing the intensely hot plasma are to surround it with a powerful magnetic field or to use lasers. The method most favoured by British scientists is generally the Tokomak. The word 'Tokomak' is derived from the Russian *toroidal'naya kamera v magnitnykh katushkakh*) – a toroidal chamber with magnetic coils. In this machine, the plasma is heated in a ring-shaped vessel and the powerful magnetic field keeps it away from the walls of the vessel. Heating the plasma requires massive amounts of electricity – typically around 5 million amperes. Further heating is obtained by injecting beams of deuterium or tritium ions into the plasma. Although vast amounts of power are required to generate the heat and the magnetic fields, the fusion process generates far more. The latest machine being tested is ITER in Caderache in France, run by an international collaboration. It should be able to produce around 500 megawatts – something approaching ten times the power needed to heat the plasma to the temperature required to get the protons to collide and fuse.

The beauty of fusion is that it requires very small amounts of fuels – lithium, deuterium and tritium – that are quite plentiful. No greenhouse gases are produced. The power plant becomes radioactive during the process, but the decay thereafter is relatively rapid, taking around fifty years – a much shorter time than the radioactive waste produced by fission takes to decay. The process is safe because if there is any deviation from the conditions required to produce proton fusion, the plasma rapidly loses its high temperature. It all sounds a little too good to be true. But the engineering and the physics are immensely complex, and it is likely to be at least thirty or forty years before we can expect any usable power from a fusion machine. So, for the moment, nuclear fission, with its waste production and its use of uranium, which is also in rather limited supply, will have to keep us going.

At this point, therefore, it is worth recalling the untimely death of

William Huskisson MP on the opening of the Liverpool to
Manchester railway line. It sums up much of the point of this book
– namely, that contained within every act of creation and innovation
there exists the potential, also, for our undoing. This point is acutely
relevant to the matter of energy. Deaths have resulted from the use
of nuclear power – thirty-one in the explosion at Chernobyl, some
thousands from radiation poisoning afterwards. But once we accept
that every act of human creation has a kickback of destruction,
then the matter becomes one of maths. In one year alone, 2007, the
number of deaths from the Chinese coal industry was 3,786. Factor
in all the other places where coal is mined, the deaths from
pollution, the possible consequences of the emission of carbon
dioxide, and the number of years we've been mining coal, and
nuclear power, quite clearly, looks safe. So why isn't that more
widely understood? Why does fear guide us more than the facts?
Another aspect of the energy issue may provide the answer.

THE CHANGING PLANET

In Auckland, a somewhat unappealing city of flyovers and shopping
malls, a group of men meets once a week to play rugby. They are on
low incomes, most of them, and these games are a way to let off
steam and socialize. They also provide a welcome opportunity for
networking, because these men belong to New Zealand's newest
immigrant community. They hail from Tuvalu, a group of six atolls
and nine islands far out in the Pacific Ocean. Separated from one
another by large tracts of ocean, the various island peoples of
Tuvalu have had few opportunities to interact. Flung together in
Auckland, they are discovering vital links, but also new and un-
anticipated tensions. Some of those tensions can only increase as
New Zealand continues to accept migration from Tuvalu over the
next thirty years. It has little choice. Tuvalu is a Pacific neighbour,
and is facing an emergency.

In September 2003 the Prime Minister of Tuvalu, Saufatu
Sapo'aga, addressed the 58th session of the UN General Assembly
with these portentous words: 'We live in constant fear of the adverse

effects of climate change. For a coral atoll nation, sea level rise and more severe weather events loom as a growing threat.'

Tuvalu's plight – as well as its bid, in 2003, to sue the Australian government for refusing to ratify the Kyoto Protocol – is well publicized. For the Sunday supplements, images of the island's eroding shoreline, its leaf-huts flooded, its coconut trees and taro plants poisoned by seepage of salt water, are a gift. So, too, are images of the Auckland migrants – a picture-perfect demonstration of the miseries of man-made climate change: noble tribesmen forced into the concrete jungle. The planet is warming as a result of the excessive accumulation of greenhouse gases in the atmosphere – a direct consequence of our overindulgence in fossil fuels. The ice-caps are melting, the seas are rising, and the Tuvaluans have to move. That, in a soundbite, is the media version of climate change.

But not everybody accepts this picture *in toto*. University of Auckland geographer Paul Kench thinks Tuvalu is far more resilient than it suggests. Atoll groups like this are made up of the remnants of other islands which have been swept away during other periods of high sea level. These places do not so much drown as regroup. In an article published in *Geology* in 2005, Kench pointed out how the Maldives fared during the 2005 tsunami. Some 9 per cent of their land area was swept away in the giant waves, but much of it was deposited back on the far side of each island.

It is also not entirely clear how much of the dramatic tidal seepage in Tuvalu is a direct consequence of climate change. The recent gravelling-over of the tiny archipelago state's road network has blocked the ability of the underground water table to rise and fall, forcing it outwards, up and over the shoreline. Much irrevocable damage, including the digging of vast, unfilled holes, was wrought by the US military, who used Tuvalu as a base during the Second World War.

Nor is it entirely decided whether climate change is the *only* factor driving the rising of the seas, or indeed whether they can be said, uniformly, to be rising. In parts of the world such as Alaska and Japan, shifts in the tectonic plates beneath the continents are so pronounced that the level of the land itself rises and falls, and tidal gauges – still the main means of measuring sea level – cannot be relied upon. In other areas, such as the Gulf of Bothnia between

Sweden and Finland, melting glacial ice is freeing up land that has been held down under its grip for centuries, resulting in a rebound of around 10 millimetres per year. At the same time, globally there is something of a mismatch between the total rise in sea levels and the contributions we would expect from individual factors such as the expansion of the oceans as a result of rising temperatures or the melting of the planet's ice-caps and glaciers.

None of the above information is top secret – it is there for any-one interested in the subject to see. But in today's hyper-fast world, where news is sent by text message, and stories have to be reduced to their baldest, starkest and most dramatic elements, such un-certainties grab few headlines. Flooding islands do. So – to refer back a few pages – do dangerous nuclear power stations, to an infinitely greater extent than clean, safe, boring ones. And so, of course, do the melting ice-caps – another poster-child for the journalist or TV documentary crew seeking to do a piece on climate change. We look at these images and feel an understandable sense of guilty shock. Was Mr Al Gore really right? Have we really done this, we fret, just by driving to Sainsbury's? We may overlook the fact that the Arctic ice-caps have tended to go through a regular cycle of expansion and contraction. Chunks are always falling off into the sea, always have been – but just possibly we are more aware of the process because of increasingly sophisticated satellite imaging.

There is no doubt that our planet is currently enduring a climatic high, and one which is most likely to have disastrous consequences for many people. But as they are presented to the public, the issues of global warming and human-generated greenhouse gases are almost always conflated. One is created by the other, so the reason-ing goes, and there is no doubt about it. But science is inevitably *about* doubt and the weighing of evidence – or it is meant to be.

Within the scientific community, a number of voices have argued for a more balanced view of climate change, human-generated greenhouse gases and the relationship between the two. It has been argued, for example, that our climate has gone through periodic warming and cooling cycles before. The very notion of a stable climate may be an oxymoron. Global temperatures have been warmer than they are now – for example during the Holocene Maximum, some 8,000 years ago – and the polar bears, quite clearly, survived. The role of greenhouse gases – most notably carbon dioxide – in this fluctuation, and their contribution to the process, are complex. Between 1940 and 1970 there was a global dipping of temperature, at the same time as the output of human-generated carbon dioxide reached a peak. Conversely, during the Ordovician period, roughly 433–488 million years ago, carbon dioxide levels were far higher than they are today, and yet the earth was in the grip of an ice age.

So just possibly we need to be more reserved about the causal relationship between carbon dioxide and global warming. Dr Ian Clark, Arctic palaeoclimatologist and Professor of Earth Sciences at the University of Ottawa, does not doubt that the current temperature trend is heading upwards, or that our activities have boosted the amount of carbon dioxide in the atmosphere. He does, however, think that carbon dioxide emissions may be caused by warmer temperatures, rather than the reverse. Samples of ice excavated from the Arctic may indicate a gap of at least 800 years between a hike in heat and a rise in levels of the gas. This could be because the oceans store a vast amount of carbon dioxide, which is released in periods of high temperature. But they are so vast and so deep that it takes centuries for the warming to have its gas-releasing effect.

Science tells us that if the land and the seas are warming up because of the accumulation of greenhouse gases, then the troposphere – the lowest level of the earth's atmosphere, in a band about 10 kilometres above us – should be experiencing a parallel rise. But this has not been definitively established – in fact, it may be that the surface of the planet is warming more than the atmosphere. In turn, some controversy hangs over the basic premise of whether things really are hotting up in the way they are said to be. It has long been mooted by several scientists that the picture may be distorted by what is known as the 'urban heat island' (UHI) effect, in which expanding cities and populations produce disproportionate amounts of heat.[23]

It is worth looking at the recent history of this particular idea. In 2003 Thomas C. Peterson, of the US National Climatic Data Center, published a report dismissing the theory of UHIs, and its conclusions subsequently became incorporated into the official standpoint of the UN's Intergovernmental Panel on Climate Change

(IPCC). Subsequently, the Canadian geologist Steven McIntyre, founder of the climate sceptics' website http://www.climateaudit.org, challenged Peterson's findings on a number of grounds, not least the quality and location of a number of the weather stations whose data was used, and found an urban heat rise difference of 0.7 degrees Celsius. Dr David Parker of Britain's Hadley Centre for Climate Prediction and Research then countered this by looking at temperature differences on calm and windy nights. Since more of a city's heat is blown away in strong winds, we would expect that, if the UHI exists, there would be a bigger temperature rise on calm nights. Finding no such correlation, he argued that the effect was of minimal consequence. The mildly sceptical climatologist Roger A. Pielke responded in 2007 with a study indicating that minimum temperatures are highly sensitive to the height at which they are taken and the wind speed.

I summarize here the academic to-ing and fro-ing over this one minor tributary of the subject because it is important to show that there is a robust debate going on among some scientists. The public impression is sometimes very different – one of uniform agreement on the issue of human-generated climate change, with just a few heretic voices at the fringe. There are certainly flaws in the arguments of the sceptics, as there are inadequacies in those who totally endorse the idea of human-generated climate change. Some of those who deny it have few qualifications in the field upon which they speak, or have published very little in the peer-reviewed literature. Some are undeniably in the pay of the fossil fuels industry. None of that, however, undermines the fact that science should be founded on evidence and debate about the evidence and presented to the public as such – not as truth, but as enquiry.

That engagement of the public is more important now than at any time in the history of science and technology. In spite of increasingly intensive negotiation involving the leaders of all industrialized countries, there is as yet no agreement on how much we shall need to reduce harmful emissions beyond 2012. This is surely where public attitudes and opinions need to have much stronger influence. We shall need to influence our politicians because, if the most careful scientific assessments are correct, our need to do more to control greenhouse gases is extremely urgent. According to Dr Michael

Raupach and colleagues at the Commonwealth Scientific and Research Organization in Australia, global emissions have continued to increase by about 3 per cent per year.[24] This is a faster rate than that predicted by the IPCC in its most pessimistic assessment of the world's current fossil fuel usage. And this limited estimate suggests that there will be an a massive increase in global mean temperature of about 4 degrees Celsius (2.4–6.4°C) by 2100,[25] which very few people doubt would have the most appalling consequences.

But the way the media sometimes packages a false consensus and sells it on is particularly dangerous. The louder the claim, the more alarming the images, the bigger the controversy – the greater the interest. This distorts our understanding of the evidence. Think back to the SARS epidemic of 2002–3, and how we were shown images of Asian people going about their daily business in face-masks. The effect was alarming but disingenuously achieved because, as any traveller to those windy and dusty regions knows, face-masks are quite commonly worn. Or think, indeed, of the alarmist reports in the *Daily Mail* about the purported side-effects of the triple vaccine against mumps, measles and rubella, which we now are almost certain have nothing to do with the triple vaccine. To some extent, the same may happen with climate change. All extreme weather, from twister storms to hurricanes to heavy floods, now tends to be presented as a consequence of human-generated global warming. Virtually any kind of strong wind tends to be labelled as a 'hurricane', because that is a powerful word, with overtones of irrevocable disaster. Yet any balanced reading of climate science forces us to realize that the true picture is hugely complex. Even acknowledged experts within the field admit that factors like the effect of sun spots on cloud cover, or changes in ocean currents, are barely understood. Climate is, in the words of Philip Stott, Professor Emeritus of Biogeography at the University of London, 'complex, coupled, non-linear, chaotic'. Our emissions of a gas that makes up some 0.03 per cent of the atmosphere may have some palpable effect on the hair-trigger of this system, but they are not the only factor, and it remains in doubt whether changing them can change the climate.

Many of the more balanced – and consequently less frequently heard – voices are calling for adaptation to a changing world, rather than costly and futile attempts to halt or reverse that change. If sea levels are rising, then we need to build defences, relocate people and stop the expansion of cities on to the marshland that can absorb overflows. If the oil is running out, we need to invest money in alternative energy, but also accept that many attributes of today's overconsuming global society will simply have to go. Whatever we do, we shall certainly have to change our behaviour. Given that human-generated greenhouse gases have driven the recent spurt in temperature, then the trend may be reversed as industry declines in an oil-poor world; but the problem is, of course, that by then it will probably be too late to save much of what is valuable on this planet. But if temperatures are rising for other reasons, then we need to adapt to them by growing food in those northerly areas being thawed. If temperatures are rising and fuel sources are becoming more scarce, we need to build buildings that use wind and sunlight to cool and warm themselves.

Historically, as this book has shown, that is what humans have always done. Temperature changes forced us down from the trees. Later, they compelled us to farm. Within every act of invention, as we have seen – and our use of carbon fuels is no exception – there lies the potential for destruction. But the reverse is also true. As the people of Tuvalu are finding new opportunities and challenges, so the whole planet may have a chance to renew itself.

Chapter 10

Physician, Heal Thyself

PEOPLE WERE EXCITED when a certain weight-control drug became available over the counter in British pharmacies. Every chemist's shop in the capital soon had a proud sign claiming that it stocked Alli (or Orlistat), a drug which assists weight loss by blocking the absorption of dietary fat. Though relatively expensive, a few months' supply is less painful than living on soup or panting around the local park. Those buying some of these products, however, may have been given pause by the accompanying leaflet. The listed side-effects of Alli include 'wind with or without oily spotting, sudden bowel motions, fatty or oily stools'. In a bid to sound positive about a stream of wind and oily diarrhoea, the makers add: 'Many users have told us that the effects were a signal that the capsules were working and helped them maintain healthier eating patterns.' In other words, the side-effects of the drug are so unpleasant they force people to cure themselves by other means.

It is unfair to single out Alli as the remedy with the least attractive side-effects. Accutane, an acne remedy, can, in rare instances, cause conjunctivitis, visual disturbances and vomiting, and Requip, a formula for combating Restless Leg Syndrome (I cannot recall any lectures on this subject at my medical school), removes the twitching in limbs but may replace this harrowing symptom with hallucinations, difficulty in moving or walking, problems with breathing, chest pain and a sudden irresistible urge to sleep. Its users are also warned to tell their doctor if they experience new or increased gambling, or sexual or other intense urges while taking the drug. Considering the mild nature of the conditions these drugs are meant to treat, one wonders if they're worth it.

It may seem perverse to include the healing arts in a book about the perils of technology. But I am not attempting to condemn technology, or to undermine its value. Medicine is a great human achievement, helping not merely our own species but some other creatures too, even plants. However, my object throughout this book is to focus carefully on various aspects of human ingenuity, pointing out that even the most seemingly valuable ideas need to be considered with wisdom and discretion. And while from a distance medicine may seem a glittering jewel, on closer inspection serious tarnish undoubtedly shows.

This ambiguity around medicine arises from three factors in particular. In the first place, in spite of vigorous medical traditions across the world from the third millennium BCE right up to the end of the nineteenth century, medicine's history has been for the most part a number of (painful) shots in the dark, and what easement of suffering it brought often had more to do with luck than human ingenuity. Second, many of medicine's greatest achievements have emerged to patch up problems which our so-called civilizations created – such as warfare, the malnutrition that came from dependency on agriculture, and the epidemics nurtured in our overcrowded cities. Third, medicine's undoubted achievements can produce, both in physicians and in their patients, an illusion of the doctor's omnipotence which diminishes the quality of care, lessens our capacity to deal with factors like pain, ageing and death, spreads public unease about lifestyle choices, and inhibits debate about how human societies will manage to foot the bill.

A HASTY HISTORY OF HEALING, 2500 BCE–1922 CE

Spiritual healing: medicine in the ancient world

Compress medical history into an hour, and it is only in the last two minutes – that is, about 100 years – that doctors have, in any useful sense, been able to cure people. Earliest references to the profession were decidedly guarded. The Code of Hammurabi, produced in Babylon in the middle of the second millennium BCE, included rules of conduct for physicians, and a sliding scale of rewards and punishments relating to those they treated. A doctor saving the life of a lord with a bronze lancet would receive ten silver shekels: a handsome – not to say improbable – reward. If, alternatively, he caused the lord to lose his life, he would have his hand cut off. And if he inadvertently killed a slave with his ministrations, he would have to supply a replacement.

These texts refer to a threefold division of the medical classes. Seers specialized in divining the spiritual origins of sickness, usually by examining the livers of sacrificed birds. Priests performed incantations and exorcisms – necessary since most maladies were believed to be the work of bad spirits. A third class – who knows whether their status was high or low? – prescribed drugs and performed basic surgery. But the fact that the Code of Hammurabi included fines for incompetent doctors and an astronomical reward for someone who did manage to save a life suggests that medicine didn't do the Babylonians much good. They used their first law codes to protect themselves against bungling physicians.

That is not to say that those ancient physicians didn't try. Although the diagnosis and treatment of illness was primarily a spiritual undertaking, the ancient Babylonians had a strong, proto-scientific tradition of observation. The 'Treatise of Medical Diagnosis and Prognosis', etched in painstaking cuneiform on forty tablets, is a list of ailments, among which we may recognize tuberculosis and malnutrition. There was no lack of curiosity; but it is noteworthy that, even at the dawn of civilization, all this intellectual energy was put into trying to counter the very problems which we had unwittingly begun to cause: epidemic diseases from living in cities and dietary deficiencies from the adoption of agriculture. Whether the Babylonian doctors alleviated suffering or not, it is hard to see this as progress – it was more of a strip of sticking plaster slapped on a self-inflicted wound.

The Edwin Smith papyrus from ancient Egypt, dated to around 1600 BCE, gives an inventory of forty-eight conditions and methods for treating them. Other papyri include advice on pregnancy and childbirth, and contraception – for which pessaries of crocodile dung were recommended. Incidentally, if you walk into the historical collection of the Royal College of Obstetricians and Gynaecologists in Regent's Park, you may boggle at the sight of one of these dung pessaries: it is twice the size of a tennis ball. The *swnu* – as physicians were called – could draw on an elaborate pharmacopoeia, including goose fat, ox spleen, tortoise gall, garlic, opium and cannabis, administered as pills, potions, lotions and suppositories – they could even be blown up the urethra through a tube.

While Babylon left no records of its doctors, the Egyptians elevated some of them to high status. Records mention Iri – illustriously titled the Keeper of the King's Rectum – and Imhotep, lieutenant to Pharaoh Zozer. So powerful a figure was Imhotep – also famous as an astrologer, priest and architect of the pyramids – that he became deified, his shrines associated with healing.

The sacred tradition was maintained in ancient Greece, where the sick spent a night in the shrines of the half-god Asclepius. Perhaps our ancestors had some awareness of the curative properties of a good night's sleep, but the practice also had a mystical edge: the god was supposed to give the patients therapeutic dreams whose message could be interpreted by a temple priest. This had some genuine, if minor, benefit; after all, listening to patients can be just as beneficial as scribbling something on a prescription pad. Those who felt themselves helped in the Asclepian shrines would sometimes commission memorials to the helpful divinity. One such inscription reads: 'Hermodikes of Lampsakos was paralysed in body. In his sleep he was healed by the god.'

Oaths and boasts: medicine in the classical world
The Greeks also produced a scientific tradition with Hippocrates (*c.*460–377 BCE) and those who followed him. Hippocratic doctors

scorned the dream-diviners and argued for an empirical science of healing, based on the observation of symptoms and treatment with naturally occurring remedies. The real Hippocrates shares something with his fellow Greek, Homer, in that the works attributed to him were actually written by several people over a long period of time – but what we call the 'Hippocratic corpus' nevertheless presents a consistent view.

The Hippocratic view was that health and sickness depended on the shifting balance of four key elements within the body, each associated with bodily fluids, tangible physical phenomena, colours, seasons and temperaments. The four elements were the *humours*. Blood, making the body hot and wet, was the source of vitality. Choler, or yellow bile, aided digestion and made the body hot and dry. Phlegm – all colourless secretions – made the body cold and wet, functioning as lubricant and coolant. Black bile, meanwhile, lent its dark colour to other bodily secretions, and made the body cold and dry. It is a masterly system of classification, but totally useless.

The balance of these four elements determined body shape, temperament, and the sorts of diseases from which a person suffered. For example, too much blood made the body overheated and feverish. Treatment could involve blood-letting, exercise and change in diet. Certain sicknesses were common when the climate favoured an excess or a shortage of one element or another: the cold, wet wintertime was associated with phlegm and lassitude, and some diseases of the brain, digestive system or lungs; the warm, wet spring was associated with blood and fevers.

The humoral framework linked many observable phenomena in an overarching scheme and must have reassured people whose doctors, in reality, had no power to cure them. The Hippocratic Oath gives more emphasis to this role of the medical profession, as reassurer and comforter, than it does to its curative potential. The oath consists mainly of vows *not* to do things – not to do harm, not to assist suicide, not to perform abortions, not to divulge secrets nor abuse one's power. In its way, like those snippets from the Code of Hammurabi, it perhaps represents an admission that doctors could do little to cure but had the potential to do great harm. It was therefore important that they presented themselves to the public as men of honour and integrity. And a modern form of the Hippocratic Oath is still sworn by many graduating medical students around the world.

How many of these Hippocratic values mattered to the next figure in our history? If Galen (129–216 CE) is really famous for one thing, it is a powerful sense of his own importance. 'As Trajan did for the Roman Empire,' this Latin physician declared, 'when he built bridges and roads through Italy, it is I, and I alone, who have revealed the true path of Medicine.' As well as acting as physician to the emperors, Galen performed useful dissections on apes, sheep, pigs and goats. But, like his Hippocratic forebears – whose work he acknowledged grudgingly – Galen was culturally opposed to dissecting human corpses, so his contribution to medical science was limited. And Roman writers like Cicero maintained a bluff, hearty approach to health, regarding medicine as a luxury entertained by the effete Greeks and doctors as the enemies of well-being, and extolling the virtues of cabbage soup and exercise. Perhaps they had a point.

The baton passes: medicine in the Christian and Islamic worlds

> And behold, there was a man which had his hand withered. And they [the Pharisees] asked him, saying, Is it lawful to heal on the Sabbath days? That they might accuse him. And he said unto them, What man shall there be among you, that shall have one sheep, and if it fall into a pit on the Sabbath day will he not lay hold on it and lift it out? Then saith he to the man, Stretch forth thine hand. And he stretched it forth; and it was restored whole, like as the other. (Matt. 12: 10–13)

The biblical account of Jesus' life contains twenty-eight separate incidents of healing. From reanimating the dead to curing more trivial ailments like shrivelled hands, dropsy and epilepsy, the miracles in the New Testament set Christ firmly within an older Jewish tradition of respect for healing. This inspired certain religious orders in the early Christian era to continue Christ's mission as healers – although it must be assumed that they lacked access to the miracle-cupboard.

Different Christian saints became associated with responsibility for various organs or diseases: St Valentine of Terni for epilepsy, St Blaise for sore throats, St Fiacre for venereal disease. Old Asclepian shrines were reworked as Christian sites of pilgrimage, and the sick invested considerable amounts of time, money and hope of obtaining a cure in journeys to touch their walls, sleep in their precincts, drink their waters or purchase associated relics. This, in itself, tells us how useless most doctors' treatments were. The religious élite, meanwhile, viewed the flesh as a base cloak for the immortal soul. It was beneath a true Christian to pamper his body by curing it – indeed, a virtue lay in doing the opposite.[1] With any scientific progress confined to the closed world of the monasteries, Europe entered a 'dark age' of medicine, while Middle Eastern scholars blazed a trail.

In his early career, Muhammad ibn al-Zakariya al-Razi, also known as Rhazes (865–925), practised medicine at the leading *bimaristan* (house of the sick) in Baghdad. The activities that went on inside such places were encouraged by the new Islamic faith. Belief in Allah as a just, all-powerful and merciful God meant that sickness could not be, as previous generations had decreed, caused by the *jinni* – evil spirits. The Prophet had ordained a remedy for every predicament in Allah's creation, and therefore medical intervention had divine sanction. In addition to this, *zakat* – charity – was a central tenet of Islam, meaning that both poor and wealthy could receive aid.

Rhazes left as much written material pertaining to the harmful aspects of medicine as on its benefits, including such works as *The Mistakes in the Purpose of Physicians*, *On Purging Fever Patients Before the Time is Ripe* and the incomparable *The Reason Why the Ignorant Physicians, the Common People and the Women in Cities Are More Successful Than Men of Science in Treating Certain Diseases and the Excuses Which Physicians Make for This*. A cautious observer, without the vaingloriousness of Galen, Rhazes left meticulous case records, setting down not just the symptoms of the various conditions he saw and how he treated them, but the ages, sexes and occupations of his patients. He recorded treating sore throats with a gargle of vinegar, and swollen testicles (his own, as it happens) with purging. He was also the first physician to distinguish between different skin rashes, treating smallpox, measles and others as separate entities, whereas previously they were thought to be a single complaint. His work *Secret of Secrets*, meanwhile, provided later physicians with a manual of basic chemical practice, describing the processes of purification, distillation and calcination.[2]

Hospitals thrived throughout the Islamic world – in Baghdad, Damascus, Cairo and Cordova. By the twelfth century, every major town under the Crescent boasted some centralized site for medical care, research and training. Physicians were public figures, with civic duties stretching beyond their roles as healers of the sick and administrators of hospitals. Some of them, like Moses Maimonides, were Jewish and treated favourably by their Muslim rulers. Al-Mansur Qalawun's hospital at Cairo, built in 1283, boasted separate wards for physical and mental illnesses, a surgery, pharmacy, library and lecture rooms, and worship facilities for Christians alongside its own mosque. European travellers to Middle Eastern lands marvelled at the standards of care – in particular at the humanity shown to the mentally ill. But while the quality of care reached new levels, Islamic medicine was certainly no better at finding effective cures for most diseases.

Plagued by the poor: medicine in medieval Europe

> In order that we might make a concise study of the treatment of women it ought to be noted that certain women are hot, while some are cold. In order to determine which, we should perform this test. We anoint a piece of lint with oil of pennyroyal or laurel and another hot oil and we insert a piece of it the size of a little finger into the vagina at night when she goes to bed and it should be tied around the thighs with strong string. And if it is drawn inside, this is an indication to us that she labours from frigidity. If however it is expelled, we know she labours from heat. In either case, assistance should be given.

Trotula is claimed by some feminists as an islet of womanly achievement in a man's world.[3] According to some accounts she was an eleventh-century female physician, trained at the famous medical school of Salerno in southern Italy. Recently it has been suggested that the work identified with her, known as *The Trotula*, was in fact written by three different authors, one or two of whom may have been female. Trotula opposes the widely held view of the time that suffering in childbirth is divinely ordained, and recommends giving opiates for pain relief. Perhaps it took the personal experience of the pangs of childbirth to generate an understanding of them. Whoever she (or they) may have been, the legend of Trotula would not have been passed down were it not for the vigorous medical tradition emerging in southern Europe at this period.

Islamic hospitals may have influenced Crusaders, who brought their ideas back to the West. But, as Roy Porter points out in his concise account *Blood and Guts*, the rise of European medicine was probably more strongly related to the rising importance of three other entities: cities, universities and the bubonic plague.[4] As learning emerged from the monasteries and passed into the newly founded universities, cities like Salerno became centres of a fresh medical tradition. A new interest in the classical and Arab worlds, inspired partly by trade, resulted in the mass translation, editing and reworking of previous works of medical scholarship. Training doctors became a lengthy process – seven years of lectures and oral examinations – and the profession saw medicine as a 'science' rather than a 'craft'. The true practitioner was someone who understood the reasons behind conditions; people who simply cured them with medicines or surgery were lowly artisans. To protect their trade and reputation, doctors began to band together in guilds from the thirteenth century.

Larger towns attracted many healers. At the beginning of the fifteenth century Florence boasted not only recent graduates of its own prestigious medical school, but also bone-setters from Rome and families specializing in the cure of eye disease, hernia and kidney stones. As well as prescribing herbal tinctures of dubious efficacy, specialists let blood, applied leeches, administered enemas and dressed wounds – the last function, perhaps, being their most solid contribution to popular health. By this time, French surgeons such as de Mondeville and de Chauliac had challenged the ancient Greek idea that 'wet healing', allowing the formation of pus, was beneficial, and recommended that wounds be drained and allowed to dry in fresh air. However, nonsense cures still abounded – for example, de Chauliac himself counselled epileptics to write the names of the Three Wise Men in their own blood, and recite three Pater Nosters and three Ave Marias daily for three months.

As cities grew in importance, the need to provide for their less fortunate citizens became apparent to the ruling classes. Peasants in the remote countryside could be ignored; but in the city, the wealthy lived cheek by jowl with the stinking, scrofulous masses. It therefore became a sideline of one's civic and Christian duty to make provision for the poor in the form of almshouses and hospices. Initially established to provide food and shelter for pilgrims and the most destitute, these small institutions of necessity performed a medical function because so many of their clients were unwell. The duty of care passed to monks and nuns, many of whom saw tending the sick – and, in particular, giving them a dignified passage into the afterlife – as a religious duty. But in France and Italy, throughout the thirteenth century, these poor-houses became increasingly medicalized, with doctors and pharmacists, backed up by their newly formed guilds, taking on the functions previously performed by religious orders. At the same time, the religious brotherhoods formed during the Crusades occupied themselves in peacetime by setting up proto-hospitals from Poland to Spain, and eventually in the New World.

It was necessity, more than Christian charity, that drove the foundation of hospitals, the emergence of people to run them, and a concern with the health of the populace. Rising urban populations contributed to worsening sanitation, as water supplies were contaminated with human excrement and living conditions became ever more cramped, with livestock kept among human settlements. And international trade, for example with Africa and Asia, made for a two-way traffic in deadly bacteria. The prevalence of widespread disease forced European authorities to take a new approach. From the twelfth century, provisions were made to house and bury those suffering from leprosy in separate quarters. Leprosaria – offering confinement and basic care – sprang up across Europe, some providing the framework for later hospitals such as London's St-Giles-in-the-Fields. Once again we see medicine emerging, not as some life-enhancing bolt of genius, but as a (mostly inadequate) response to problems humankind was complicit in creating.

Leprosy ceased to be a threat from the mid-fourteenth century. It was not eradicated by blood-letting, purging and hocus-pocus cures offered by the medical establishment; possibly we just developed better immunological resistance to the leprosy bacillus. But bubonic plague, though it had probably been around since biblical times (the Philistines being smitten with 'emorods in their secret parts'),[5] was first clearly documented in 1347. Its origins were said to result from biological warfare practised by Tartar tribes fighting Italian merchants in the Crimea. Laying siege to the Italians in the city of Caffa (modern-day Feodosya), the Tartars were forced by illness to retreat – but in (as they say) a parting shot, they catapulted numbers of their dead over the city walls, and on impact these decaying bodies fragmented, infecting the Italians with the plague. International trade – that supposed glory of the Middle Ages and Renaissance – was the vector for the spread of death: the Tartars had become exposed to the infection through their own contacts with China; Europe fell when the infected merchants returned by ship to Genoa.

Seen variously as the wrath of God and a Jewish plot, bubonic plague created a new role for doctors as agents of public health. Protected by a bizarre-looking costume made of leather, including a long snout containing aromatic herbs, doctors strode the affected areas, administering strong-smelling potions, ordering the fumigation of rooms and probably spreading the plague further as they made their rounds. To be fair, their altruism would have meant that many succumbed themselves. Quarantine measures were enforced throughout Italy and France,[6] and when the first plague subsided the mechanisms of medical authority remained. In Milan, a physician, a surgeon and two gravediggers formed the first permanent 'Board of Health' in 1410 (note that gravediggers made up 50 per cent of the board, indicating how effective medicine was in warding off the Grim Reaper). Similar bodies throughout Europe – generally using doctors as advisers and relying on local noblemen for executive clout – took over the regulation of borders during epidemics, and the day-to-day administration of cemeteries, noxious industries and poor-houses. Still functionally useless where the curing of sickness was concerned, medicine had begun its slow creep of involvement with daily life. Initially the promoters of hygienic

childbirth and of best practice in child-rearing, in the late Middle
Ages midwives became agents of orthodoxy, reporting illegitimate
births and pressing unmarried mothers for the names of the fathers.

In spite of the elaborate measures undertaken to avoid the plague,
and the various theories concerning its origins, no one suspected
that it was carried by rat fleas, nor that, because rats were prone to
die within six days of becoming infected, humans were ideal for the
fleas' next meal. Dung-heaps, decaying corpses and the effluent of
tan-houses and butchers were identified as sources of ill health – but
not through any understanding of bacteria or the processes of
putrefaction; instead it was thought that bad smells created bad
vapours, which poisoned the air.

Drawing aside the veil

In Christian history, the name of the Council of Tours is synonymous with unpopular legislation. It was in this French city in 567
that the Church decreed that any priest spending the night in bed
with his wife would be excommunicated for one year; moreover,
monks were now forbidden to sleep two to a bed. In 755 the
Council proposed beginning the calendar year at Easter – a move
which was vigorously opposed. And in 1163, as well as ordering
that the heretical Cathar sect be deprived of all their goods, the
Council vetoed the deliberate dismembering of dead Crusaders.

Western medicine did not remain in the doldrums until the
Reformation simply because the Catholic Church forbade the dissection of bodies. Throughout the Middle Ages, Western doctors
had access, in their own languages, to accounts of the ancient
Greeks Herophilus and Erasistratus,[7] who conducted public
dissections. By this means, for example, they identified the
duodenum and traced the roots of the nervous system to the brain.
And – until the Council of Tours outlawed it – this knowledge was
supplemented when, in order to bring back slain Crusaders to their
homelands, their bodies were cut up and boiled, leaving just the
bones. Moreover, with the emergence of the physician as a state
functionary, his role came to include the performance of autopsies
on murder victims – under which guise anatomical study certainly
took place.

The first recorded public human dissection was conducted by

Mondino de'Luzzi, author of a definitive anatomical guidebook, in Bologna in January 1315.[8] In his book he reports that he conducted a dissection on the body of a woman, giving him an opportunity to examine the anatomy of the uterus. In 1375, a dissection conducted in Montpellier was declared obscene and no further such procedures were allowed for some years. Eventually, however, the study of anatomy became crucial to a physician's training, and dissection finally received official sanction some thirty years before the first stirrings of Protestantism, when Pope Sixtus IV permitted the dissections of executed criminals. These often took place in impressive theatres with public galleries – indicating the cooperation of medicine with the judicial arm of the Church – and were conducted partly for education and partly to inspire a sense of awe. Heavily ritualized, a dissection was directed by a robed physician intoning from an anatomical guidebook, while a surgeon slit the corpse open and an assistant indicated features with a pointer.

Conducted like this, dissections replicated the errors of previous generations until, in the iconoclastic climate of sixteenth-century Europe, scholars began to challenge Galen and Hippocrates. Falloppio published mappings of the skull, ear and female genitals in 1561. Discoverer of the clitoris – an organ whose location and purpose might be said to elude some men to this day – Falloppio also described the conduits leading from the ovaries to the uterus, although he failed to realize that it was through these (which now bear his name: Fallopian tubes) that all of us once passed into the womb as fertilized eggs.

Advances in anatomical understanding aided the English physician William Harvey in his brilliant work on the circulation of the blood.[9] For over a thousand years, the prevailing view had been that the liver created the blood and then 'washed' it throughout the tributary veins of the body, in the fashion of a flood spreading through a field. The veins, it was thought, carried the blood along by virtue of a unique pulsating action inside their walls.

Harvey, influenced by Hieronymus Fabricius, whose dissections he had attended in Padua, recognized that the heart was a muscle, driving blood around the body through contractions. He realized that the amount of blood forced out by the heart in a single hour was greater than that contained in an entire body. Hundreds of gallons passed through the organ in a day – far more than could be absorbed by the body, or indeed manufactured by the liver. This led him to conclude in *Exercitatio anatomica de motu cordis et sanguinis* (1628) that 'blood in the animal body is impelled in a circle, and is in a state of ceaseless motion'.

Following Harvey, bright young physicians researched the heart, lungs and respiration. Their work coincided with the foundation of the Royal Society (1660), a body that gathered together the men calling themselves 'natural philosophers' and doctors. Within the precincts of the Royal Society, Richard Lower joined forces with physicist Robert Hooke to investigate the differences between arterial and venous blood, conducting the first blood transfusions.

Natural philosophy – expounded by early scientists like Hooke, Boyle and Descartes – had a profound influence on medicine, encouraging the view of the body as a machine. Scholars rebelled against the Hippocratic framework, with its vague talk of vapours, humours and essences, and put their effort and expertise into studying the structures of the body. What they found they expressed in terms of a mechanized system in which every part had a purpose, often with some analogy in engineering or hydraulics. With new instruments – the microscope and the thermometer – they took detailed readings of muscle behaviour, gland secretions, heart action and respiration. In Italy, men like Galvani and Volta demonstrated how muscles could be made to move with the application of electrical current. Others, like the radical Swiss doctor Theophrastus Philippus Aureolus Bombastus von Hohenheim – Paracelsus, for short – analysed the body in chemical rather than physical terms. Paracelsus argued that all bodily processes were chemical, due to fermentation or the action of a gas.

The application of real science through measurement and analysis brought significant benefits. But, as so often in this history of ideas, every benefit was achieved at a cost. The marriage of medicine and natural science created a culture in which practitioners became more interested in sickness than in the sick. This trend was encapsulated in the work of the eighteenth-century anatomist Giovanni Battista Morgagni. His mammoth work *De sedibus et causis morborum* (remarkably, published when its author had reached the age of 79) related the findings of 700 autopsies, describing in particular how the causes of death could be seen within the organs. His work was invaluable – especially for the understanding and later treatment of heart disease – but it contributed to a decline in the old, Hippocratic, caring role. One of Morgagni's followers, the eighteenth-century pathologist François Xavier Bichat, made the following comment – his words contrasting sharply with those of Hippocrates quoted beneath them:

> You may take notes for twenty years from morning to night at the bedside of the sick, and all will be to you only a confusion of symptoms.

> . . . when you examine the patient, enquire into all particulars . . .

Bichat, like Morgagni before him, believed physical observation and measurement of the body, alive or dead, were the paths to truth. Once seen as a friend to the sick, trained to take the 'whole person' into account in his diagnoses, the physician became a scientist, studying diseases, not people.

A revolution in care

Estimates vary, but between 18,000 and 40,000 people lost their lives as a result of the French Revolution. Though it was a movement against the power of the ruling classes, only 8 per cent of those guillotined during the infamous 'Reign of Terror' were noblemen. Six per cent were clergy, some of whom escaped the blade by abjuring their priesthood and accepting forced marriages. They then stood by as the civic authorities celebrated masses to the 'Goddess of Reason' in Notre Dame Cathedral. Meanwhile, although the guillotine claimed such gilded victims as King Louis XVI and Marie Antoinette, 70 per cent of those who lost their lives were poor workers and peasants, accused of hoarding grain or dodging the draft.

Two doctors, Joseph Ignace Guillotin, Deputy of Paris, and Antoine Louis, the Secretary of the Academy of Surgery, devised this means of dispatch. At the outset of the Revolution, the official executioner Sanson maintained the sword was too inefficient to terminate lives in the numbers the authorities required. Guillotin consulted with Louis, who invited a German engineer to construct a machine similar to ones depicted in drawings dating from the thirteenth century. The engineer perfected his prototype using dead bodies from a nearby hospital, and once it was completed, Guillotin impressed upon the Constituent Assembly that this was the most humane means of execution.

As doctors perfected the art of decapitation, the blood-drenched Revolution improved medical care. Those who had stormed the Bastille prison in the name of liberty, equality and fraternity in 1789 considered sickness endemic to a corrupt, unjust society. In 1792, France's newly established National Assembly made bold pronouncements about replacing the clergy with professionals who cared for bodies, not souls. Expropriated Church funds would be channelled into care for the sick. Medical officers would encourage frugal living and hygienic orthodoxy.

During the post-revolutionary years Paris, with its ready stock of poor-houses, former leprosaria and religious refuges, became preeminent in medical care. Under central control, large public hospitals like the Sâlpetrière and the Hôtel de Dieu ceased to be merely places where the sick went for a bearable death. Physicians like René Théophile Hyacinthe Laënnec – the inventor of the stethoscope who, ironically, died of tuberculosis, the disease of many of his patients – came to these new gathering-grounds of the sick. From the United States and Britain, would-be physicians flocked to Paris, returning with a stethoscope and a zeal for medicine-as-science. It became common to discount *symptoms* and look for *signs*; to disregard the patient's own account of how he or she felt, in favour of what the attending physician could observe. With that, yet another conceptual mile separated the doctor from the people he treated.

In the wake of the French Revolution, nervous monarchs and ruling élites examined their consciences. Building and endowing hospitals became a superb way for rulers and other wealthy folk to assuage their guilt. Vienna's Allgemeine Krankenhaus, Moscow's Obuchov Hospital and the Charité in Berlin reflected the concern of rulers to care for their subjects, and to be seen doing it. In Britain, it was the affluent public who took charge, endowing five new major London hospitals in the eighteenth century –

including the institution where I trained, the London Hospital.

But at the same time as Paris fell into a veritable fever of facility-building, some radical former revolutionaries were calling for all the hospitals to be burned down as breeding grounds for sickness and death. They had a point. Until the next century hospitals would remain atrociously prone to fatal infections, while offering for the most part little more than blood-letting, purges and mostly doomed amputations. Those with money stayed at home, to be treated by their own physicians, whose armoury of medicaments was no more effective but had the benefit of containing amounts of opium and alcohol. Seventeenth- and eighteenth-century pharmacopoeias included extracts of human cranium, oil of earthworms, spider webs, woodlice, hyssop, vipers and coral, none of which probably had any curative effects whatsoever – except hyssop, which modern research shows is apparently a good source for *Penicillium* mould.

In the midst of this rather gloomy review of eighteenth-century medicine, we should not ignore the Reverend Edmund Stone's discovery in 1763 that willow bark relieved fever. The Royal Society chose to overlook it, though, and it was not until the end of the nineteenth century that its active ingredient, salicin, was investigated for its analgesic and fever-relieving qualities. Today we know its derivative as aspirin.

Stone's contemporary William Withering had better luck. Having heard of a widow whose special herbal teas were sought after by people suffering from swollen legs, he discovered that the tea contained foxglove – whose leaves yield the drug digitalis. The botanist Withering administered some to a fifty-year-old builder suffering from asthma and swelling of the abdomen, with dramatic results. Eight years later, in 1783, digitalis began to be included in pharmacopoeias as an effective treatment for cardiac oedema. Digitalis strengthens the beating of a failing heart and, by increasing blood flow, reduces the build-up of fluid in the lower limbs; it is still used today. But if digitalis and aspirin were now launched as new products, safety restrictions and drug regulation might make it impossible for either to get a licence. Though they have been immensely valuable over many years, their potential side-effects would probably be considered too risky to allow them to be made available even on prescription.

As the nineteenth century approached, chinks of light began to appear in the prevailing gloom. Hospitals teeming with disease and ailments persisting in the face of ineffectual remedies did force some physicians to think carefully about why their patients weren't getting better. The science of bacteriology emerged as a by-product of surgery, which was taken to a new level thanks to the first effective use of chloroform as an anaesthetic. Used for the first time to dull labour pains in 1847, it was opposed by some on the ground that God intended women to give birth in torment. Such arguments swiftly subsided when Queen Victoria made ready use of the chloroform-mask to introduce young Prince Leopold to the world in 1853. But while anaesthesia greatly advanced surgery, it initially resulted in limited benefit, because now surgeons did not have to rush their operations to minimize the pain. Extended exposure of the tissues gave more opportunity for fatal infection to take hold.

The deadliness of hospitals in general exercised the mind of Ignaz Semmelweiss, a physician at Vienna's Allgemeines Krankenhaus, who was appalled by the numbers of women who died after giving birth. The illness was called 'puerperal fever' because of the high temperature these women had after delivery, but the symptoms were due to septicaemia, usually as a result of severe infection of the uterus or vagina. A prolonged labour, repeated vaginal examinations before birth, a traumatic delivery, or the use of forceps by a doctor were all frequent causes of infection and, before the antibiotic era, puerperal fever was often fatal. In 1848 Semmelweiss noted that the first obstetric clinic, run by male doctors, had a higher mortality rate than the second, run by midwives. He concluded this was because the male medical staff carried infections on their hands from the mortuary – indeed, from everywhere in the hospital – into the maternity wards. The midwives, who stayed in one place, did not. So Semmelweiss recommended hand-washing and sterilization of instruments; but to his authoritarian colleagues the idea that doctors might spread infection proved highly unpalatable, so they laughed him out of his profession, and eventually committed him to a lunatic asylum. His death was as ironic as that of Vavilov, related in chapter 2. While protesting his sanity to the asylum's guards, he was badly beaten, bound in a straitjacket and placed in a dark cell, where he died of septicaemia from his open wounds two weeks later.

The French chemist Louis Pasteur (1822–95) was undeterred by criticism. Pasteur's elaborate experiments showed that maggots were the larvae of flies feeding upon decaying matter and not the result of spontaneous generation. In proving this he strayed on to contentious ground, for in nineteenth-century Europe the cities were plagued with epidemics. Pasteur did not invent the idea that microscopic living organisms caused infections, but he popularized it with clever public experiments.

Pasteur identified links between specific organisms and specific diseases – a process continued by Robert Koch in Berlin, who established a bacteriologist's credo known as Koch's Postulates. To prove that an organism caused a particular condition, all four postulates must be satisfied: the organism is found in every instance of the disease; the organism can be grown in culture; inoculating an animal with the organism produces the disease; and the organism can be recovered from such an animal and grown again in pure culture. Using the postulates as their guide, Koch and his students successfully identified the microbes responsible for typhoid, diphtheria, leprosy, tetanus, whooping cough and meningitis.

Victory at last: medicine on the edge of modernity

Syphilis has been known variously as the French Disease, the Spanish Disease, the Polish Disease, the Russian Disease, the Portuguese Disease, the Castilian Disease, the British Disease and the Disease of the Christians. Although there are disputed claims for an outbreak in thirteenth-century Hull, the first clear epidemic came in 1493, as French troops stormed into Spanish-occupied Naples. When the victorious soldiers returned home, some found they had genital sores, then painful rashes and ulceration, and eventually a plague of seeping abscesses which ate away flesh and bone. Madness followed, then death. Contemporary accounts tended to blame the immediate neighbours – hence the plethora of pan-European nicknames for the disease – but the origin of this condition remains obscure. Some say it came with Spanish troops from the New World; others that it was an entirely European super-bug, mutated out of pre-existing venereal diseases. At any rate, it was ghastly, and the treatment – with mercury, or guaiac wood from the Caribbean – rarely much use. Syphilis, known as the Great Imitator on account of the resemblance of its symptoms to those of many other conditions, is an infection that can affect every organ in the body in a multitude of different ways. Even though it had become a relatively rare disease by the time I entered medical school, several of our teachers would still say, 'know syphilis and you know the whole of medicine'. Nowadays, the same might be said of diabetes or AIDS, as syphilis has largely been eradicated, and these two diseases manifest themselves in numerous different ways.

It was a microbiologist, Fritz Schaudinn,[10] and a dermatologist, Erich Hoffman, who in 1905 identified the spirochaete *Treponema pallidum* as the cause of syphilis. Blood-screening tests were then developed, and it fell to the German chemist and Nobel Prize winner Paul Ehrlich to work on a cure. Noting the occasional use of an arsenic-based remedy in medieval times, Ehrlich synthesized over 600 compounds of the element, testing them on patients in the brutal end-stages of the disease. The results were spectacular: by 1910, some 10,000 syphilitics had been treated with compound number 606, otherwise known as Salvarsan. The success of this drug proved a major boost to the medical profession at the outset of the twentieth century.

Medicine's reputation was further advanced when researchers came to understand the role of the pancreatic hormone insulin in the often fatal condition of diabetes. Two Canadian scientists, Fred Banting and his assistant Charles Best, were given the run of their boss's laboratory while he went fishing on holiday in 1922. After treating diabetic dogs with pancreatic extracts, Banting and Best eventually took their findings to Toronto General Hospital, where a fourteen-year-old lay dying. Within minutes of receiving an injection of the extract – called *insuline*, because of its origins in a part of the pancreas known as the Islets of Langerhans – the boy's blood-sugar levels dropped and he survived.[11] Banting was awarded the Nobel Prize; so incensed was he that Charles Best, his assistant, was not also recognized that he gave him half his prize money.

At this time, after centuries of limited therapies, medicine entered a new epoch. An understanding of the processes of infection resulted in improved hygiene, contributing to better rates of survival after surgery and for mothers and new-borns. Aspirin and digoxin could temper fevers, cure simple aches and pains, and relieve the

symptoms of heart disease. There were several forms of immuniz-
ation against infectious diseases, and the effects of syphilis and
diabetes had been ameliorated by the human intellect. At long last,
the arrogant boasts of Galen, even the semi-divine status of the
Asclepian physician, seemed justified.

DRIVEN BY CONFLICT

Of all historical events of the past 200 years, perhaps the Crimean
War (1853–6) most aptly expresses the paradoxes and ironies of
human history. With naval force backing him up, the French
emperor Napoleon III coerced the Ottoman Turks into relieving
Russia of its long-standing role as the protector of Christianity in the
Holy Land and passing the job to the French. Particularly symbolic
were the keys to the Church of the Holy Nativity, in Bethlehem,
which were withdrawn from the safe keeping of an Orthodox
patriarch and placed in French Catholic hands. Russia responded to
this theological affront at first with diplomacy and then by seizing
those western extremities of the Ottoman empire where Orthodox
Christians were in the majority. Matters reached a head when Tsar
Nicholas I attacked Ottoman ships off the coast of Turkey in 1853,
prompting Britain and France to go to the Turks' aid. It was like a
rerun of the Crusades, albeit with the Crusaders attacking one
another on Saracen soil.

Rooted in ancient history, this war was a relatively modern
conflict. Its events were immediately reported to the populations of
Europe and the New World, using the technology of the telegraph.
Railways were used for the first time for the mass movement of
troops. Disasters like the Charge of the Light Brigade – caused, in
part, by a dispute between two headstrong aristocrats, Lord
Cardigan and his brother-in-law, the earl of Lucan – led to a rethink
of the policy of selling army commissions.

Perhaps the greatest irony of this conflict – which resulted in
20,000 deaths – is its contribution to medical care. Nursing the sick
was an integral part of the role of Catholic religious orders, but
Protestant Europe had lagged behind. We gain an idea from the

pen-portraits of Dickens of how nurses were seen in early
nineteenth-century Britain. Sairey Gamp, whose society it was
'difficult to enjoy without becoming conscious of a smell of spirits',
and the slatternly Betsy Prig of *Martin Chuzzlewit* were hostile,
brutal women on the fringe of the criminal classes. A resurgence of
the concept of *noblesse oblige* did something to change things, as did
a notion of the value of education for women: in consequence, by
the 1840s nursing began to be seen as a worthwhile endeavour for
well-bred young ladies. The Quaker social reformer Elizabeth Fry
established standards for training English nurses after visiting a
renowned institute run by a Lutheran pastor near Düsseldorf.

The Crimean War created nursing as a profession open to all, a
move spearheaded by the well-connected Florence Nightingale. She
had studied at Düsseldorf, and with a religious order in Paris, before
news reports of the appalling treatment of wounded British soldiers
– zapped at high speed to the world's presses – reached those in
charge of the war. Asked by a high-ranking family friend to help out,
Florence took thirty-eight nurses to the barracks at Scutari on the
Black Sea. In just six months they reduced the death rate among
the injured from 40 per cent to 2 per cent. Wired home to a
populace eager for news, accounts of the 'lady with the lamp'
created public outrage at the treatment of our soldiers, and
adoration for the 'angels' caring for them. Donations poured in to
fund the formal training of nurses, which began at St Thomas's
Hospital, London, in 1860. So-called Nightingale schools carried
the credo of hygiene, discipline and fresh air throughout the empire.

Waging war has resulted in more medical advances than any other
activity.[12] The battlefields of Europe were proving grounds for
surgical techniques which were compelled to greater levels of
sophistication as gunpowder and bullets multiplied the complexity
and severity of wounds. Surgeons became advanced in the skills of
amputation, learning to remove more bone than flesh, so that
healthy skin could grow over the stumps and form a useful base for
fitting of artificial limbs. The enhanced technique of leading
surgeons like the Frenchman Ambroise Paré and the Englishman
Richard Wiseman – both of whom had perfected their arts at war –
elevated their status. By the early eighteenth century, every sizeable
vessel of the British fleet had its own surgeon. Acceptance enabled

surgeons to distance themselves from the blood-letting street-barbers and set up their own professional bodies, which were grudgingly admitted into the wider medical profession in the mid-nineteenth century.

As with improvements in surgery, so with improved hygiene: it is telling that Lister's anti-bacterial practices were first given a whole-sale testing in the Franco-Prussian war. German military doctors followed the London-born Quaker's advice about cleaning wounds with carbolic acid and were rewarded with far fewer losses than the French, who ignored his well-publicized findings. In the following century, the need to restore some dignity and quality of life to those suffering horrific facial injuries during the two world wars drove Sir Harold Gillies and his cousin, Archibald Hector McIndoe, to pioneer treatments in plastic and reconstructive surgery. Between those two conflicts, the Spanish Civil War provided the arena in which those blood transfusions first attempted by Hooke and Lower in the seventeenth century were perfected.

The British National Health Service – still rightly regarded as the envy of the world – also emerged out of war. At the close of the nine-teenth century, the military had alerted our government to the shockingly poor health of those conscripted to fight the Boer War. This prompted concern about public health, in particular that of the industrial poor – whose meagre diets, poor working conditions and cramped accommodation made them prey to debilitating, even fatal conditions. In the newly unified Germany, a similar awareness had prompted Chancellor von Bismarck to establish the first state-run system of medical insurance in 1880, turning health care into a government duty. In 1911, Liberal leader David Lloyd George launched a similar plan for Britain, with tax-supported 'panel doctors' – forerunners of the modern GPs – taking over the primary care of the family unit. But it was two world wars that finally brought the NHS into being.

During both those wars, the vast numbers of returning casualties forced the government to commandeer the country's diverse collection of health resources. Professionals gained huge experience from treating the injured and from working as part of a centralized mission. After the Second World War, it became clear that many of Britain's 1,143 voluntary and 1,545 municipal hospitals might collapse if the state did not continue to support them. In the face of this looming threat, and also of the visible strides in provision being made as the communist world delivered free health care to its 200 million or so citizens, nationalized health care seemed the proper response.

The marriage of state and medicine – first brought about when our teeming cities and busy trading ports fell prey to the plague in the Middle Ages – also spawned deadly offspring. Both fascist and communist governments dangled improving public health as a carrot in front of their citizens, distracting their attention from the lack of freedom and the abuse of human rights. Although the Nazi government in Germany was one of the first to institute a form of nationalized health care with free treatments for its citizens, many of its doctors and scientists who wished to 'improve' society participated in the forcible sterilization of the mentally ill, the disabled and the alcoholic. Loyal physicians rationalized the racism and mass murder of that era with pseudo-science, 'proving' the genetic inferiority of Jews, Slavs and Gypsies, and the biological superiority of the Aryans. Dr Josef Mengele, performing horrifying experiments on inmates of Auschwitz concentration camp, was the polar opposite of the Hippocratic healer. So, incidentally, were those who administered the Japanese research centre at Pingfan, in occupied Manchuria. Under the leadership of Dr Shiro Ishii, and with government blessing, this unit produced enough anthrax, typhoid and bubonic plague to wipe out the planet's population, and tested these infections on local people.

We can see the murky and unwelcome message emerge. Human medicine could solve problems, of course, but humankind had created some of these problems in the first place. One example could be said to be tropical medicine, a discipline reaching its apogee in the Age of Empire. Somerset Maugham's 1934 novel *The Painted Veil* encapsulates the values of this period in a tragic love story, whose hero is a bacteriologist fighting a cholera epidemic in a remote province of China. Saddled with an unfaithful wife, Walter Fane struggles to improve sanitation conditions for the town, while gangs of nationalist thugs paint 'Foreigners Out' slogans. The ensuing chaos forces Fane and his wife, Kitty, to cooperate with, to respect and finally to love one another, until their happy future is cut short by the epidemic they've devoted themselves to fighting.

When Kipling wrote that colonialism's duty was to 'fill full the mouth of famine and bid the sickness cease', he summed up the feelings of many a leader. For the founder of France's colonial medical service, the general and fascist sympathizer Hubert Lyautey, it sufficed to say, 'La seule excuse de la colonisation, c'est la médicine.' From developing DDT to combat malarial mosquitoes to providing vaccines and clean water, the imperial age assisted the warmer parts of the world in their battles against the killer diseases of malaria, yellow fever, sleeping sickness and cholera.

But imperialism also created some health problems. The occupation of foreign lands was followed by the mass movement of people along improved networks of communication. As people spread, so did diseases. Meanwhile, opportunities to earn wages on plantations and in factories prised agricultural peoples off the land that could have nourished them and into a more precarious existence, buffeted by the exigencies of the global economy. Colonized territories also suffered from Western infections to which native populations had no immunity. As a Catholic missionary remarked to the anthropologist Bernard Deacon on Malakula in the early twentieth century, 'We're not civilising them. We're syphilating them.'

The mathematics is suggestive. The sum of lives saved or improved by hospitals, surgery and improved hygiene, blood transfusions, tropical medicine and the increased involvement of the state is truly remarkable. But subtract from that the number of lives lost to war and imperialism, and the total is less impressive. Of course, medicine in itself is not such a 'bad idea'; but factor in the loss of life caused by cities teeming with sickness, malnourished populations overdependent on agriculture and diseases blown in on the winds of trade, and it starts to seem more like a brake on humankind's own self-destructive tendencies. As we entered the twentieth century, that brake did begin to work with increasing efficiency – but, as we shall see, the credit for that lies only partly with inventive, creative human beings.

STUMBLING ON A CURE

In December 1940, Police Constable Albert Alexander scratched his cheek on a rose-bush during the execution of his duties in Oxford. This trivial injury failed to heal over Christmas, and he was eventually admitted to the Radcliffe Infirmary with septicaemia. By February 1941 his left eye had had to be removed because of the infection, the sight of his remaining eye was under threat, his right arm was swollen with pus, and he was suffocating in his own phlegm. On 12 February, Constable Alexander was given an intravenous infusion of 200 milligrams of penicillin – by modern standards a tiny dose. Within twenty-four hours Alexander's temperature had dropped, his appetite had returned and the infection had begun to heal. But the penicillin the doctors had was unstable and it was wartime – so restrictions were placed on the laboratory. Only a small quantity of penicillin could be extracted from Alexander's urine and reused, and by the fifth day the supply had run out. Constable Alexander died on 15 March.

Freakish as PC Alexander's fate may sound, this sort of life-threatening sepsis was once commonplace. The discovery of penicillin dates back to 1928, when Alexander Fleming was observing various strains of bacteria in his lab at St Mary's Hospital, London. Having taken a holiday without washing some Petri dishes, Fleming returned to find an unfamiliar mould growing on one of them. He discovered it inhibited the growth of micro-organisms around it. Other scientists could not replicate his findings, so the research was temporarily dropped. Fleming's discovery was fortunate. The mould – *Penicillium notatum* – had grown on Fleming's dish because of an unseasonally cool bout of weather, and the spores it came from were not to be found in the average atmosphere – they were rare, and had been wafted up from the adjacent lab of a fungus expert.

Fleming's discovery lay ignored until 1940 when, in Oxford, Howard Florey, professor of pathology, and Ernst Chain, a biochemist and refugee from Nazi Germany, tested it on mice infected with streptococcus. They had high hopes that the major pharmaceutical companies would show interest. But Britain was at war and had just escaped massive defeat at great cost at Dunkirk. In spite of the possible benefits of a drug like penicillin in battle, there was little cash for research. In a scene reminiscent of the best Ealing comedies, Florey and Chain ingeniously decided to manufacture the drug themselves for testing on humans and organized a human chain of medical students, armed with bedpans and milk-jugs. By February 1941 they had just enough for a trial on a human subject – the unfortunate PC Alexander.

After the war, in 1945, Florey, Chain and Fleming shared a well-deserved Nobel Prize for work arming humanity against tuberculosis, meningitis and many deadly infections. And yet, revolutionary though it was, the discovery was not entirely to do with science. It is unlikely any expert would have come up with antibiotics from first principles. Finding new antibiotics is still a relatively obscure process today, after decades of research. Presented to the world as a triumph of scientific reasoning, this great discovery was to some extent a fine example of not taking a chance finding for granted, and of ingenuity combined with healthy collaboration.

The writer and doctor James Le Fanu asserts that a number of major medical 'breakthroughs' of the twentieth century came about through serendipity rather than science.[13] He is too critical of medicine, but he has a point. For example, it was the machinations of military intelligence during the Second World War that led us to understand the nature of adrenal hormones. The US authorities learned that Luftwaffe pilots were being given injections of adrenal gland extracts to help their performance at higher altitudes. So there was an injection of cash into hormone research, which eventually furnished two Minnesota doctors with a small quantity of a synthetic compound that would revolutionize many treatments.

The tale starts in 1928, when Dr Philip Showalter Hench was treating a colleague with rheumatoid arthritis. He had been admitted to Hench's care at the prestigious Mayo Clinic with jaundice and had noticed that, as he turned increasingly yellow, his arthritis improved. The painful swellings in his hands and feet remained at ease for several months after the jaundice was resolved. Over the following years, Hench observed a similar phenomenon in other patients. Arthritic swelling also decreased when women were pregnant, suggesting that the effect was not caused by jaundice itself but by a change in hormonal state. Nor was arthritis the only condition which improved when people were jaundiced or pregnant – hay fever, asthma and the neurological disorder myasthenia gravis all tended to get better. Hench thought he should try to isolate the elusive Substance X causing all this.

By chance, his friend and colleague Edward Kendall was working just down the corridor on hormone deficiencies. As Hench pondered over Substance X, Kendall isolated a number of interesting chemicals – human hormones – and dubbed them Compounds A, B, E and F. The two men compared notes, and that was the end of it. In any case, Hench's researches were put on hold when he was called up to serve in the Second World War. It was at this point that whispers began to be heard that Germany was bulk-buying bovine adrenal glands taken from cattle in Argentina. As Canadian research suggested these extracts might alleviate the stresses of oxygen deprivation, this military intelligence was worrying. The US government commissioned a major research programme, which continued after the Luftwaffe rumour had faded. In 1948 these researchers synthesized a few grams of Kendall's Compound E and Hench, making a wild stab at suitable doses, gave some to patients crippled with arthritis. The effect was remarkable, and for their contribution to the discovery of cortisone, Hench and Kendall received the Nobel Prize in 1950.

Cortisone, of which the adrenal glands are powerful producers, relieved the symptoms of rheumatoid arthritis, but its side-effects were considerable and there was a hunt for alternatives. Its discovery had led to advances in the treatment of many conditions, from asthma to conjunctivitis and lupus. The application of steroids – as they came to be known – to a range of inflammatory conditions was a medical watershed. Casting aside decades of science-based enquiry, advocates of steroids admitted that they didn't fully understand the body's inflammatory responses, or the mechanism by which a naturally occurring hormone kept them in check. The point was: steroids helped people with a wide variety of medical conditions. Like the discovery of antibiotics, this advance would have been hard to make from first principles. When Hench first had his conversation with the jaundiced, arthritic doctor, arthritis was thought to be caused by infection. He had no theoretical grounds for considering the role of hormones. And even if he had had, synthesized compounds might not have come his way without some groundless wartime rumours, or the presence of a friend doing related research in the same hospital.

The first two-thirds of the twentieth century saw a number of stunning medical advances – some linked to the two I have just sketched out. Steroids provided the inspiration for immunosuppressant drugs which, combined with improved techniques in surgery and microsurgery, paved the way for the first effective organ transplants in the 1960s. Antibiotics provided the source for several of the first anti-cancer drugs, as well as, incidentally, an early antidepressant, imipramine. More broadly, the synthesis of hormones led to the availability of beta-blockers to lower blood pressure and prevent strokes. An understanding of the body's own responses to damage, meanwhile – sparked by the cortisone discoveries – led to the anti-histamines, one of which, in an equally bizarre stroke of luck, was then found to relieve the symptoms of schizophrenia. This cascade of new drug treatments was complemented by solid advances in technology: the dialysis machine, heart bypass pumps, the operating microscope, the laparoscope, operative catheters which afforded access to the furthest recesses of the body. At the same time, particularly sophisticated engineering techniques provided major advances such as the artificial hip and the pacemaker, as well as powerful diagnostic tools such as ultrasound, computerized tomography and magnetic resonance imaging. All of these are of the greatest value and depended on huge scientific and engineering progress. But they sometimes had the disadvantage that they encouraged doctors increasingly to look at the results of tests, rather than at the patient.

It would be ludicrous to claim that we would be better off without antibiotics, or immunosuppressants, or hormones, or anti-cancer drugs, or treatments to alleviate the harrowing symptoms of mental illness; but to some extent, having stumbled upon many of them serendipitously, humankind is winning a series of battles with perhaps less hope of being victorious in a longer war. There is also some risk of our luck running out. In tending to mistake our good fortune for mastery, both the medical profession and those it treats may be led into a potentially unhealthy state.

Over forty years ago, as a very junior house-surgeon, I saw at first hand and in a practical fashion the state of mind which our sense of mastery can induce. My chief, a famous authority with a renowned international reputation – less than two decades earlier (before the NHS was founded), house-surgeons had flocked to pay him for the privilege of being unpaid juniors on his team – was expert in many advanced operations. So highly skilled and extremely eminent was he that there were few operations he was not prepared to attempt. Occasionally he would try his hand at removal of the adrenal glands, together with castration (by removal of the ovaries) for women with advanced cancer of the breast. This operation was based on the belief that some breast tumours were hormone-dependent, the corollary being that if all steroid hormones were reduced, the cancer would fail to grow. There were few convincing randomized studies supporting the use of this heroic surgery – heroic, that is, for the patient, though my chief was regarded as a hero by visiting international colleagues who witnessed his surgery.

At this early stage of my career, just after I had assisted at my first of these operations, I had a brief discussion in the operating theatre with my chief about how much cortisone the patient would now require to support her vital functions. The anaesthetist gave a dose in theatre, but my boss adamantly vetoed my prescribing further doses to support the patient overnight. He asserted that his great experience proved this was quite unnecessary and a waste of drugs. I was worried that without sufficient circulating steroids a woman in this situation was liable to go into shock, becoming collapsed with a dangerously low blood pressure. But I also knew that to argue further with that particular consultant could mean the end of my job. (I should add that nowadays patients would always be nursed in an intensive care unit after adrenalectomy because of this risk of collapse, but at that time these patients were nursed in a routine surgical ward.)

Doing my midnight rounds before bed, I found our post-operative patient with a dangerously lowered blood pressure. I rang my registrar, who grumbled for being troubled so late.

'If we don't give cortisone,' I said, 'Mrs Smith could die.'

'Well – don't inform me about it – and if you give it, don't write anything on the drug chart,' was his answer.

Needless to say, within minutes of my trembling injection – illicit and unrecorded – the blood pressure came up and all the patient's vital signs improved dramatically.

The following morning the chief came on his ward round. Standing at the bottom of Mrs Smith's bed and examining her charts, he turned to me and said: 'See, Winston – that outrageous fuss you made yesterday. I told you that extra cortisone was quite unnecessary for my patients.'

MEDICINE'S CUPBOARD BARE

In the second half of the twentieth century medicine faced a dilemma. Better treatments resulted in high accolades for the science. The few dozen effective remedies available to the doctor of the inter-war years had become some 2,000 by the time Harold Wilson (prophet of the 'white heat of the [technological] revolution') came to power. Unfortunately, extension of medicine's prowess on this scale couldn't continue. Many drugs that were developed were offshoots of the great discoveries noted above. To some extent, beyond tinkering with dosage and absorption and reducing side-effects, for a while there was a limited amount that pharmaceutical science could add, not least because, as we have seen, great discoveries often come about, not through scientific research, but by chance. 'Breakthroughs' continued to be trumpeted, partly because of the sheer volume of drugs being churned out by the pharmaceutical industry. This is by no means to criticize the industry, which, with a few highly publicized exceptions, has largely conducted its research and its business with increasingly high ethical standards. But the tendency to prescribe drugs for any condition increased the likelihood of their having unexpectedly beneficial effects for other conditions.

By the 1970s, however, there was a marked reduction in both the number of new drugs and the speed with which they made their way into general use. In the 1940s, two Parisian psychiatrists had begun treating schizophrenics with the antihistamine chlorpromazine within months of its being synthesized by the Poulenc company.

Twenty years later, it was no longer possible to market drugs for clinical use without a lengthy and costly process of trial and review. This radical change in policy was in large part the result of the Thalidomide scandal. Women who during pregnancy in the late 1950s and early 1960s had taken an apparently harmless pill to combat morning sickness or insomnia found themselves giving birth to infants with severe defects in limbs and internal organs.

Accordingly, the flow of new drugs dwindled from seventy a year in the early 1960s to less than thirty a year by 1971. Attempts were made to return to the pharmaceutical drawing board, engineering drugs on the basis of our understanding of bodily processes, and these have resulted in, among other things, really valuable vaccines, including one against hepatitis B, and combination drug therapy for HIV. But many of the best-selling drugs today are derivatives of the antibiotics and anti-inflammatories developed at the start of this golden age. Pharmacology had, to some extent, started to return to the 'suck it and see' approach of previous eras, using new technologies to screen millions of chemical compounds for their possible biological benefits. So it is only relatively recently that we have started to learn how to design new drugs more rationally – advances in X-ray crystallography, nuclear magnetic resonance spectroscopy, molecular biology and immunological techniques being important examples of how science is being applied to improving therapeutics.

The medical profession embarked on strategies to retain its heroic Asclepian status. Envisioning its mission as one of conquering disease with science, it has tended to foster a new breed of medical men who have often approached human suffering somewhat as if it were a problem in the laboratory. In the age of 'clinical science' and the preoccupation of universities with the Research Assessment Exercise, research has tended increasingly to take precedence over cure. One very early critic of this trend was Maurice Pappworth, who argued vigorously against what he called 'the sin of diagnostic greed'.[14] Pappworth graduated in the 1930s in Liverpool, where he had been informed that he would never get a consultant appointment because he was a Jew. After military service in the war, during which he rose to the rank of lieutenant-colonel, he held out for a teaching post in London. When spurned, he turned to private teaching, preparing medical graduates for the difficult exam for

Membership of the Royal College of Physicians (MRCP). The pass rate for this exam was around 7 per cent at the time, but his clinical teaching was so outstanding that it was not unusual for 50 per cent of successful candidates to have gone through his courses.

During the 1950s and 1960s Pappworth became increasingly concerned by the apparently unethical experiments that his students told him they had witnessed. He collected numerous examples of ethically dubious research and eventually published them in his book *Human Guinea Pigs*.[15] It described 205 different experiments in all, including examples of procedures carried out on children, the mentally defective and prison inmates. The book was particularly critical of Britain's leading research hospital, the Hammersmith Hospital, at the Royal Postgraduate Medical School, where some of the earliest cardiac catheterizations and liver biopsies had been carried out, and where I was subsequently a researcher myself. My senior colleagues furiously condemned Pappworth, who was totally ostracized by the establishment. (He held one unenviable record: because he fell foul of colleagues who loathe criticism, it took him fifty-seven years to attain election as a Fellow of the Royal College of Physicians.) But his criticisms had a profoundly positive effect, and within months of his book's publication Hammersmith set up excellent ethical review committees to ensure good clinical conduct and proper patient consent for all research.

Pappworth's key argument was that medicine was forgetting its duty to alleviate suffering. There is no doubt that his bravely made criticisms were massively valuable at Hammersmith and elsewhere; but even this timely corrective had its costs, for the research carried out at Hammersmith was possibly never quite as clinically innovative afterwards because the emphasis tended to turn towards laboratory-based research rather than clinical trials which might have been thought to be ethically questionable in some way. This hospital remains a very important centre for basic research, but it is perhaps questionable whether this more rarefied activity always translates quickly to the benefit of quite so many patients.

For Pappworth, a horrifying example of doctors' slavish devotion to research would have been the notorious forty-year-long Tuskegee Study, carried out between 1932 and 1972, in which many underprivileged black men in the United States were not only denied effective treatment (penicillin) for syphilis, but were not even informed that they had the disease, being used as controls in an experiment. Many of these men died unnecessarily, and often their wives or partners contracted the disease as well. As a result of the failure to disclose information, a number of children were born with congenital syphilis and consequently maimed for life, if they survived.

During the 1970s and 1980s it gradually became clear that clinical science wasn't always delivering the goods. The national beacon for clinical science in Britain was the costly Clinical Research Institute at Northwick Park, opened by HM The Queen in a fanfare of publicity in 1970. Sadly, it was shut down only sixteen years later,[16] just a couple of months after Sir Christopher Booth, its director, had published an article extolling its reputation for research in 'psychiatry, dermatology, allergic asthma, anaesthesia and communicable diseases'.[17]

Inevitably, some branches of medicine gained real impetus from burgeoning medical technology. Much excellent work was done, and is still being done. For example, there is now a huge array of monitors, incubators, sophisticated respirators and sensors for premature babies, all designed to perform a vital function in maintaining life. Nowadays, a tiny baby weighing just under 500 grams – less than half a bag of sugar – has close to an evens chance of survival, something which was unthinkable when I was training as an obstetrician. But the quality of life of these children, if they survive and grow into childhood (many with brain damage), is frequently a vexed issue. Only about 20 per cent of babies born before twenty-six weeks, by which time they usually weigh more than 500 grams, will be free of long-term health problems, and at least half of all these surviving babies will have significant or severe disability.

In certain other areas of medical practice, however, the consequences of the accent on technology are even less clear. Too many of us will die in hospital, of conditions a hospital is unable to cure, surrounded by and uncomfortably attached to catheters, drip sets and bleeping machines, under harsh strip lighting, with the additional risk of being carried off by some ward-borne bacterium. This is a dehumanized, impersonal, undignified and rotten end. It is a predicament disturbingly portrayed in the final movement of Shostakovich's last symphony, where the composer pictures his own demise.

This kind of terminal care is also hugely expensive. By 1976, around half of all medical expenditure in the United States was incurred by patients during the final sixty days of their lives. There is no question that prompt action, perhaps with cardiac resuscitation and intensive care, has saved the lives of many people after coronary thrombosis; but that also means that it is common practice to place all heart attack and stroke victims in the bewildering setting of intensive care units, where sometimes the associated technology is unnecessarily intrusive and may not help their recovery, occasionally even hampering it. In an American study of 150 severely ill cancer patients who had spent time in ICUs, more than three-quarters of those who had survived to go home had promptly died within three months. Of course, buying time for patients and their loved ones can be immensely valuable, but one may question the use of intensive care facilities for this purpose.

Medicine, having made such prominent and well-publicized advances, seems to have suffered from a compulsion to do something, and to be seen doing it. This trend has been particularly marked in America, where medical care is largely paid for by insurance schemes. To pack a patient off home with no more than an injunction to get some rest could be seen as no fair return for his investment: hence the battery of tests, scans and questionable treatments on offer in increasingly complex and costly hospitals.

In his polemical *Limits to Medicine*, the American philosopher Ivan Illich quotes a 1934 study in which 1,000 eleven-year-olds attending public (equivalent to British state) schools in New York were examined.[18] Over half (610) of them had had their tonsils removed. The remaining 39 per cent were examined by a group of doctors, who selected 45 per cent for tonsillectomy. The 55 per cent spared the operation were examined by a fresh group of doctors, who selected 46 per cent for tonsillectomy. And so the experiment continued until, eventually, virtually every child was recommended for tonsillectomy. Incidentally, the trial took place at a free clinic, so financial considerations can't have been behind the repeated desire to operate. So what was? The desire to seem capable, argued Illich.

Illich was so hostile to the medical profession that he left his own facial cancer untreated, eventually dying from it. So he is likely to be biased and selective in the studies he chooses to cite. But his point cannot be ignored entirely. For him, medicine's greatest crime lay in teaching humankind to hide from its own mortality. In claiming limitless powers over the physical, he argued, doctors made us spiritually sick. His critique of unnecessary tonsillectomy haunts me. Although we may dismiss it by observing that this kind of thing could only happen in the overmedicalized United States, I cannot forget that when I was a house-surgeon in the earliest stages of training I must have done close to 100 tonsillectomies in small children, after cursory consultation and examination (usually by another doctor) in the outpatients' department. I am fairly sure now that many if not most of these operations were entirely unnecessary. I don't think I was 'spiritually sick', but I do think now that my young patients were victims of fashion. Of course, these tonsillectomies were done largely because we could not think of another way of treating recurring sore throats and probably did not want to admit there was not much else we could offer by way of intervention. We should have been more prepared to tell the parents of these children that they would eventually get fully better without any treatment.

Living in more 'sophisticated' societies, many of us are increasingly seduced by a futile obsession with physical perfection. Some doctors trained with public funds to serve the public good in the NHS are sucked into a lucrative beauty and cosmetic surgery industry that holds out the paper-thin promise of beauty to those with a healthy bank balance. For a fee you can now opt for removal of the signs of ageing or (rather than wear spectacles) sharpen up your vision; you can have fat sucked away from your belly and thighs, your lips plumped, your balding pate permanently re-sown with lifelike hair. But somewhat ironically, the treatments offered are seldom perfect – you can nearly always spot which of your acquaintances has had plastic surgery.

I fear that my own field, reproductive and infertility medicine, has done badly in this respect. Infertility causes truly devastating grief in some people – having conducted a number of gynaecological cancer clinics, I can affirm that the distress and pain I saw experienced by those attending infertility clinics were usually far more profound and corrosive. The reasons for this are complex – but my cancer patients always tended to believe they would get better; many infertile ones felt they had no hope. A significant proportion of sterile couples are so badly affected that they lose the will to work, to socialize, to have sex, and they become increasingly depressed and alienated from those around them. Many infertile men and women suffer so much they will leave no stone unturned in attempts, no matter how futile, to conceive.

In vitro fertilization is a wonderful invention, and has brought into the world nearly one million babies who might otherwise not have existed. But the delivery of IVF is an extreme example of my concern about the drift of medical practice. Prospective patients can now queue up at an IVF clinic complaining of infertility, and, after a very limited conversation about their previous clinical history and the most cursory physical examination, be mechanistically subjected to a few basic tests – to assess in some limited respects their suitability for IVF, but not in an attempt to diagnose the underlying condition causing the infertility. One consultant at a famous teaching hospital in London has openly admitted that he sees no point in investigating the infertile women who come to his clinic requesting IVF, because they are going to be offered IVF anyway. What I find upsetting about this is the pride with which he makes this statement – after all, it saves money! But infertility is not a disease; it is merely a symptom of something wrong. If you or I went to our doctor complaining of pain in the chest and requested a heart bypass operation, we would be laughed at. The chest pain might be due to angina and therefore a symptom of heart disease, but it might equally be pneumonia – in which case we would probably need an antibiotic. Equally, the chest pain might be caused by indigestion, in which case an antacid might do the trick. It could be due to gallstones, in which case removal of the gall-bladder might be indicated. Or the pain might be due to a fractured rib, a viral infection, spinal disease, inflammation of the tissues around the heart, lung cancer or an attack of herpes. Very often, the pain could be simply due to mere bruising of the chest.

But a patient requesting IVF at an IVF clinic will almost certainly get just that, and only that – IVF. And if this treatment doesn't produce a pregnancy she almost certainly will be offered another cycle of IVF and, even if the same outcome is repeated again, she and her partner are most likely still to be left completely in the dark about why they have been unable to conceive. And it turns out that for many patients failure to establish a cause of infertility is one of the most distressing aspects of the problem.

This accent on IVF has wider consequences. It means many doctors specializing in reproductive medicine are poorly trained, because IVF is the only treatment they get properly familiar with and really understand. It also means that there is a dearth of good research in Britain in this field. It has recently been found very difficult to fill senior academic posts in this specialism in Britain, because there are so few good candidates coming up through the ranks. But there is another issue, too – a further reason why I believe IVF is an example of what is increasingly happening in medical practice in developed countries. Unfortunately, when IVF first became available, infertility was not recognized as a serious clinical problem and it was relatively poorly funded. So many pioneers involved in IVF were tempted to set up in private practice, in lucrative clinics. Nowadays, these clinics have a very big turnover, often making huge profits; some have an annual income of many millions. One international health-care company recently bid for a particular IVF clinic in London, offering an undisclosed sum which was certainly in excess of $100 million. I am told of another clinic in Australia that was sold last year to a health-care consortium for around $200 million.

One might expect things to be better inside the NHS. But the NHS, always under financial pressure, has followed suit. IVF clinics in NHS hospital trusts are working in the NHS 'internal market', and hospital trusts sell their services on a contract basis to primary health-care trusts, who send them patients. The prices that are charged are based not so much on any precise analysis of what it actually costs to deliver the care offered as on what the hospital trust's administrators consider the market will bear. Practically speaking, the private sector has a big impact on how the NHS prices the delivery of this service. This means, of course, that many NHS trusts make substantial profits from reproductive medicine. I believe that the IVF clinic in the trust where I once worked must have produced a useful income of at least £1–2 million for the hospital managers. Of course, this money does not go into the pocket of any individual; it is laudably used to meet the needs of NHS patients suffering from a wide range of illnesses. But the fairy-tale costing is an example of our inability to establish a mature, accurate accounting system in our National Health Service. Although the NHS is massively expensive, we still do not know what different treatments in a wide number of medical and surgical areas actually cost.

In the private sector, the readiness of patients to pay substantial fees and their desperation makes a very unhealthy combination. Sadly, many infertile women who attend these independent clinics – I know this from the copious flow of emails I receive daily – are too ready to be persuaded, or to persuade themselves, to undergo IVF even though the chance of success may be virtually zero. This is not usually evil exploitation on the part of the doctors, though it may sound like it. But it has the same effect. The practitioners conducting these forlorn treatments are probably not trying to make a massive financial killing – in fact, they do not need to, as they are wealthy enough from the overflowing waiting rooms where they work. No; the reason why these patients are not turned away is that it is very difficult for the doctor to accept that he or she is impotent, and cannot provide clinical success. So treatment is offered 'to give hope'.[19] How shocking that, in 'giving hope' to these patients, my profession feels it necessary to 'give hope' to itself. Our medical care system has helped reinforce our feelings of invincibility, and failure becomes nearly as difficult to countenance for the doctor as for his or her patient.

As I say, IVF treatment is an extreme example, but it is a revealing one. Because of the wider availability of these 'improving' practices, we have come to believe that there is – or should be – a solution for everything, that no state of affairs should simply be endured. This insulates us from reality. How can we handle such insoluble, unavoidable, endemic hardships as chronic pain, or terminally ill children, if we dwell in a fantasy world where white-coated magicians will take every niggle and blight away? Far from making us bolder and stronger in the face of nature and its constant exigencies, medicine has made us more afraid. And nothing makes us more afraid than the medical profession's current obsession with prevention rather than cure.

CLINICAL GOVERNANCE

Oddly, while the finances of the health service are poorly regulated, bureaucracy and micromanagement have increasingly resulted in over-regulation of NHS medical care, an issue brilliantly exposed by Professor Raymond Tallis in his book *Hippocratic Oaths*.[20] Some of this regulation is a result of the reaction of politicians to the mass murders committed by Harold Shipman, the general practitioner who gave many ageing patients lethal injections, to the Alder Hey 'scandal', and to the apparent 'cover-up' of mortalities after heart operations on small children in Bristol's Royal Infirmary.[21] At Alder Hey Hospital, in Manchester, a number of post-mortems were done on infants to ascertain the cause of death. Unfortunately, after many of these autopsies organs – for example, brains, hearts, livers and kidneys – were retained for further pathological examination without informing the next of kin, and therefore without their permission. This meant that a number of burials were conducted with relatives unaware that the bodies of their loved children were incomplete.

Obviously, I am not condoning the actions of those pathologists who did not get proper consent for the conduct of autopsies – nor, of course, the shocking crimes of Dr Shipman. But the government response seemed hasty and ill-considered as well as disproportionate. We now have a cumbersome system of clinical governance that is often unnecessarily bureaucratic. It greatly increases the mound of paper which medical and nursing staff are required to produce while deflecting them from spending time on patient care, an imbalance which often inhibits good medicine.

An example is provided by the way the NHS bureaucracy handled an issue in a hospital where I once worked. A senior consultant, the most outstanding surgeon in the hospital trust, renowned for his excellent skills, was consulted by a woman of child-bearing age because of increasingly severe abdominal pain which was so incapacitating that normal life, even proper sleep, was impossible. Thorough investigation revealed a benign tumour involving many blood vessels; various tests confirmed that, although this was not a cancer, surgical removal was the only way of curing her symptoms.

But the consultant recognized that surgery would be dangerous. The tumour had grown in a difficult area close to many vital structures, and the large blood vessels around it presented a serious hazard.

The consultant explained the situation to the patient and her husband clearly. Surgery put important anatomical structures at risk, and if the blood vessels could not be sealed, there was a risk of fatal haemorrhage. In effect, he said, this surgery was quite experimental and he could not predict how he would go about things once the operation was in progress, nor precisely how he would deal with the tissues around the tumour. But he was convinced that its removal was her best chance of gaining a pain-free, normal existence. The matter was carefully and repeatedly discussed with the patient and her spouse, and both, recognizing the serious risk involved, readily gave consent to an exploratory laparotomy.

Some days later, the laparotomy was done and the tumour removed without any damage to the patient's vital organs. Unfortunately, haemorrhage set in and was so profuse and rapid that this brave woman died. The conscientious consultant was devastated but felt he had taken the right decision. The husband was remarkably supportive – indeed, he thanked my surgical colleague profusely for his care and for the quality of the information he had provided. Ironically, in spite of his own deep distress at the loss of his wife, he attempted to offer comfort to the doctor.

A day after the death, my colleague was summoned to the Medical Director's office. The Medical Director – a post set up by the new NHS clinical governance procedures – was critical about the conduct of the case. Though he had no surgical experience (actually his total medical practice was limited to two medical clinics a week) he told the consultant that, in his view, he appeared to have been negligent, and suspended him from all duties. Conforming to standard NHS governance procedure, he was able to insist on immediate suspension without hearing any legal representations on the consultant's behalf. In my experience it is sometimes the case that the professionals appointed to posts supporting the burgeoning bureaucracy of the NHS are filled by those whose duty it is to follow procedure and who do not always appear to appreciate that good medicine – and especially good surgery – requires skilled people to improvise sometimes to achieve the best treatments. It is interesting to consider that Florey and Chain's use of bedpans – filled with urine enriched with penicillin salts – would not be countenanced in the modern NHS.

Repeated protests by the husband and many senior fellow consultants eventually resulted in this consultant being reinstated. But the damage done to this good, caring surgeon was immense, and the defensive conduct of surgery by anxious colleagues in this hospital trust thereafter certainly did nothing to improve patient care or surgical technique.

PREVENTION: THE LAST REFUGE

I awoke this morning to the news that the mother of a small boy – aged five, weighing in at a reasonable 3 stone 5 pounds (21 kilograms) – had received a letter from his local health authority pointing out that, during a recent screening programme, this young person's weight had been decreed to be 500 grams over the desirable limits for his height and age group. According to the letter sent to his mother, this put him at risk of developing diabetes and cancer in later life.

Many similar stories have made news over recent months. One council threatened mobile burger and kebab vendors with the withdrawal of their licences unless they offered at least one healthy eating option on their traditionally grease-heavy menus. Elsewhere, chip shops have been ordered to replace their many-holed salt-shakers with a standardized healthy appliance that allows customers a minimal amount of the blood-pressure-elevating condiment. Certain sections of the media make too much of these tales – seeing in them the meddling fingers of the Big Brother state, a dangerous new tendency towards state control over the minutiae of daily life. In fact, public health has long been a concern of governments: in the nineteenth century, concerns over squalid living conditions in the inner cities led to the construction of sewers; in the twentieth century, the poor health of recruits to the fighting forces in the two world wars convinced governments of the need to play a greater role in health care. Arguably, living in any society involves a mixture of rights and responsibilities, and the free medical care we citizens of the United Kingdom enjoy can continue to exist only if we all accept some responsibility for our well-being.

But these councils are not acting alone: they make edicts on the basis of guidelines originally created by doctors. It is as the result of medical research that people are advised to lose weight, cut down on fatty foods or reduce the salt in their diets, medical research that has drawn a link between obesity and certain types of cancers or stroke. This may seem an obvious point, but the likes of Hippocrates or Rhazes would probably have been astonished to hear it. Since when, they might want to know, have doctors been telling the healthy what to do, instead of curing the sick?

Perhaps since 1976, when an influential report by Professor Thomas McKeown of Birmingham University argued that the enormous improvement in human health over the previous 100 years had been due to improving social conditions rather than advances in medical science.[22] This idea was popular with Western governments, facing ever larger bills for the costs of health care. Shift the emphasis to prevention, rather than cure – so the thinking went – and health costs diminish. Not long afterwards two major studies appeared suggesting, among other things, that many cancers could be directly linked to eating habits and the environment. Each of these studies was slightly flawed, as was the original Birmingham report; but together they created a climate in which medicine could influence the conduct of the healthy alongside the sick, and in which people began to imagine themselves assailed on all sides from unseen threats lurking in their food, the air they breathed and the lifestyles they had little choice but to lead.

The Birmingham study was faulty partly because it chose tuberculosis statistics as its benchmark. Noting that TB levels dropped more sharply between 1838 and 1945 – that is, before the advent of effective medication – than thereafter, it concluded that living standards had played a greater part in this improvement than drugs. They certainly did play a part; but so did the practice of isolating the infected in sanatoria, an innovation brought about by medical understanding of the processes of infection.

The study linking 70 per cent of cancers to diet was undertaken by the eminent Oxford Professor of Medicine Sir Richard Doll.[23] His findings were influenced by those of two other scientists: first, his erstwhile collaborator Sir Austin Bradford Hill, who in the 1950s had definitively proved the link between cancer and smoking, and second, the University of Minnesota's Ancel Keys, who had concluded that the rising incidence of heart disease was related to the amount of fat in the diet.[24] This latter theory had actually been rejected when it first appeared in 1957: distinguished critics, such as the American Medical Association, pointed out that the numbers of those affected appeared to be rising steadily, while patterns of food consumption were stable. They also argued that the fatal element in most instances of coronary thrombosis was not the furring of the arteries with cholesterol, but the presence of a blood clot. Keys' theory, however, became mainstream in the sixties, because the anti-clotting drugs didn't always work, and in the absence of any other credible alternatives his was the one that remained.

Two large studies – one in the United States, another in four European countries – also showed tiny decreases in the rate of heart disease when people decided to behave differently, eschewing fatty foods and taking moderate exercise.[25] In a nutshell, forty-one out of every thousand people who made no changes died of coronary thrombosis. In the group who adopted the lean, ascetic lifestyle, the figure was forty in every thousand. It stands to reason, really. The human body has been built to withstand extreme variations in climate and diet. Human groups survive and thrive from the Arctic to the Gobi, so it's unlikely that minor variations in diet – such as those seen in the West over the past fifty years – could account for major killer conditions like heart disease and cancer.

Nonetheless, the link between heart disease and diet had been forged in the public imagination. Media bombardment entrenched a popular terror of heart attack and a widespread belief that coronary problems could be prevented by diet. Food became a threat, doctors had a duty to tell people what they should eat – and these two perceptions carried over into Sir Richard Doll's study of cancer.

That report, too, had some flaws. Contrasting the diets and disease patterns of the West with those in India, the study concludes that meat and dairy products must play a role in the prevalence of certain specific types of cancer found here. But the findings are not consistent. Take the example of Mormons and Seventh-Day Adventists, Western sects who lead similarly sober lives, but only one of which abstains from meat. If a meat diet is linked to cancer, one would expect to see lower rates in Seventh-Day Adventists than in Mormons, but there is no evidence for this. It seems 'being Western' has more of a cancer-causing impact than eating meat. In spite of this, because his publication glossed over the link between cancer and age in an increasingly long-lived population, Doll's report became widely accepted.

Take a trip to even the smallest supermarket in 2009 and you will find an impressive array of 'organic' products. As I note in chapter 3, concern about the adulteration of our food may be justified. But it may also be excessive. The fear of man-made pesticides ignores the quantity of naturally occurring bug-killers present in our daily diet, and the extent to which these may be carcinogenic. Dr Bruce Ames of the University of California estimates that some 99.9 per cent of the cancer-causing substances in our food come from the foods themselves, and just 0.01 per cent from the substances sprayed on to them. Similar observations can be made about the 'harmful' radiation emanating from nuclear power plants, electricity pylons and mobile phone masts – none of it dangerous ionizing radiation, the bulk of which comes from the earth itself. (The only other way we are most likely to be exposed to it is through repeated X-ray examinations.)

Based on what seems, in some cases, to be the wobbly interpretation of research, there has been a shift in the relationship of medicine to society. Where once we sought out a doctor's advice in sickness, now this advice tends to guide the very lives we lead. Not all this counsel is excessive. The advice to avoid tobacco was perhaps the most valuable advance in public health of the twentieth century. Cigarette smoking – widely promoted as healthy right up until the Second World War – is now rightly seen as a real killer (and in this area Richard Doll and his colleagues did important epidemiological research studies). It is also undoubtedly true that antenatal care, vaccination programmes and some interventions to limit the spread of HIV have been really important. But it is not nearly so clear whether many other public health campaigns are quite so valuable.

For example, as I write we have just had a ludicrous campaign conducted by senior members of the Royal College of Obstetricians and Gynaecologists advising women not to leave child-bearing too late because of the risks of infertility and the high risks to women who get pregnant after the age of forty. Does this kind of advice have any serious value? The extra medical risk of having a baby later in life is very slight indeed. It is true that the chance of miscarriage is greater and there are slightly increased risks to the baby – but I am very dubious whether this well-meant advice has a positive effect. It certainly has a negative one, as evinced by the surge of well-to-do women seeking to freeze their eggs in private clinics in a vain attempt to preserve their fertility until they meet 'Mr Right'. Few experts point out that egg freezing is highly inefficient and also very expensive. It also may carry much greater risks to any child than a 'natural' conception at age forty-three or forty-four, because insufficient animal or human research has been conducted to determine whether the process of freezing may impair the way genes are 'switched on' during development of the foetus.

We should also be more sceptical about, for example, the vast amounts of public money being spent on chlamydia screening. This micro-organism is indeed very common and is spread by sexual contact. It may well cause low-grade inflammation in the genital tract in some (by no means all) people who are infected with it. But the evidence that chlamydial infection really causes infertility – in particular, inflammation which leads to blocked Fallopian tubes – is not so good. I know of no studies that clearly show that the £80 million being spent annually on chlamydia screening is playing a useful role in protecting reproductive function. But apart from any waste of public money, the government's campaign has had a very serious effect on the self-esteem of many anxious, infertile women who feel guilty that they have infected themselves and believe that it is their behaviour which has led to their childlessness.

Campaigns against obesity are well-meaning, but it is not clear whether they are effective in either getting people to reduce their weight or improving public health in the longer term. In some schools certain foods are increasingly labelled as unhealthy and are banned – not just from being sold on their premises, but even from being brought on to the premises by pupils. On medical advice, manufacturers remove salt, fat and sugar from the foods they produce, and boost others with fish oils and vitamins. Despite the fact that most of us will die of some complication of old age, we are made to fret about the quantity or quality of bread, or butter, or jam we consume, assailed each week with conflicting claims about the consequences of a pint of bitter or a glass of burgundy. Should we risk all and go out to a public house for one or two, we may have to pick our way through a clutch of smokers, who have been, on doctors' orders, legally excluded from the premises. Medicine has won many battles, most notably against the killer infections, a battle which has reduced mortality rates over the last fifty years. But its most recent campaign is somewhat questionable: less about curing the sick and caring for the elderly, more to do with almost harassing the healthy – with unproven efficacy.

I was once a student on Lord Evans' team – though in those days, it wasn't called a 'team' but a 'firm'. Lord Evans was a formidable diagnostician, physician to the Queen, and utterly brilliant. We were quite frightened of him because he was so eminent, but actually he was gentle and considerate, a good old-fashioned doctor in the best sense. One summer, he was on holiday as a guest of the monarch at Balmoral. During his absence we admitted a 27-year-old porter to one of his beds with a fever. Extensive history-taking and examination did not give us a clue why this man had a high and fluctuating temperature; during the next three weeks we threw every known laboratory test at our patient – but still without arriving at a clear-cut diagnosis. When we did not get a positive answer, we repeated the tests. Some keen wannabe doctors opined he must have a rare cancer, the more learned among us asserted that he had a lymphoma, those interested in conspiracy argued he was malingering by deliberately putting his temperature up by drinking hot tea before ward rounds, others said that this was a rare tropical infestation (though he had never travelled east of Wapping) and the rugby players muttered darkly that he probably had some strange venereal infection, even though this man was unmarried and claimed he had never had sex with anybody.

On the day after Lord Evans returned from Scotland, he strode ponderously into his ward and we waited delicately by the entrance, attentive acolytes in a gentle semicircle. Being told of this immense clinical puzzle, and our total failure to fix a useful label to this patient, he shook his head, strolled towards the bottom of the patient's bed, picked up the temperature chart and gazed lengthily at it. In his characteristic deep growl he finally pronounced just five words, loudly and slowly. 'This man has a fever,' he said, and firmly hung the chart back on the bed-rail. Then, very slowly, he silently walked to the top of the bed and, from behind the patient, placed the flat of either hand on the man's neck. 'This man has glands,' he announced. He took his time to walk to the bottom of the bed, whence he looked around at his assembled audience while we all hovered in breathless anticipation. He paused. Then: 'This man has glandular fever,' he pronounced with absolute certainty, and without any fuss strolled towards the next bed.

You could always argue with Horace Evans. His senior registrar was not having this. 'But, sir,' he objected, 'we've done the Paul Bunnell test three times and it's been negative; and there are no circulating mononucleocytes in the blood films either – which we have also repeated.'

'Lab is wrong,' asserted God. 'Do the tests again.'

A short discussion followed, but nobody had the heart to upset this delightful but apparently senile peer of the realm for long. So, most reluctantly, another blood sample was collected.

Two days later, the laboratory diagnosis came through. Glandular fever.

To this day, I am not sure how he did it – swollen glands in the neck can be caused by at least 200 diseases, and fever accompanies just as many. But it seems the shape of the temperature chart and the precise nature of its rise and fall, and the curious consistency of the slightly rubbery glands in his neck, were all that Horace Evans needed.

Lord Evans generally used to listen hard to what his patients told him – but on this occasion he wanted to show us the value of *looking* at the patient. I really doubt whether any of today's doctors would have made that diagnosis – because in many medical schools, and in the teaching hospitals where we train the physicians of tomorrow, they are all 'scientists'. As a profession we rely increasingly on the laboratory test and the computer, we talk less and less to our patients, and we have largely forgotten how to examine them or really to understand how they feel. Medicine is becoming so mechanized that we are even starting to use robots to do the more menial medical history-taking, and recently even some of the surgery – which separates us increasingly from those we are supposed to be treating and risks alienating both the doctor and the patient.

To be fair, a real change does now seem to be happening – at least in training. In many medical schools students are being more actively encouraged to understand that communication with a patient is really important and is a key, two-way process; that it involves listening and needs some better understanding of how the patient feels about his medical problem and the effect on his nearest and dearest. But after qualification, the pressure on young doctors is so great that this important aspect of medical care gets pushed to one side. The NHS, with its governance systems, its targets, its bureaucracy, its paperwork and its limited staffing, easily snuffs out this most important aspect of health care.

GENES, DRUGS AND MONEY: DECISION TIME

The latest medical focus (dare I say fad?) is the genome. As we will see in the next chapter, there is a belief that understanding how a sick person's genes are printed will revolutionize medical treatment and improve what we can offer. Forgive me if I seem slightly sceptical. At this stage there is undoubtedly a potential for improving public health through knowledge of the genome: it will almost certainly help in the treatment of some cancers, and may possibly improve drug treatments by enabling us to tailor drugs to a person's or an organism's genetic make-up. It might improve diagnosis in some cardiac conditions, certain forms of blindness, and possibly diabetes. Almost certainly it will help obstetricians to identify foetuses suffering from genetic diseases in their mission to search and destroy.

But, just as with the supposed links between heart disease and a high-fat diet, nobody has shown that changes in behaviour because of a perceived higher risk of certain conditions associated with variations in the DNA will make a statistically valid difference. And even if this information can be shown to be seriously useful, a massive investment in genetic services will be needed to make it widely available. Centralized genetics laboratories will have to be established; every pathology service will need complex equipment to identify gene sequences; substantial training of doctors and nurses, including those in primary care, will be essential; complex assessments of the validity of various tests will be needed; and complex clinical trials will be mandatory. Considerable investment in genetic counselling will also be essential. But above all, the most complex computing system known in any government service will be a precondition, and its security must be assured. Given the difficulties besetting the NHS's current attempt to establish a multi-billion-pound system for simple medical record-keeping, this may be a very tall order.

Against this background, of course, people's expectations that they will continue to have the benefit of increasingly effective medicine are rising. But the subject of the rapidly increasing costs of the NHS tends to be ignored. It seems unthinkable that these costs will fall – indeed, with the increase, for example, in numbers of elderly patients and the costs of inpatient care, and potential advances in areas such as diagnostics and genetic medicine, one must question whether our therapeutic aspirations will continue to be sustainable. What we have never had is serious, mature consultation with the public about how we are to pay for these improvements in medical care.

One example of public ignorance concerns the costs that pharmaceutical companies face. Big pharma often gets a bad press, and we recognize that, for example, the costs of new drugs given to cancer sufferers are spiralling because rationing within the health service has received much recent publicity. But most people are quite unaware of how much it costs to launch new drugs on the market. Typically, to launch just one new drug, after ten years or so of research that will include screening perhaps 5,000 compounds and undertaking comprehensive testing under stringent regulation, costs around half a billion dollars. And the patents, the intellectual property which makes it worthwhile for a company to develop new pharmaceuticals, are granted for surprisingly short periods, making it very hard for drug companies to recover their development costs before the patent expires.

Like the majority of doctors, I believe in the NHS and take pride in the fact that this system was established in England. If we accept that the protection of human life is a central tenet of our moral system, then it follows that privatized medicine is likely to be fundamentally unjust. A system of free health care, with equal access for everybody, is one of Britain's finest aspirations. But, of course, health care is not free; it is paid for out of the taxation system to which we all contribute.

Every political party claims the NHS as its own, and every political party uses it to attract voters in the polling booths. Surely it is time for politicians of all persuasions to agree that the health service is not the plaything of any particular party, and that NHS structure and policy should not be changed repeatedly at the whim of whoever happens to be in government or for reasons of political expediency. A very strong case can be made for taking the NHS out of the present system of government and having it managed independently by a body on which all political interests are represented. Once parliament had decided on its core grant, there might be some consensus on sources from which extra funding could be derived. Independent governance of the NHS could also improve its accounting system and its continuity of management. The constant and costly management and administrative changes of recent decades have caused much damage and do not seem to have greatly improved the delivery of health care in the UK.

The political parties claim they want to 'empower' patients by involving them in decision-making, but this vague, well-meaning aspiration seems to miss the point. Perhaps we need to reflect on our history. Is it not time to recognize that we still cannot cure many human illnesses – such as glandular fever, for which the basic treatment has changed little in sixty, if not one hundred, years? And that, this being so, it is just possible that our patients might feel rather better if we spent less time pursuing a kind of diagnostic holy grail by performing batteries of (increasingly complex and expensive) tests *on* them, and more time finding out how they feel, and how we might help them, by talking – and listening – *to* them?

It seems extremely unlikely that politicians will readily lead the debate about how to balance rising health costs against the aspiration to provide free health care for all. Yet, if we are to manage expectations better, more effective public discussion – both *in* public and *with* the public – is going to be needed. If politicians are not prepared to lay out the agenda for the future, perhaps we doctors owe it to our patients to commence this important area of public engagement.

Chapter 11

Genetics: Potent Art or Rough Magic?

A SHORT, yellow-flowered bush with remarkable properties grows in equatorial Africa. The Fang people who occupy this territory tell of its origins. Their ancestor Bitama was gathering fruit from the top of a tree when the creator god, Zame, struck him down. Bitama fell and was killed, upon which his god cut off his fingers and toes and planted them. They germinated, producing the eboka bush, and when Bitama's widow came searching for her spouse, Zame instructed her to eat its roots. When she did so, she acquired the ability to see the dead.

Consumption of the eboka bush is a central part of the Fang people's Bwiti cult, a religious movement which emerged in response to colonial oppression. Bwiti adherents reinforce their ethnic identity through regular contact with their ancestors. They achieve this by ingesting large quantities of eboka, which has hallucinogenic and euphoria-inducing qualities.

Western scientists have been interested in this bush for a long time. Its active ingredient, ibogaine, was first isolated in 1901, and by the 1930s it was on sale in French pharmacies. In 1962 a former heroin addict, Howard Lotsof, obtained a US patent covering the

administration of ibogaine to conquer addiction – and today a
number of private clinics and self-help groups still offer ibogaine
treatment. It is thought that ibogaine interferes with neuro-
transmitters, inhibiting the effects of serotonin or dopamine, so that
users cease to experience the pleasurable sensations associated with
taking their drug of choice.[1]

For the Fang, ibogaine remains the route of contact with their
ancestors. Although they are now settled farmers, this tribe was
itinerant – a way of life that creates cultural anxiety about continuity and tradition. If outward proof of your identity cannot be
found in a piece of land, it must be sought elsewhere, and for the
Fang this end was achieved by carrying around the bones of important tribal leaders in cylindrical bark boxes. Keeping in such
close contact with their forebears gives the Fang continuity over
time, if not space, and this is emphasized by the nature of their art.
To protect themselves the Fang make offerings to wooden ancestor
sculptures, with large heads, long, folded bodies and short limbs.
These curled, foetus-like effigies express their belief that their
ancestors are remade constantly; there are no 'new' people, only an
eternal, rolling cycle of the original Fang given new flesh. As they
say: our bones are the ancestors' bones.

We can interrogate our past in a different way, using it to shape
our future. As I write, we stand upon what seems to be the shore of
a new, largely unexplored continent in scientific achievement. Each
month experts announce the identification of genes implicated in a
range of human conditions – from breast cancer to Alzheimer's
disease to multiple sclerosis. Advances in the manipulation of stem
cells seem to imply that the cultivation of spare organs and the
repair of badly damaged tissues will become a possibility. Concerned
parents can now take advantage of sophisticated tests to ensure their
offspring are born free of hereditary conditions. Amid a blaze of dis-
graceful publicity, a pseudo-scientist claims to have produced several
human clones, and another biologist to have cloned himself. And,
intriguingly, a few scientists have started using the science of genetics
to create 'synthetic' life in the test tube.

None of this would be possible without the realization that – to
borrow the Fang's poetic phrase – our bones are our ancestors'
bones. This knowledge – first glimpsed by breeders of crops and

animal herds, enthralling many scientists like Vavilov (see chapter 2), later guiding two researchers in Cambridge to the beautiful spiralling structure of DNA, and now expanded into the Human Genome Project's road-map of human variety – may be one of humankind's most significant advances. Logic dictates that it should be the most dangerous, too – as well-educated and respected voices have asserted. But must it be? Can we learn the lessons from our millions of years of inventing and discovering, to ensure that this discovery is applied for benefit, and not our downfall?

GARDENERS' QUESTION TIME

Those familiar with the Old Testament will know the story in the Book of Genesis about Jacob and Laban. It is the earliest reference I can think of to genetic manipulation, as well as to fraught family relationships. After fourteen years of servitude, stockman Jacob, who has married in succession two of Laban's daughters, asks for his freedom. As a reward for faithful service, he requests a few animals to establish his own herd: just the unwanted black lambs, and the spotted, speckled or striped animals. That way there can be no accusation of theft when he moves on. His father-in-law, Laban, agrees – but double-crosses Jacob by removing all the black, spotted, speckled and striped beasts overnight. Undaunted, Jacob waits for the lambing season, when Laban's snow-white flock, as expected, still produces a quotient of black, spotted, speckled and striped young. Jacob takes the healthiest specimens and fences them off from the remainder of the herd, ensuring that they give birth to the strongest, fattest, most vigorous progeny.[2] The biblical account also describes how Jacob shows the sheep mottled black-and-white sticks when they are drinking or copulating, in accordance with the ancient notion that experiences at conception influence the characteristics of our children.[3] Gaining a huge flock as a result, he is enriched by genetic engineering.

The notion of heredity has always been obvious to humans: obvious when children grow up to resemble their parents; obvious when animals carry traits across their generations. It was, for example, apparent to English horse-breeders in the twelfth century, who crossed local fillies with Arab stallions in order to engineer the perfect racing model. It was equally apparent to Henry VIII, who in 1541 banned any horse under 15 hands from grazing on common land so that only the larger beasts should have the opportunity of mating – even though by today's standards 15 hands is a small nag indeed. In the eighteenth century, English pig farmers were eager to mate their sows with super-sized boars brought in from China, because it made for fattier bacon – a popular delicacy during the 'Little Ice Age' of the time.

Although this primitive understanding of heredity was widespread, it fell to a relatively obscure nineteenth-century monk in Austria-Hungary to elucidate its mechanism. There had been increasing interest in the matter throughout the previous century, particularly from the Herefordshire horticulturalist Thomas Andrew Knight.[4] Knight was fortunate in having inherited wealth, with some 10,000 acres of land at his disposal, and like Vavilov was fascinated by heredity. He found the common pea to be of more value to his enquiries than spotted goats or extra-fat pigs. Unremarkable though it is, this common accompaniment to Sunday roasts comes in an enormous variety of sizes, shapes and colours. Accordingly, crossing members of this plethora of sub-types has provided invaluable insights for those interested in genetics. It led Knight to observe an interesting phenomenon. When the offspring of two dissimilar pea parents emerged, they usually looked fairly similar. When these hybrids were mated with other hybrids, however, the next generation was a riot of different characteristics.

John Goss took up this observation in 1824 when he presented a paper to the Royal Horticultural Society, outlining his theory of dominance.[5] He had noticed from his gardens on the edge of Dartmoor that certain characteristics from one parent plant could outweigh characteristics from the other parent, although those latter characteristics would still reside, seemingly dormant, in the offspring. In this way, a yellow-flowered pea might cross with a white-flowered type, the yellow trait proving so strong that the offspring inherited it. But then that offspring, depending on its choice of mate, might go on to produce a white-flowered descendant. Unlike the monastic gardener Mendel who followed him, though,

Goss appears to have been uninterested in the important mathematical relationships in inheritance, preferring to grow peas for their comestible qualities. 'The edible qualities of this Pea I have not tried, having but few,' he wrote.

Gregor Mendel was well aware of these kinds of observation and they exercised his intellect. History tends to present Mendel as a simple cleric in a remote central European backwater, stumbling on the central truths of genetics while tending his vines. This charming picture doesn't do credit to Mendel, his era or his country. He was born in Brno (now part of the Czech Republic) in 1843 and his poor background led him, as it did so many would-be scientists, into monastic life as a means of pursuing an education. This took him to the University of Vienna, where he studied under many leading scientists of the day. He returned to the Augustinian friary of St Thomas, eventually becoming abbot in 1867; here he fought and won many battles against the region's conservative authorities and became patron to the composer Leoš Janáček. Far from being a solitary gardener, Mendel was an urbane, well-connected scientist at the heart of an intellectually curious Europe.

Nor did he arrive at his realizations by chance. With a solid grounding in experimental techniques and statistical analysis, Mendel concentrated on seven characteristics, including colour of flowers, appearance of seeds and height of plants, in twenty-two different types of pea over an eight-year period. It was a formidable undertaking – but, as is so often the case in science, Mendel did have one huge stroke of luck. The pea plant he studied has seven chromosomes. Each of the seven characteristics he decided to study was determined by genes on a different chromosome (or, in one case, at the opposite end of a long chromosome). Consequently, the pattern of inheritance of each trait was pretty well independent from the others and easier to unravel than it might have been. Unfortunately, not much of Mendel's work is now available – a thorough colleague burned the bulk of his written records after his death. Happily, though, the text of his 1866 presentation to the Brno Natural Science Society, 'Experiments in Plant Hybridisation', survived.

Building on John Goss's observations about strong and weak traits, Mendel coined the terms 'dominant' and 'recessive' to explain how traits were passed down. But he wasn't content merely to note that it happened. With his superior mathematical skills, he observed something important about the ratios of these intergenerational changes.

When two different pea plants were crossed, the offspring with the dominant characteristic – for example, round seeds – outnumbered the offspring with the recessive characteristic – wrinkled seeds – by 3: 1. If a plant that had two dominant characteristics – like round seeds and green pods – was crossed with one that had two recessive characteristics – such as wrinkled seeds and yellow pods – all the offspring had round seeds and green pods. If two of those were then crossed, a precise arrangement could be observed. For every sixteen offspring, an average of nine would have round seeds and green pods, three would have wrinkled seeds and green pods, three would have round seeds and yellow pods, and one would have wrinkled seeds and yellow pods. Mendel also noticed something extremely important: there was no blending of characteristics. Seeds were either wrinkled or not, pods either green or yellow.

None of this may seem so very remarkable now, but these findings were the result of years of patient work. First, Mendel chose a plant, *Pisum sativum*, whose reproductive organs are largely enclosed – greatly reducing the risk of accidental cross-fertilization. Second, he spent some years establishing that the plants he studied were true breeding.[6] He also made sure that any cross-fertilization was under his strict control by removing all immature pollen-bearing stamens from all true-breeding plants that displayed a particular trait he wanted to study. And his researches involved tens of thousands of plants as he meticulously recorded the results for each trait.

It became clear to him that each characteristic was created by some factor within the plant. Each plant contained a set of these factors, which must be mixed up and recombined in the act of reproduction. So, if a plant needed to inherit just one dominant factor in order for its offspring to express a corresponding characteristic, but two recessive factors in order for its offspring to express a recessive characteristic, it was easy to see why the ratio of 3: 1 applied.

In a nutshell – or a peapod – Mendel explained how genes work, before the word existed. It was a remarkable intellectual achievement – but even more remarkable was the resounding silence that greeted it. Charles Darwin allegedly had a copy of Mendel's paper, but apparently never read it. The distinguished Swiss botanist Carl Nägeli said Mendel's experiments must be wrong. It was another thirty-four years before Mendel's research received the recognition it deserved.

In the meantime a Swiss biologist, Johann Friedrich Miescher from the University of Basel, developed an interest in the composition of white blood cells.[7] He obtained these initially from the pus-soaked bandages of patients at a nearby hospital, but later moved on to the more readily accessible resource of trout gonads. In 1869 Miescher announced that he had isolated a compound that was found inside the nucleus of every living cell. It was rich in phosphorus and composed of very large molecules. He called it, on the suggestion of a student, 'nucleic acid'.

Physicians were now, slowly, taking the exciting implications of Darwin's theories into their consulting rooms. One very fashionable English doctor, Sir Archibald Edward Garrod, was surprisingly fascinated by coloured urine. Some of his patients produced deep red urine, and eventually he established that they had the disease haematoporphyrinuria. Then, in 1897, he changed colour, getting interested in people who, as he put it, 'advertised their condition' by producing urine which turned black after it had been left standing. These people, whose condition he called 'alkaptonuria',[8] were often well – but he noted that alkaptonuria occurred in families, with brothers and first cousins who were affected frequently suffering from arthritis.[9] As the condition had emerged in the children of unaffected parents, who were in each case first cousins, Garrod recognized that susceptibility to alkaptonuria could be a recessive trait of the kind Mendel had described forty years earlier. But he felt it might be a bit problematic to conduct the definitive experiment, writing to William Bateson (see chapter 2) in Cambridge: 'I do not see any way of introducing any marriageable alkaptonurics to each other with a view to matrimony.'

While some were developing an understanding of heredity, and others groping towards an apprehension of DNA, it took a long time for the sides to meet. It became clear that genes, located on the chromosomes within the cell, were the source of proteins, the versatile molecules that could be said to run the entire human body.

But, precisely because we had started to begin to understand the importance of proteins, we were inclined to overlook the significance of Miescher's nucleic acid. Its chemical structure seemed very simple – too simple to have anything to do with the complex array of proteins necessary for life. It was assumed that genes were composed of protein, and that Miescher's nucleic acid must just serve as the glue.

This substance – properly termed deoxyribose nucleic acid, or DNA – was found to be more significant in 1944 by American scientist Oswald Avery. Like many distinguished scientists before and after him, Avery never won a much-deserved Nobel Prize, but he devised an important experiment to look at the chemistry of the genetic information that was transmitted between bacteria. By targeting this information with an enzyme that broke down proteins, Avery was able to assess how well the genetic 'message' was being conveyed. He discovered that, even when proteins were knocked out of the equation, the genes still got through. Only when the DNA was targeted did the bacteria prove unable to pass on their load. This was clearly more than glue; it must, in fact, be the very stuff of genetic information. One remarkable thing about Avery was his rare modesty – a feature not shared by all the geneticists who preceded and followed him. In 1943 he wrote:

> If we are right, and of course that is not yet proven, then it means that nucleic acids are not merely structurally important but functionally active substances in determining the biochemical activities and specific characteristics of cells and that by means of a known chemical substance it is possible to induce predictable and hereditary changes in cells. This is something that has long been the dreams of geneticists.

The race now began to discover the structure of DNA, and how it worked. The tale of Crick and Watson, the two Cambridge scientists who loudly celebrated their discovery of the 'secret of life' in the town's Eagle pub, is well known. It is, however, worth reminding readers that they would not have succeeded but for the brilliant work of the X-ray crystallographer Rosalind Franklin, of King's College, London. Controversy still surrounds the fact that Crick and

Watson were shown a particularly telling snapshot of the DNA molecule's structure, and used it without Franklin's express permission. When the pair published their results on 25 April 1953 in the premier scientific periodical *Nature*,[10] Rosalind Franklin received acknowledgement for her contribution. She did not, however, share the Nobel Prize – and in 1958 she died of cancer, still unaware of her major role in the discovery. Although there were personality clashes, and some knotty issues of scientific etiquette between King's and Cambridge, there is no evidence that Crick and Watson knowingly cheated Franklin out of her rewards, and many believe it was just an unfortunate turn of events. But she serves as a reminder that many discoveries have largely forgotten, or unknown, authors. Some, like the early farmers, and the clerks who first grasped the significance of writing, had the misfortune to make their marks before we had the means to record or retrieve them. And some, because of chance and birth and the behavioural quirks so peculiar to our species, simply get edited out of the picture.

MIRACLE IN MINIATURE

In the Jewish religion, most details of daily life are sanctified by some prayer, blessing or ritual. So concerned have generations of my co-religionists been to link every experience to God that there are specific words of thanks to be said upon eating fruit, smelling odorous plants, hearing thunderstorms and seeing the sea. One favourite is the blessing to be recited when one sees a gentile king or meets with a wise man: 'Blessed art thou, O Lord Our God, who hast given of His glory to thy flesh and blood.' It almost seems like a thank-you for DNA, genes and chromosomes, enzymes and proteins.

No account of this astonishing discovery would be complete without some explanation of how it all works, but before that it is worth stepping back to consider its magnitude. Picture a cell. Its contents will be much the same whether it comes from an elephant or an East End builder. But for the moment consider one belonging to a human. Inside the nucleus of this tiny object you will find long sections of

DNA, coiled up and spooled on to chromosomes. Closer examination reveals that within each section there are thousands of
subsections composed of groups of molecules. Stretched out, they
would measure about 25 millimetres. Instead, they are compressed
into the centre of a cell no more than 0.00067 millimetre in
diameter. This remarkably compacted unit of information
determines, among other things, the formation of limbs, the workings of the central nervous system, the body's metabolism, the
growth of the fine neural networks of the brain, the cells in
the reproductive system – indeed, the growth of the entire organism
and susceptibility to some of the diseases that may kill it. It is the
most complex item in the known universe – and if anything might
make you catch your breath at the awesome intricacies of that
universe, then it has to be the DNA molecule.

When discussing its workings, I am reminded of the quantum
physicist who quipped: 'If you think it's simple, then you haven't
understood it.' But it should be possible to describe with relative
accuracy how this molecule does its job. Picture a spiral staircase –
the classic double helix featured in virtually every popular book
on the subject. The two banisters of this spiral stairway are strands
of the sugar deoxyribose. The steps of the staircase are formed by
horizontal bars of four different nucleic acid molecules or
'nucleotides' – adenine (A), guanine (G), cytosine (C) and thymine
(T) – assembled in pairs and arranged in varying sequences. These
steps, called 'base pairs', are effectively the equivalent of the
alphabet of four letters for the chemical language of life, ideally
suited to the task of replicating information. The English language
has twenty-six letters, enabling precise descriptions to be rendered
comparatively briefly. But with just four letters to play with, repeated
in various combinations, rather more printing is required to give precision – which is why each cell in the human body has a DNA message
containing roughly 3 billion base pairs in various combinations.

When cells replicate, a complex chemical ballet takes place within
the DNA. Enzymes are choreographed up the staircase, breaking the
hydrogen bonds that keep it together. The staircase splits down
the middle and each 'banister' with its base-steps now acts as a
template for the formation of a new DNA molecule, a new staircase.
And this can only happen in a certain way. The process can be

thought of as resembling how the teeth of a zip mesh, or how the ridges of a key fit a lock: the nucleotide molecules can only pair up in a predetermined pattern, adenine always pairing with thymine, guanine with cytosine. 'Free' nucleotides made from available chemicals in the cell can now rush in and line up in the correct formation alongside the separated strands.

The way this molecular staircase creates the thousands of proteins necessary for life is a little more complicated. All proteins are cocktails of twenty different amino acids. To create them, the four nucleotides of DNA work in groups of three – adenine–guanine–cytosine, for example – each triplet making up a single amino acid. Since there are four different nucleotides, a total of sixty-four combinations is possible (four to the power of three) – which, given that there are only twenty amino acids, is more than sufficient. In order to create a protein, a section of DNA unravels, creating a parallel, slightly modified form of itself known as 'messenger RNA' or simply mRNA. This mRNA passes out of the nucleus into the body of the cell, where it is absorbed by the ribosomes – minute protein-manufacturing sites. The ribosomes 'read' the coded information from the mRNA, and then begin to construct the proteins by drawing in the necessary amino acids from their surroundings.

Another much used, but still appropriate, metaphor for the workings of DNA is provided by mighty, multi-volume reference books like the *Encyclopædia Britannica*. If you can imagine the 'steps' of the DNA inside a single cell, printed out in normal 12-point font in the format of one of these weighty tomes, you would need 300 volumes to express all the information. One volume may be said to represent a chromosome. The entries inside the volumes are the genes. So a single gene can be regarded as just one entry in the encyclopaedia – for example, 'Naples during the Risorgimento' or 'Industries: Extraction and Processing'. I suspect the entry about 'Naples' would be quite short; that about 'Industry' quite long. Some of these entries or genes might be 'read' once during a lifetime – for example, at a particular moment in development, perhaps during fertilization or organ formation in the embryo. Others might be 'read' on a daily basis – such as while making red blood cells or during digestion.

While some members of my research team at Hammersmith seem

to spend hours happily gazing at computer screens displaying thousands of these letters written in columns, for most people these 'entries' are very boring to look at: line upon line of, for example, AAG CCG ACG CCT – not ideal bedtime reading and certainly not *Strictly Come Dancing*. But imagine the effects that can occasionally occur if just a single letter is misprinted. Unlike an average encyclopaedia, chromosomes contain a good deal of meaningless 'paragraphs' of printing between genes. This material is sometimes referred to as 'junk DNA'. Because there is so much junk DNA most misprints probably will not matter much, because they occur in these meaningless paragraphs. But if a misprint – even a single letter in one million – occurs in a gene, the results may be fatal. There may be misprints in this book – and if you come across one, you will almost certainly be able to work out my meaning using your knowledge of English and the surrounding text. But, in general, the cells in the human body cannot be sentient in this way, cannot compensate for misprinting.

FINDING GENES, SEEKING CURES

We use the term a 'blue-eyed boy' to describe a favourite person, someone who attracts an enviable, perhaps even undeserved, amount of affection. But why should a biological feature like eye colour have such significance? It may be something to do with its relative rarity. A recent report suggests that blue eyes are derived from a mutation in a gene which arose in one individual living near the Black Sea some 8,000 years ago. The team responsible for this report, led by Dr Hans Eiberg of Copenhagen University, believe the culprit is a gene called OCA2, which doesn't so much 'make' blue eyes as switch off the mechanism responsible for creating the more usual brown.[11] A mutation is simply the result of a change in the DNA lettering – a 'misprint' – and this mutation is thought to be due to a misprint in just one nucleotide – just one letter out of over 344,000 letters which make that particular gene. Quite why this mutation became so valued is a matter for speculation. An oddity like blue eyes, while not intrinsically linked to keener vision, higher intelligence or better health, became a marker of 'good genes', in much the same way as a peacock's tail, though useless, became the gold standard by which peahens judged their mates. Maybe blue eyes were an add-on of lighter skins, which were better adapted to extracting Vitamin D from the relatively weak sunlight available in northern and western Europe.

Whatever its causes, blue eye colour is undoubtedly still highly valued. In recent years, it has become possible for people to change their eye colour temporarily using pigmented contact lenses. The advertisements for these products show people changing from brown to blue or green – never the other way round. More recently, people have begun to speak of another, more permanent way of securing the eye pigment of their heart's desire – if not for themselves, then for their offspring.

Inside some tabloid newspapers and in a certain genre of non-fiction book sometimes discovered undeservedly in the science section of your bookstore, you may find a common myth. It is a new type of story – a myth about the future, as opposed to some heroic past. In it, two prospective parents visit a clinic for IVF. The embryos they have created are screened and their genetic profiles built into a detailed dossier. The parents then have a choice as to the embryos they wish to nurture into life. Do they want this one, with the potential for developing Alzheimer's disease in his eighties, but also a good chance of being a genius? Or this one – lithe, beautiful and athletic, but sadly none too bright? This talented, blue-eyed piano player, with a tendency to bipolar depression? Or this utterly ordinary, unremarkable, brown-eyed boy, who'll probably die of old age? How do they want their baby to be designed?

This myth is fed by the strides in understanding of genetics that have been made since Watson and Crick's discovery and the hyperbole that we scientists have used to describe it. The science is complex, but it boils down to perhaps three major advances. In the 1960s it became apparent that DNA was served by a host of specialist enzymes, which helped it to repair and replicate itself. One of their functions is to disable viruses, by chopping up their genes into such tiny fragments that they can't insert themselves into new cells. In 1972 a pair of Californian scientists used one of these chopper-enzymes (termed restriction enzymes) to cut a section of DNA from a bacterial cell and to insert that section into the DNA of another bacterium. To return to our book analogy, it became possible to cut pages out of the 300-volume encyclopaedia and paste them in again elsewhere.

Bacteria were also vital in the next advance, because of the way they reproduce. Despite being asexual, they can nevertheless share genetic information by means of an extra loop or piece of DNA which sits outside their own genes, and which can be passed around. In 1973 scientists first perfected the means of isolating these bits of DNA (known as plasmids), cutting them open and inserting strands of human DNA inside them. Re-attached to bacteria, these strands would then be replicated every time there was a round of plasmid-sharing. In effect, we found a way of photocopying mini-sequences of human DNA thousands of times over.

Then two scientists at the University of Wisconsin–Madison discovered that some viruses could produce a very special enzyme, which effectively sent proteins into a rewind. As we saw above, the sequence of DNA letters in a cell's nucleus sends out messenger RNA into the main part of the cell, in order to pass its instructions on to the ribosome protein-making factory. The enzyme in a certain kind of tumour-causing virus, however, possessed the ability to reverse the process. Exposed to mRNA, it converted it back into the DNA sequence which formed it. This meant that, instead of rooting around in our 300 volumes for a solitary strand of letters, we could select from a far smaller pool of specialized proteins. Mixing them with the rewind-enzyme – known as 'reverse transcriptase' – we could then find the genes that created them. This advance would prove important in detecting genes that might, for example, be associated with diabetes or sickle-cell anaemia.

By the 1980s a number of scientists were translating the fruits of this knowledge into reproductive medicine. Those wishing to know more about this may read my book *A Child Against All Odds*.[12] This knowledge helped equip us with a tool to prevent suffering. It was possible to identify some genes responsible for a range of disorders, not just in adult humans or in the unborn foetus, but in fertilized eggs. Armed with this knowledge, we were able to prevent the birth of children with life-shortening and incurable problems like haemophilia, muscular dystrophy and cystic fibrosis.

The understandable mood of public optimism arising from these advances was bolstered from 1990 onwards by the Human Genome

Project (HGP), initially funded by the Department of Energy and the National Institutes of Health in the United States and the Wellcome Trust in London. The HGP had the mission of deciphering the entire order of the base pairs which make up human DNA and identifying all the genes. Initially it was expected that humans would have at least 100,000 genes, but it transpires that in fact there are only about 25,000. The Project completed its goal of mapping the human genome in 2003, although some areas still need to be explored.

Using the findings of the HGP, scientists across the globe have captured media attention. In 2007 the *Guardian* triumphantly announced that scientists had 'unlocked the secrets of diseases afflicting millions'.[13] The piece went on to say that a research programme funded by the Wellcome Trust had studied the DNA of 17,000 people. Its fifty separate research groups, coordinated by Peter Donnelly, Director of the Wellcome Trust Centre for Human Genetics and Professor of Statistics at Oxford University, had identified twenty-four new genetic links for common problems like bipolar disorder, Crohn's disease, types 1 and 2 diabetes, rheumatoid arthritis and high blood pressure, effectively doubling the number of genes known to be implicated in these conditions.

This kind of report makes headlines all the time – regardless of the limitations of the new knowledge. It is not that treatments have been found for these conditions – merely that more of the genes involved in their origin are known. In some ways, the prospect of a cure becomes more distant with every new gene we find, for a large number of the conditions which afflict us may depend on the action of dozens of genes in tandem. For every gene which codes for a protein, there are other genes which regulate that action, switching it on or off (as the OCA2 gene does in brown-eyed babies), or intensifying or minimizing its outputs. Professor Donnelly pointed out that, 'By identifying the genes underlying these conditions, our study should enable scientists to understand better how disease occurs, which people are most at risk and, in time, to produce more effective, more personalised treatments.' That certainly is the hope; but we shall have to wait to see if it is borne out in the reality. As the *Guardian* writer said, such discoveries 'pave the way' for treatments, but the extreme length of the road, and its varied terrain, are not always made completely clear by journalists.

In 1881 a London-based ophthalmologist inadvertently identified
a major genetic disease while examining children's eyes. By spotting
the characteristic red spot on the retina of a few of his Jewish
patients, Warren Tay acquired a certain kind of immortality, because
his name lives on in the condition he helped to identify – Tay–Sachs
disease. It is a grim legacy. Because of a mutation in the HEXA gene,
the body produces insufficient levels of an enzyme, hexosaminadase
A, which is responsible for breaking down lipids. As these build up
in the brain, babies experience a range of terrible symptoms at about
six months old, while their parents can only look on helplessly. The
children cannot sit up because of muscle weakness; they suffer
progressive blindness and paralysis, and inexorable mental
deterioration. Most of these children die before the age of four.

Tay–Sachs is a recessive condition. The parents are healthy, but
both carry the mutation causing the illness – so there is a one in four
chance of each child suffering in this way. Some families can be very
unlucky: the probability may be only one in four, but the nature of
random chance means that in some cases each pregnancy is affected.
Tay–Sachs is thought to be a result of inbreeding in small Orthodox
Jewish communities. But it also occurs in a few other contexts, for
example in the Cajun communities of Louisiana and the Québecois
of Canada – both of whom, of course, trace their origins back to
sixteenth- and seventeenth-century France. People have been so keen
to label Tay–Sachs a Jewish disease that its appearance in North
America was once explained by the idea that travelling Jewish fur
traders indulged in extra-curricular activities with the wives and
daughters of their customers. This 'fur trader' hypothesis has been
discounted, the Cajun and Quebec strains of the disease having
been traced back to two French families, with no apparent Jewish
link. We just do not know why Tay–Sachs should have occurred
within these three communities.

We do know we can't cure it, though programmes of mass
education, screening and counselling have drastically reduced the
incidence of Tay–Sachs among all three affected groups. For strictly
observant Jews, the organization Dor Yeshorim offers a form of
anonymous screening, so that concerned couples can be told of the
risks of marrying and having children, without the stigma of a definite
Tay–Sachs label falling on one or the other party. This has been of

far more value than the fairly useless attempts to create a genetic cure by injecting the missing enzyme into the cerebrospinal fluid.

Tay–Sachs is fairly typical of illnesses caused by a mutation in a single gene. There are about 6,000 diseases resulting from single gene defects. Many of them are very serious, causing the deaths of babies or children. This is because they cause disturbances of basic bodily functions and therefore show themselves early in life. In technical language, the genes 'express' early. And the great majority are recessive. If they were dominant, they would rapidly die out – because any carrier would become ill before being old enough to have children. So the few dominant genetic disorders tend to be illnesses which only start late in life. One such disease is Huntington's chorea, affecting adults from around the age of thirty to forty-five – by which time they may well have had children and unknowingly passed their condition on to the next generation.

There is also a third category of genetic disorder – so-called 'sex-linked' gene defects. These are caused by mutations in genes on the X-chromosome – the 'female' chromosome. Because every normal woman has two X-chromosomes, the fact that one of these chromosomes carries a mutation does not matter immediately: the second, normal chromosome compensates for it. For this reason, a woman with a mutation on one of her X-chromosomes generally does not recognize any problem until she gives birth to a son. Males have only one X-chromosome, so if the boy inherits the 'wrong' one from his mother, he will suffer from the disease caused by the mutation. As one of two X-chromosomes will be inherited, the chance of any son being affected will be 50 per cent. The most infamous of these diseases is haemophilia – carried by Queen Victoria and a number of her female relatives, and passed on to Tsar Nicholas II's son the Tsarevitch, who suffered from the disease.[14] There are roughly 300 sex-linked diseases, of which the commonest is Duchenne muscular dystrophy.

Duchenne muscular dystrophy causes paralysis in boys who have the defect in the gene which produces dystrophin. Dystrophin is a protein, and although it actually accounts for only about 0.002 per cent of the total protein in muscle, a defect in it can be a death sentence. These children have slowly developing muscle paralysis: initially this makes it hard for them to walk, and most are confined to wheelchairs by the time they are adolescent, but as the weakness gets worse they have increasing difficulty in breathing. Because the muscles in the chest don't work at the end of their short lives, they often risk suffocation if they do not die of a chest infection first.

Duchenne muscular dystrophy has been studied extensively by one of my great heroines, Professor Kay Davies at the University of Oxford.[15] Her work brings an effective treatment much closer – a degree of success fairly unusual in the panoply of diseases caused by gene defects. The dystrophin-producing gene is an interesting example of a problem in the DNA analysis of gene disorders. The gene is one of the largest in the body – it is 2.3 million base pairs long – and different families often have different mutations. There are around 500 different misprints which may occur, any of which may cause a defect in the protein, and this variability in the mutations means that the severity of the disease may vary from person to person. It also makes diagnosis and the prediction of its severity a bit more complicated.

There may be a source of hope for people with Duchenne muscular dystrophy. Drs Qi Long Lu and Terrence Partridge have experimented with a 'gene patch' that has been tested on mice at my own institution, Imperial College at Hammersmith Hospital. The patch involves injecting small pieces of genetic material into the ribosomes – the protein-making factories of the cells – so that they are blind to the mutation, allowing the protein to be made in almost working order. When tested on mice, the 'patch' successfully restored dystrophin production, improving muscle function.

For a long time it seemed that the exciting discoveries in genetics would be of very limited practical value. But recently there have been some notable successes. Ashanti de Silva was born with another rare condition, in which her genes were unable to create an enzyme called adenoside deaminase (ADA). This led to her lacking the T-cells in her blood which fight off bacteria and viruses. With her immune system severely compromised, Ashanti was confined to living under sterile conditions. Without strict sterility, the slightest mild infection – something that might cause you or me to have a bit of a sore throat or a pimple – would be a killer.

The treatment for ADA deficiency, before gene therapy, was for affected children to have injections of a synthetic form of the enzyme. But this, as well as being very costly and not as effective as the naturally produced enzyme, produces diminishing returns – and by the age of four, Ashanti's condition worsened. In 1990 her despairing parents approached the geneticist William French Anderson of the University of Southern California School of Medicine, then working at the Children's Hospital in Los Angeles, who was seeking permission for the first gene therapy trials on humans. Using some of the techniques discussed above, this team inserted functional copies of the ADA gene into Ashanti's blood cells, and put them back into her body. There was no rejection – as there would have been in the case of a transplant from another person – because Ashanti's body recognized the cells as her own. Within six months, her T-cell count had shot up. Within two years, she could go to school to live the life of any normal six-year-old.

This treatment was not an unqualified success. For ethical reasons, Ashanti continued to receive injections of the synthetic ADA throughout the gene therapy trial, and when these were stopped, her condition worsened. For that reason, she is still on a regular dose – albeit a low one. But doctors took the lessons they had learned into the next phase. In 2002 they tried out a new technique, using modified stem cells to stimulate the body into creating its own supplies of ADA. Crucially, this was tried in children who hadn't been treated with the synthetic version of the enzyme, and its outcome was positive. The first was a two-year-old Palestinian girl called Salsabil, whose life is now so ordinary that she even managed to fight off chickenpox unaided, a virus which would certainly have killed her before.

Occasionally, then, our understanding of genes can be turned into treatments. There are three basic ways of doing this: by replacing faulty genes with working copies; by inserting modified genes which can inhibit the actions of others; or by the use of 'suicide genes', which can switch on self-destruct mechanisms, or make dangerous cells more vulnerable to attack. Between them, these techniques create exciting prospects.

Some possibilities being thrown up by gene therapy are truly remarkable. A team at the University of Rochester Medical Center in New York found a way of reanimating dead bone tissue with a gene graft. Treatment for bone cancer typically entails removing the affected tissue and replacing it with a graft from a corpse. But the bones inside us are actually living and breathing, constantly acquiring tiny cracks, constantly reknitting new tissue for repair. Dead bone can't do this, so fitted grafts usually have a short shelf-life before they begin to crumble and a new transplant is needed.

The Rochester team, led by Professor Edward Schwarz, identified two of the genes which enabled the production of repair proteins in mice.[16] They then engineered a harmless virus and attached the DNA of this virus to the DNA of the required genes. They developed a canny technique for freeze-drying a virus-containing paste, which was then painted directly on to a bone graft during surgery. The virus infected the tissue around the bone, switching on the repair proteins – effectively converting dead matter into living, self-regulating tissue. It sounds like science fiction and so far has only been done in mice. And where treatments for most human conditions are concerned, much of all this is still remote and even closer to the realms of science fiction.

THE NEW ALCHEMISTS? STILL SEARCHING FOR THE ELIXIR

Few sports can have changed as much over the years as cycling. New materials have made for stronger, lighter bikes. Medical advances have enabled athletes to fine-tune their bodies with diet, isotonic drinks and specialized training regimes. Small wonder, then, that whereas the first Tour de France was won, in 1903, at an average speed of little over 25 kph, today's leaders cycle at 40 kph for the best part of 4,000 kilometres.

If one aspect of cycling has remained unchanged, it is the drugs. The Paris–Roubaix race is one of the classic contests, completed in a single day over approximately 280 kilometres of rough and cobbled roads; the winner is presented with a mounted cobblestone, together with a respectable sum of money. (Rather unfairly, it seems, it is the organizers of this race who get the gold-plated cobblestone.) In April 1903 the luxuriantly moustachioed Hippolyte Aucouturier, wearing his famous blue-and-red striped vest over his barrel chest, won this race by just 100 metres. So he became favourite to win the next major event, the inaugural Tour de France to be cycled that July. But just 320 kilometres into the first stage, Aucouturier stopped cycling with stomach cramps, collapsing in sobs by the side of the road. It seems he had been drinking wine and sniffing ether, both ways of numbing the pain of long hours in the saddle.

Henri Pélissier became champion after the 1923 Tour de France. Some years later, this hugely aggressive athlete, disappointed and frustrated at being well past his best, tormented his wife Leonie, who eventually shot herself in 1933. Three years after that, Pélissier threatened his lover, Camille Tharault, with a kitchen knife and lacerated her face. Seizing the gun with which Leonie had killed herself, she fired five shots. One bullet hit Henri's carotid artery and he bled to death. It was not the first time he had hit the headlines. Back in 1924, the sport had been rocked by scandal when Henri Pélissier and his brother Francis gave an interview to a French journalist during the Tour de France.

'You have no idea what the Tour de France is,' Henri had said. 'It's a Calvary. Worse than that, because as the road to the Cross has only 14 stations and ours has 15, we suffer from the start to the end. You want to know how we keep going? Here—' (he was reported as pulling a phial from his bag) '—that's cocaine, for our eyes. This is chloroform, for our gums.'

'The truth is,' said Francis, 'that we keep going on dynamite.'

Henri Pélissier spoke of being as white as a sheet after the day's cycling and of being drained by diarrhoea. 'At night,' he said, 'we can't sleep. We twitch and dance and jig about as though we were doing St Vitus's Dance . . . '

On 14 July 1967 the *Yorkshire Post* announced the death of Tommy Simpson, Britain's best-known cyclist. The newspaper reported that he died from heat exhaustion, having cycled up Mount Ventoux 1,800 metres above sea level during the Tour de France. Within a few metres of the summit he slumped over the handlebars and fell off the bike. He was helped back on to his machine, but immediately collapsed again. Whether he was suffering from heat exhaustion or not, the reason for the cardiac arrest that killed him was a large dosage of amphetamines used as a stimulant. The four remaining members of the British team decided to continue the Tour, the team manager, Alec Taylor, saying: 'I knew Tommy very well and I am sure that this is what he would have wanted.'

The evidence suggests that doping continues today, with even greater efficiency. Admittedly, the ability to detect drugs has also improved – and this is in part thanks to advances in genetics. During the Tour de France of 1998, customs officials on the Belgian–French border stopped Willi Voet, the massage therapist for the Festina Team. They found 400 vials of erythropoietin (EPO) in his possession. EPO was first isolated in 1985 by an American company seeking to boost the body's production of red blood cells. It is prescribed as a successful but costly treatment for the severe anaemia caused by kidney failure or HIV treatments. It became clear that EPO injections boosted the red blood cell count, whether that person was anaemic or not. In healthy individuals, more red cells increases oxygen capacity and that enhances endurance. In events like cycling, where victories are secured by fractions of a second, the use of EPO can offer an edge.

Treatments like this are risky. Increasing the number of red cells can thicken the blood and may make clots more likely, particularly in narrow-bore vessels. And if the blood becomes more viscous, the heart may not be able to pump it so efficiently. Heart attacks and death can follow – the fate suffered by a number of Dutch cyclists who took the drug in the early nineties. But what if the human body could be made to manufacture its own extra servings of EPO and also regulate the process through its own internal mechanisms? Suppose our genes could be modified to ramp up our endurance – not necessarily to win prestigious cycling events, but to cope better with daily life, or, for example, to lessen the effects of getting old?

This might just be possible. In 1997, at the University of Chicago, Professor Jeffrey Leiden and his colleagues injected monkeys and mice with a virus containing an extra EPO gene.[17] Yet again, the virus was a kind of Trojan horse, used to gain entry into the nucleus of cells where its DNA is incorporated into the genome along with the relevant DNA from the EPO gene. After just one treatment, the numbers of circulating red blood cells soared in both mice and monkeys. It may not have improved their cycling, but monkeys that had had EPO gene therapy still had increased red blood cell-counts four years later. This does raise the possibility of people taking one-off or very infrequent treatments to boost their strength. It also demonstrates the blurred line between using gene therapy as a means to a cure and using it as a form of enhancement.

The proposals of Dr Aubrey de Grey, chairman and chief science officer of the Methuselah Foundation, are more extreme. This foundation is a puzzling affair. Boasting many pledges of substantial sums, it seems in fact to be funded for the most part by very small donations from individuals wanting to live eternally. Perhaps, for those with such ambitions, small donations are the best strategy. Although his BA degree was in computer science rather than biology, de Grey has a PhD from Cambridge University in the field of ageing research. He has suggested that eternal life is possible and that in the future anyone wishing to have children will have to find someone else willing to make room for them by dying. Comments like this – and his striking appearance, with a long, reddish, straggly beard reaching almost down to his navel – make Dr de Grey a gift to journalists, most of whom refer to him as a 'Cambridge University' gerontologist or geneticist, even though he does not seem to have a substantive appointment at the university.

He is, however, an author of a number of papers: I counted ninety on the definitive medical index in Washington, PubMed, where virtually all biological research papers are registered. Number of publications is not itself any indication of their merits or demerits, of course. As far as I can gauge, none are reports of any serious research of his own – none demonstrate that de Grey has any new experimental evidence that he can prolong human life; they appear to be merely reviews or expressions of ideas that seem plausible, and most of them are published in de Grey's own journal, of which he is editor-in-chief. He is also co-author of a paper published in the *Annals of the New York Academy of Science*, on which his name appears alongside those of a number of respected researchers – not the first time that a controversial figure has benefited from the co-authorship of people of recognized status within a field. In this paper – which is a statement of interests, rather than an account of completed work – de Grey and his co-authors set out a number of intriguing techniques for prolonging life.[18] One is to relocate the mitochondrial DNA inside the nucleus, where it would be safe from the ravages of time, and from the actions of free radicals – charged molecules that build up and cause mutations. Other proposals include the removal of the 'junk' material which accumulates inside and outside cells and leads to the formation of plaques like the amyloid that clogs the brains of Alzheimer's sufferers. But it is noteworthy that de Grey has criticized other work in the field as pseudoscience,[19] when his own assertions that humans could live for several hundred years seem to me just that – pseudoscience.

De Grey is certainly a controversial figure. But some of his co-authors on this review paper in the *Annals* have done sound research into the link between genes and ageing. One of them is the respected Andrzej Bartke, of Southern Illinois University, who was a colleague of mine years ago when he and I were professors in the same department at University of Texas at San Antonio. Professor Bartke has genetically modified mice so that they produce less growth hormone and are less sensitive to insulin – and these mice have a lifespan 66 per cent longer than normal laboratory mice.[20] That is the equivalent of you or I living to be 130 years old.

Research on species as diverse as nematode worms, mice and fruit flies has aimed at discovering how tiny genetic changes might extend lifespan dramatically. Some hundred or so relevant genes have been identified. Some of them, when modified as in Andrzej Bartke's mice, work by reducing sensitivity to insulin. Others protect cells against the bruising, electron-stripping effects of the free radicals. Some mimic the effects of a low-calorie diet, which, for reasons still not clearly understood, lessens the damage wrought by factors like time, stress, heat and toxins on the DNA in our cells. None of these have been tried in human subjects yet. But the fact that similar genetic modifications produce related results in species as different as mice and fruit flies suggests they might be of some use.

Another (surrogate) way of 'living for ever' is to have oneself cloned. And when, in January 2008, a scientist called Samuel Wood announced that he had done it, the world's press was quick to comment. Dr Wood, employed by a Californian stem cell company called Stemagem, took unfertilized eggs donated by women having IVF treatment. He removed their nuclei, replacing each nucleus with the nucleus from one of his skin cells. Cell division was jump-started with a tiny electrical charge, and at least three embryos survived for a total of five days.[21] They were then destroyed by Dr Wood, who said that to continue with their growth would be unethical.

To some of those enthusiastic about the potential of stem cell research and concerned by the legal, ethical and technical difficulties of obtaining embryonic cells, his work was welcome. To others, it was a moral outrage. It is difficult to justify taking eggs from human volunteers unless there is a serious chance of using them to make a significant improvement in medical therapy. But 'moral outrage' is a strand which runs through history in responses to human innovations of many varied kinds. When Ottoman sultans first heard of printing presses, they instituted the death penalty for anyone using them. In the nineteenth and twentieth centuries, medical voices were raised in concern about what would happen to the human body when steam and petrol engines transported it at speeds exceeding 20 miles per hour. To fear for the consequences of an invention seems as human as to invent in the first place. It falls to each new generation to consider which of its fears are valid – and why.

DANGEROUS CURES

A photograph taken on 1 July 2007 shows Jolee Mohr on the sundeck of a motor-cruiser, a picture of health and beauty. At thirty-six, she was the mother of a five-year-old daughter and lived a full, active life, despite having rheumatoid arthritis. A few hours after the photograph was taken, Jolee Mohr received an injection of a gene-loaded virus into her right knee. A few days after that, she was dead.

According to lawyers acting for Jolee's family, she was encouraged to take part in a gene therapy trial by her doctor. They claim that he presented the therapy to her as a potential cure, rather than as an experiment to gauge its safety. Her death has called into question the ethics of gene therapy for any other than life-threatening conditions. When she died there were some 130 licensed gene therapy trials under way in the United States, the majority of them for terminal conditions like cancer. Around ten, however, targeted conditions which were either less severe or, as in Jolee Mohr's case, under control through existing treatments.

This was not the first time gene therapy had made bad news. In September 1999 eighteen-year-old Jesse Gelsinger died after experimental gene therapy. He had a rare disease called ornithine transcarbamylase deficiency, which causes a build-up of ammonia in the body and in severely affected individuals, particularly babies, results in brain damage and death. But Jesse was fit, active and sporty, in no great danger from the disease provided he kept to a restricted diet – though he did need to take a large number of pills each day. It was Jesse's misfortune to be offered a place on Dr Jim Wilson's gene therapy trial when another volunteer who was to undergo the treatment pulled out. It is very difficult for me to see how his inclusion in the trial was justified: even then, it was clear that gene therapy could be a risky business, and Jesse did not particularly need it to improve his health.

By many accounts, Dr Jim Wilson is a highly focused, go-getting physician participating in the highly competitive race to be among the first to produce a successful gene therapy treatment. In such a high-profile field the rewards were likely to be high. So it seems it was also Jesse's misfortune that Jim Wilson was involved with a trial where informed consent did not appear to be taken seriously. Neither Jesse nor his parents appeared fully to understand the risks involved; nor had they been told that several monkeys treated by the method had died shortly afterwards. Jesse gave his consent to take part because he knew this disease killed babies and he wanted to do his bit towards improving human knowledge.

Jesse's death was horrific. He had been given a particularly large dose of the adenovirus which was designed to carry the DNA into his cells. He developed a fever of 40 degrees Celsius and became jaundiced; the blood in his vessels gradually clotted, and little by little he swelled up, waterlogged, as his lungs and kidneys failed and all his vital organs ground to a halt. He took four days to die. Many people find it difficult to escape the feeling that in Jesse's case the researchers may have been prepared to take life-and-death risks for glory. But it was Jesse's life or death for the researchers' glory.

X-SCID is another rare sex-linked gene defect, caused by mutations in the IL2RG gene that inhibit formation of a particular protein. A nasty disease affecting baby boys, it results in an in-adequate immune system. These children are so prone to infections that they have to live in a germ-free environment and can have no normal human contact. Without the normal protein, the white cells in the blood cannot develop properly, and so cannot protect the body. The outlook for these babies is, to say the least, very bleak unless they are lucky enough to get a bone-marrow transplant – in itself a risky, expensive treatment.

Gene therapy for this 'boy in the bubble' disease was first applied in France, in 2002, when eleven babies were treated. Within a few years, four of these boys developed leukaemia and one died. It appears that the virus used to transport the corrected gene into cells had inserted itself randomly in different parts of the patients' DNA – so that in some cases many copies of the gene had been inserted into the babies' genome. And in at least one case the insertion occurred in a cancer-controlling gene, disrupting its function. This may have increased that child's risk of developing leukaemia.

This cautionary outcome did not stop doctors at Great Ormond Street Children's Hospital in London from trying a similar treatment. Ten children with X-SCID have been treated so far there, and all appeared to have benefited: one boy did develop leukaemia, but it was treated successfully.

The difference between attempts to undertake gene therapy on infants with X-SCID and Jesse Gelsinger's case is stark. It seems reasonable to have tried gene therapy for these children, because without it they were likely to die. While it is always problematic for a parent to give consent for any experimental treatment on a child, all these parents knew of the potential risk and, after being provided with full information, decided in what they believed was the best interests of their children. Consequently, the decision to allow this trial to go ahead seems ethical even though it involved treatment with a potentially dangerous viral vector.

Scientists are working on the problems that may arise from 'Trojan horse' viruses. One possible 'vehicle' is called AAV, a cold/flu/sore-throat virus that is carried by some 80 per cent of the world's population with very few symptoms. Another technique avoids viruses altogether, instead placing the genes inside packages of fatty lipid molecules and delivering them to specific sites by triggering their release with ultrasound. A third, and equally experimental, method involves using a part of the 'junk DNA': its role is still only partially understood, but we know that it has the ability to home in on specific parts of a cell's gene sequence.

All or none of these techniques may deliver a solution. It is likely that someone will come up with an answer, of course – but only if public and government attitudes to gene therapy remain fair and open-minded. The responsibility for that rests mainly with the scientists. Proceeding without proper caution simply to be first to make a 'breakthrough' may prevent improved treatments.

DNA SCREENING, DNA EVIDENCE: THE PROMISE AND THE THREAT

When the human genome was finally sequenced, British scientists were understandably proud that much work had been done on this side of the Atlantic. That remarkable institution the Wellcome Trust had financed the British effort, and the announcement of the sequencing was greeted with acclaim by its director, Mike Dexter, who enthused that this advance was 'more important than the invention of the wheel'.[22] He went on to say, 'I can well imagine technology making the wheel obsolete. But this code is the essence of mankind, and as long as humans exist this code is going to be important and will be used.' The genome may be more important than the wheel – but, nearly ten years after this announcement, it does not seem to have rotated that much. John Sulston, one of the scientists who did the work,[23] boasted perhaps unwisely that 'People may live for a thousand years, and we may arrive at a total understanding of not only human beings but all of life.' One might be pardoned for observing that it should be possible to get a great understanding of human beings by reading any one of Shakespeare's plays, or by glancing at the poetry of Andrew Marvell.

Of course, this knowledge is important; but it is difficult at this stage to argue that the sequencing has benefited many of the organisms on the planet. Meanwhile, enthusiastic hyperbole from scientists may raise ludicrous expectations. So it is hardly a surprise that the US President who had supported so much of the genome project, Bill Clinton, should have responded with the optimistic speculation: 'It is now conceivable that our children will know the term cancer only as a constellation of stars.' Perhaps this statement tells us even more about human nature than the information derived from sequencing our genome.

The letters of the DNA alphabet actually tell us rather little about ourselves. Yes, individual variations in the sequence may give each of us a unique 'fingerprint' – but this tells other individuals little more about me than does my signature at the back of my passport. True, some variations of spelling may give more than a hint about what diseases I might have more than average likelihood of contracting, and occasionally – in the case of a rare single-gene disorder – might fairly accurately predict what symptoms I am likely to have from one of those nasty diseases. But DNA sequencing in general is not a good way to predict the likely health of an individual because so many other confounding factors come into play. I may well have genetic variations which are associated with high blood pressure, heart disease or a greater than average risk of cancer in one of my organs, but that does not mean I will develop those illnesses, nor is it clear how a change in so-called 'lifestyle' will mitigate that risk.

The reality of genetic testing should produce more alarm. In a recent TV programme, three celebrities underwent genetic screening to discover their chances of being killed by various conditions, such as heart disease, stroke or Alzheimer's. The programme seemed for the most part to be an extended advert for the London-based company providing the screening service – and while in this instance a medically qualified doctor explained the nature of the tests and their findings quite clearly, one was left wondering where this trend might take us. One British company now offers a postal screening service for just £500, in return for which it will provide the low-down on a client's likelihood of developing various ailments from Crohn's disease to multiple sclerosis and breast cancer.

These tests can only ever be of limited value, given the complex interplay of different genes and environmental factors. The picture is likely to be changing continually, along with our state of knowledge – and that knowledge cannot be conveyed effectively using the internet or printed reports. When I have dealt with couples wanting screening for genetic defects, they have undergone a thorough process of counselling. They have been given detailed advice and

explanation by trained professionals of the implications of the information, and time to consider what they have heard. This process
will need to be just as thorough (if not more so) when screening for
more common problems with genetic causes is offered. What does it
mean if I am told I am 'four times more likely' than the average person to contract Alzheimer's disease? Could I take any preventative
steps? Should I have children? Tell my spouse, or my boss, or my
insurance company? Without better information, such postal tests
could be a minefield, delivering fear and uncertainty where they
promise to bring peace of mind.

I think the optimistic pronouncements of some members of my
profession have encouraged companies offering DNA screening
services. These companies seem to offer a kind of certainty by
exploiting the fears of mostly perfectly healthy people who are
unlikely to gain any real benefit from having their genome (actually
just a tiny fraction of it) analysed. One such company advertises
'DNA screening – Scientifically proven, unquestionable facts!' 'Our
objective with the Gene DNA Screening Analysis service', it goes on,
'is to offer clients a total lifestyle plan for health and illness
prevention, optimized to their genetic profile.' Effectively, the advert
implies that DNA sequencing is accurately predictive of good or bad
health – but this is just not true, except for the rare individuals
who carry a dominant gene defect like Huntington's chorea
which will not start to show itself until middle age. And no known
'lifestyle change' will alter that prognosis. Certain individuals may
have a greater chance of getting breast cancer or heart disease
because of their genetic background, but the controlled evidence
that a lifestyle change will make a major difference has not yet been
established.

The advert continues by stating that DNA is the genetic
instruction manual of life and that, as biologically unique individuals, we react to food and the environment in unique ways. To
achieve optimum health and performance, we need 'to synchronize
our diet and lifestyle according to our individual genetic instruction
manual'. After receiving a DNA sample, this company promises a
personalized report giving the customer's 'genetic score' (whatever
that is) and revealing 'the following important basic life elements:
Heart health, B vitamin utilization, Detoxification, Antioxidants,

Bone health, Inflammation, Insulin resistance'. It will also set out 'a personalized set of diet, nutrient, vitamin and mineral requirements tailored to your genetic profile and directed to attainment of optimum health and defense against illness'. The advert concludes: 'Our nutritionists, where appropriate, will recommend a personalized nutraceutical regimen proven to enhance health and well being.' I could not find the word 'nutraceutical' in my dictionary. As far as I am aware there is not the slightest research evidence that DNA screening of this kind – which is not cheap – is of any real health value.

It is, of course, true that genetic risk prediction may get better as we learn more about different variations in DNA and their association with the more common diseases.[24] Some estimates suggest that in three to five years from now these commercial tests, provided they are sufficiently powerful to analyse a significant amount of each person's individual DNA, may be more useful. But as Jennifer Couzin points out in a recent article, even for one of the common diseases with the strongest genetic links – breast cancer – these tests may identify a risk increased by 8–10 per cent at best.[25] The real risk then may be that of having unnecessary screening and tests, undergoing much anxiety – and even having an operation on the breast as the result of a false positive, to remove a tumour which turns out to be entirely benign when a pathologist examines it after the breast has been mutilated.

This enthusiasm, partly misinformed, for genetic science is spilling over into other aspects of our lives. One procedure which is somewhat worrying is genetic fingerprinting, now commonplace in examination of material taken from suspected crime scenes. Very tiny traces of DNA – perhaps a fingerprint on a glass, or a smear of seminal fluid on underclothes – may be used to provide 'incontrovertible' evidence of the guilt of an individual whose samples seem to provide a 'perfect' match. It must be of some concern that law courts may possibly take too seriously the evidence of 'expert' scientists on the probabilities of a match. For example, there must be an element of doubt with trace amounts of DNA which may have 'denatured' (decomposed slightly) when they are examined long after an alleged crime has been committed. Moreover, contamination of really tiny traces of DNA is a very significant problem when testing for a particular 'informative' sequence. To analyse the DNA successfully, a considerable amount of the molecule is needed. So the number of copies available needs to be artificially amplified (effectively, repeatedly photocopied) – but this risks also amplifying information from DNA which is contaminating the sample. So properly judging the credibility of such evidence greatly depends on the scientific and mathematical literacy of those in court.

It should also be of growing concern that information about stored DNA samples may be increasingly used. This is an invasion of a citizen's privacy. For some time now the police have wished to keep records of the DNA of all crime suspects – including those subsequently found by the courts to be entirely innocent and not convicted of any crime. Since April 2004, the police in England and Wales have been able to take DNA samples without consent from anyone arrested on suspicion of any recordable offence, irrespective of whether or not they are charged or cautioned. And at present the Forensic Science Service analyses one sample and retains another until that individual would be 100.

Important issues of civil liberties and equality are at stake here. For example, many of the 3 million samples currently stored are from members of Britain's black population – indeed, it is said that about one-third of all black citizens are on the record. The fallibility of any information held, the risk of misuse and abuse, and lack of security in its keeping are all causes for anxiety. If the British Forensic Service were privatized (as has been seriously suggested) this information, misused, might yield a useful profit. This is not scaremongering: instances abound of 'public' data being lost. There has to be solid justification to hold such data.

EPIGENETICS: THE EFFECT OF THE ENVIRONMENT ON INHERITANCE

The picture of inheritance is turning out to be a good deal more complex than that depicted by variations in the spelling of the DNA alone. As Marcus Pembrey, a distinguished paediatric geneticist from University College London, has said: 'What we do know is that completion of the human genome project is just a start. It has defined the notes of the piano. We must now think in terms of the music.'[26]

During the latter part of 1807 and the beginning of 1808 the Swedish government, under their feeble king Gustav IV, had ignored all reports of massive manœuvres on the Russian border. Alarm bells had been ringing at the Swedish embassy in St Petersburg, where the ambassador was all too well aware of Russian ambitions regarding Finland, which was then part of the Swedish kingdom. But the Swedish ambassador's messages back to Stockholm were ignored and, without a formal declaration of war, Russian troops crossed into Finland on 21 February 1808. The future of the 700-year-old union between Sweden and Finland hung in the balance, and the pious and incompetent King Gustav IV was hardly a suitable leader to defend it; sure enough, the resulting war, fought as a sideshow while the rest of Europe was engaged with Napoleon, was the end of the Swedish–Finnish union.

Tsar Alexander I had designs on a massive chunk of land. He wanted to extend Russia's borders as far as the River Kalix, to the north of the Gulf of Bothnia and just west of what is now the border between Sweden and Finland. Over the next year, various pitched battles were fought around the Baltic Sea. Finland was overrun and a *coup d'état* in Stockholm resulted in the removal of the Swedish king by his own people.

The precise details of this relatively obscure war are somewhat irrelevant here, but its consequences for the as yet unborn local population turned out to be quite profound. Some remarkable detective work recently done by the Swedish medical scientist Lars Bygren shows that what happened in 1809 almost certainly resulted in silent changes in the population at the time. Though the health of many local people living then was not particularly affected, these silent changes were inherited by some of their children and grandchildren. Amazingly, two generations later, some of them faced a shortened lifespan.

In 1809 the little village of Överkalix was, like most of Sweden, a poor community almost entirely dependent on its agriculture. Various years had seen crop failures, but although no battles were fought there, 1809 was particularly devastating because, in this year of indifferent harvests, marauding soldiers involved in the conflict requisitioned all the available grain.

Today Överkalix is still a small place with a population of around 2,000. Just a few kilometres from the Arctic Circle to the north and from the modern Finnish border to the west, it sits beneath its beautiful mountain, Brännaberget, and next to the salmon-filled Kalix River which flows through the pine forests to the north of the Baltic. Tourists come to admire the scenery, to bask in the midnight sun at the height of summer and to ski in winter, and to marvel at the striking displays of the Aurora Borealis, the Northern Lights.

Överkalix is now served by the E10, a well-surfaced main road, but in earlier times it was relatively cut off. For Lars Bygren, its relative isolation and the quality of its well-preserved parish records, going back well over 200 years, made it ideal. He decided to trawl the Överkalix parish for children born before 1905, and he identified their parents and their grandparents who had lived around the time of the Napoleonic and Russian wars and during the years that followed.[27] He also analysed historical data to find out in which years there was likely to have been a shortage or an abundance of food in the area, examining official records of harvests, annual food and grain prices, the records from local community meetings and estimates made by a famous nineteenth-century statistician, Johann Hellstenius. Total crop failure was common – food was very short not only in 1809, but also in 1800, 1812, 1821, 1829 and during 1831–6. But there were also particularly good years when food was unusually abundant – 1799, 1801, 1813–15, 1822, 1825–6, 1828, 1841, 1844 and 1846.

It has been surmised for centuries that dietary conditions in early childhood may affect health in later life. The Ulster radical Mary McCracken, Secretary of the Ladies' Committee of the Belfast Poorhouse, wrote about the importance for future life of proper nourishment during infancy in 1830. (Incidentally, she must have been a rather formidable woman – at the age of eighty-eight she could still be seen in Belfast docks handing out anti-slavery leaflets to those boarding ships bound for the United States, where slavery was still practised.) Back in the seventeenth century, Francis Bacon offered his thoughts about the effect of a woman's diet in pregnancy:

It hath been observed that the diet of women with child doth work much upon the infant; as if the mother eat quinces much, and coriander-seed, it will make the child ingenious; and on the contrary side, if the mother eat (much) onions or beans, or such vaporous food; or drink wine or strong drink immoderately; or fast much; or be given to much musing; (all which send or draw vapours to the head;) it endangereth the child to become lunatic, or of imperfect memory: and I make the same judgment of tobacco often taken by the mother.[28]

Bygren was interested to know what kind of environment the parents and grandparents had experienced during childhood and just before puberty, the so-called slow-growth period of a child's life. When he focused on the menfolk, he found that grandfathers who had lived through very bad harvests during infancy and early childhood were very slightly more likely to die at an earlier age – on average, 69.5 years compared with 69.9 years. But though grandfathers whose slow-growth period had occurred in a time of famine were likely to live very slightly shorter lives, they produced children who were more likely to live longer than average. And remarkably, the paternal grandsons (the sons of the sons) of those men who were aged between eight and twelve during times of plentiful harvests were four times more likely to die early from heart disease or diabetes.[29]

As Dr Bygren points out, there are problems with a study of this kind. The sample size is small: only 276 grandchildren were available for follow-up. Also, the authors had little knowledge about what happened during the adult life of the ancestors or even their grandchildren. The estimates of food availability are just that – estimates. And in a small village cut off from the outside world like Överkalix, intermarriage is likely to be more common, which could increase (or even decrease) the prevalence of inherited diseases like diabetes and heart disease. But it is striking that the pattern of inheritance seems to be so skewed towards the male line. Female grandchildren showed no differences in their susceptibility to disease and a plentiful food supply available to their mothers and grandmothers appeared to have little or no influence on longevity in this particular study.

This suggests some factor inherited via the Y-chromosome – the only part of the DNA 'handed' down from father to son. And why should being eight to twelve years old be significant? One possibility is that at this stage of life pre-pubertal boys are starting to make the precursor cells which will become their sperm. With this new testicular activity, genes which are on the Y-chromosome possibly may be programmed by the outside environment. So it is likely that although the 'printing' of the DNA is unchanged, the way the DNA actually functions may be modified.

These ideas are confirmed by animal research. Jennifer Cropley and colleagues in Sydney, Australia, were able to reproduce a rather similar mechanism in mice,[30] and by altering their diet in pregnancy could modify the colour of the yellow fur of their offspring – a trait inherited by the next generation.

Changes in susceptibility to disease are one thing, but could the parental environment change behaviour or possibly brain function? There is some evidence that the way a human mother treats her child influences the way that child behaves with its own children. Around 70 per cent of abusive parents were themselves abused when babies, and about one-third of abused babies will become abusers. Previous good maternal bonding is an indication of how a new mother will behave to her own newborn infant. Of course, the big question is whether this is just learned behaviour. Frances Champagne, from Columbia University in New York, has studied rats and mice and compared mothers that groom their young attentively and those that don't.[31] Some rodents, it seems, are not particularly caring. So she swapped litters between high-grooming and low-grooming mothers. When low-grooming rats or mice act as foster mothers, the babies tend to inherit the same trait. In a series of elegant studies she has shown that although the DNA is unchanged, the way that some genes work is altered and that this effect is inherited in a subsequent generation. The poorly nursed rodent babies show alteration of the activity of a gene in the brain which is sensitive to oestrogen. These animals have altered activity of steroid hormones in the brain, have a different reaction to stress, have less inclination to explore their surroundings, are more fearful, and have impairment of memory trying to find their way out of a maze; they also have a different sexual response after they reach sexual maturity. All these traits tend to be passed on to the next generation. Given that all mammals have very similar physiology, one wonders whether these mechanisms play a part in human development. Is what Frances Champagne has observed in mice a model for human experience? Is what we are seeing here just the tip of an iceberg?

And it may be a large iceberg. Throughout this book we see how humans are particularly vulnerable to changes in the weather – and especially when these events are sudden. On 5 January 1998 a massive ice storm struck eastern Ontario, raging through to sudden Quebec and Nova Scotia. Four inches of freezing rain produced sufficient ice to fell millions of trees and bring about 130 electricity pylons and 30,000 poles carrying electric cables crashing to the ground. The extensive damage to electrical installations resulted in power cuts for over 4 million people, in some cases lasting several weeks. The cold was so intense that over 30 people died and around 1,000 were injured; some 600,000 abandoned their homes. Reparing the damage cost about $5.4 billion.

Professor Michael Meaney and his colleagues at McGill University in Montreal had the opportunity to study many families who had been caught without power during this disaster.[32] Some pregnant women were profoundly stressed by these events and the researchers found an effect on a baby's growth and later development if the mother was between 14 and 22 weeks pregnant at the time. What is interesting is that, when they studied the digits of these children after birth, the researchers found that the babies whose mothers felt they had experienced the most stress had abnormal, asymmetric fingerprints. And this seems to be related to a rise in the stress hormone cortisol in the mothers' blood.

Abnormal fingerprints do not matter much unless one is going to be a burglar. But these minor physical abnormalities are also quite strongly associated with abnormal brain development. For example, schizophrenia is more common in people with such curious fingerprints. And in some earlier research Professor Meaney's colleagues found that such stress in pregnancy is frequently associated with intellectual impairment of children after birth and their ability to use language – the very tool that is so important for human creativity.[33]

SYNTHETIC BIOLOGY

The idea that humans might create life goes back to earliest antiquity. One of the oldest ideas is that of the golem – a humanoid creature made artificially using magic, and brought to life by invoking holy names. Around 1,500 years ago the Talmud told how Rabbah created a man whom he sent to help Rabbi Zera.[34] Rabbi Zera spoke to this golem but he did not reply – at which point Zera says to the golem: 'Thou art a creature of the magicians; return to dust.' Elsewhere in the Talmud, a golem made of clay is brought to life with an amulet – which is usually stuck on to his forehead – on which is written 'Emet' ('Truth'). When the first letter 'Aleph' or 'E' is removed, one is left with the Hebrew word 'Met' ('Death') – so one way of dealing with a golem is to scramble for his forehead (somewhat awkward, because it is rather high up given his monstrous size) and pluck off the letter Aleph.

The most famous story (which, of course, predates Mary Shelley's *Frankenstein* by over a century) is of the golem made by Rabbi Loew ben Bezalel of Prague. A visit to the medieval graveyard behind Prague's Altneuschul (the old synagogue) still causes the hairs to stand up on the back of one's neck, particularly if it is made after sunset in the evening gloom. The graveyard is dotted with thousands of gravestones placed very close to each other, apparently entirely at random and slanting crazily; when I visited it some years ago in the diminishing twilight, I hesitated to look round for fear of seeing the monstrous golem stumbling after me.

The German film director Paul Wegener made a famous Expressionist silent movie about the golem in 1920.[35] In it, Rabbi Loew makes a huge figure out of clay because the Jewish people in the ghetto are threatened by the Emperor Luhois. Initially, the golem is passive, a slave-like creature chopping wood and drawing water. But when the rabbi's beautiful daughter is seduced by the emperor's foppish messenger, Count Florian, her Jewish boyfriend commands the golem to kill Florian. Once the golem has thrown Florian's lifeless body off the highest roof of the ghetto, he runs amuck, destroying everything in his path and setting fire to the rabbi's house. Finally, the golem breaks down the ghetto's massive doors to escape into the outside world, where he sees a flimsily dressed little three-year-old girl playing innocently in a field. The resulting frame is one of the great classic shots of cinema. The golem gently bends down and the little blonde girl offers him the apple she carries. The golem picks up the child and gently holds her in his arms – whereupon the little innocent grabs the magic amulet containing the word 'Emet' and pulls it off the golem – and the golem keels over, dead.

The golem legends show our preoccupation with life created artificially. Paul Wegener's film is also interesting in depicting three key stages in the invention of the golem. First comes the wild enthusiasm at the appearance of a new, scientifically produced device which would help humans. Then the unexpected downside – wanton destruction of human life, with the golem being manipulated sometimes as a result of the evil inclination of some humans. And finally – and rather too late, of course – come the ethical implications and a better understanding of the effect on our humanity.

Until recently the idea of the artificial creation of life has been merely the stuff of fiction. But at some point this year, 2010, it is very likely that somewhere in the world a scientist will announce that he or she has finally created life in a test-tube. In October 2007 a team of scientists headed by DNA researcher Craig Venter and Nobel laureate Hamilton Smith from San Diego, California, stated that they planned to create the first artificial life form in history by injecting a synthetic chromosome into the bacterium *Mycoplasma genitalium*. This, they announced, would result in an artificial species which they intended to call *Mycobacterium laboratorium*. And in 2008 Craig Venter's team reported they had made their synthetic bacterium in their laboratories.[36]

The naturally occurring bug *Mycoplasma genitalium* is one of the most common sexually transmitted bacteria – not quite as common as chlamydia but more prevalent than gonorrhoea. It is said to cause inflammation of the uterus and fallopian tubes, and possibly male infertility too. What Venter's team had done was to recreate the entire DNA alphabet from the Mycoplasma organism by sequencing in their laboratory. Mycoplasma is an organism which is not circumscribed by its own cell wall, a slight advantage in this experiment. Because the genome of any organism is lengthy and unmanageable, they reproduced it (in total 589,000 base pairs in length) in twenty-five chunks of DNA and then 'glued' them together inside an empty yeast cell. The yeast's cell wall was dissolved in a culture dish and then allowed to regenerate around the introduced foreign DNA. Subsequent analysis confirmed that the entire new Mycoplasma genome was present. But one of the definitions of 'life' must be that a living organism can reproduce – can replicate itself. And the key problem with Venter's new bug is that his new 'organism' doesn't actually work. It shows no evidence of life, does not grow or divide, and will not infect any tissue. So the laboratory personnel working with it will not even have to worry about whether to wear a condom.

Set against the popular assertion that DNA is 'the blueprint for life', this surely puts things into perspective. Indeed, Craig Venter's experiment may seem totally pointless – but this is by no means the case. Such experiments elucidate some of the steps needed to assemble the DNA 'building blocks' in place, and his hope is that an artificial bacterium of this kind might be highly useful. Certain bacteria use carbon dioxide; if rapidly dividing organisms were created, it could be possible to use such cultures to absorb high carbon emissions from industrial processes: a highly useful prospect if we are to mitigate the effect of greenhouse gases.

Tailor-made bacteria using DNA technology could be helpful in other ways. Humans used micro-organisms to manufacture a desirable product – alcohol – before recorded history. Fermentation, which breaks sugars down to ethyl alcohol and carbon dioxide, has mainly been done using yeasts, but many bacteria have a similar process. Lactic acid may also be produced if different bacteria or fungi are used, and hydrogen production may result from the fermentation of other compounds – for example, butyric acid.

Biofuels might be manufactured like this. James Liao and his team at UCLA have developed a method for producing them by genetically modifying *Escherichia coli* bacteria. *E. coli* is a very common bacterium found in the gut and other parts of the body. Much of the time it lives happily inside its host without causing any disease process. The modified organism seems to have the potential to be an efficient biofuel synthesizer and its use could enable production of biofuels on a grand scale.[37] These bacteria do not produce ethanol, the simple molecule and the common form of alcohol with which we entertain ourselves. While this can be used as a fuel, it is not nearly as efficient as gasoline and must be mixed with gas for use in automobiles. As it tends to absorb water, it is corrosive and can damage engines, and is difficult to store in large quantities. But after genetic modification these bacteria can make bigger alcohol molecules such as isobutanol, with a much higher octane rating. Because growing enough plants to make biodiesel demands considerable agricultural resources and is not always environmentally desirable, the use of modified bacteria looks promising.

'The ability to make these branched-chain higher alcohols so efficiently is surprising,' says Dr Liao. 'Unlike ethanol, organisms are not used to producing these unusual alcohols, and the fact that they can be made by *E. coli* is even more surprising, since *E. coli* is not a promising host to tolerate alcohols. These results mean that these unusual alcohols in fact can be manufactured as efficiently as what evolved in nature for ethanol.' It remains to be seen whether a tailor-made artificial bacterium – such as that proposed by Craig Venter – or one which has been genetically modified would provide a more efficient or safer method of producing biodiesel.

Safety is, of course, a crucial issue. Obviously, any organism manufactured in a laboratory must be carefully designed so that it could not become a pathogen, capable of infecting and harming other living organisms. But there is the concern that this may be difficult to guarantee – or that the organism might change its DNA by mutation if it is allowed to replicate continually. And given that this would be a bacterium that no immune system had 'seen' before, its effects could be devastating as resistance to it might be very low.

Many people worry that the greatest risk is not that of an accidental change in any artificial bacterium, but of synthetic biology methods being used to build a human pathogen deliberately. Many governments – including those in democracies like our own – experiment with germ warfare, if only to find ways of protecting the military or the general population against the use of such weapons by a potential enemy. It would be a small step to manufacture such organisms for purely offensive purposes. And a number of commentators have pointed out that such organisms would be a terrifying threat in the hands of bioterrorists. But the Nobel Prize winning scientist David Baltimore suggests that concern about the production of an 'artificial' form of the deadly virus Ebola, for example, as a weapon largely ignores the real danger – which is from organisms that already exist. He argues that the idea of synthesizing something worse, for example taking bits of Ebola and other viruses to create something more deadly, underestimates how hard it would be for a man-made virus to survive in the natural world. In any case, because it kills its host so quickly it never spreads widely – it is actually too lethal to be in the population long enough to be propagated.

But anxieties will remain. In the case of Liao's biofuel-producing organism, it is understandable that the university has patented the process. The pressure on all academics to find ways of exploiting technology is considerable. This is not necessarily motivated by academic greed; governments see investment in science as a crucial way of driving the economy forward, and universities are invariably short of money. In the case of Liao's work, UCLA has licensed the technology to Gevo Inc., a company in Pasadena, California, which explores the manufacture of biofuels. So this technology, like all technologies, will be driven increasingly by commercial interests. This may be the best way of establishing it as a useful, viable technology. But it is not unreasonable to have worries that technology like this could just be advanced too quickly, without adequate measures to protect human and environmental safety. And if synthetic biology were used to manufacture pathogenic organisms, the cause for concern would go beyond straight commercial interests to the potential purchase and/or misuse of such organisms by a government or other organization.

The development of synthetic biology could be enormously valuable. Possibly the most successful project so far has been Jay Keasling's at the University of Berkeley in California. He has produced an anti-malarial drug, artemisinin, using modified yeast cells.[38] The WHO calculates that around 500 million people living in the tropics and subtropics become infected with malaria each year, and possibly 3 million victims die – many of them children. Artemisinin and its derivatives are very effective in killing the Plasmodium parasite which causes malaria, but it is costly to produce. Until now, artemisinin has been extracted from the wormwood tree or manufactured by chemical synthesis. But wormwood is sensitive to climate changes and both methods of production are expensive. Jay Keasling believes that he could make the drug for one-tenth of its present cost, making it more affordable in many poorer regions, such as Africa, where malaria is endemic. Another application which looks interesting but which so far has not yet been used is the production of extremely strong threads of spider silk from modified bacteria. Chris Voigt from the California Institute of Technology has modified salmonella bacteria (the microbes responsible for causing typhoid), producing a thread that could be woven to make new blood vessels for human transplantation. Because the thread is strong, is capable of stretching like naturally occurring arteries and does not cause any adverse immune reaction in the human body, Voigt believes that his tubing could be highly valuable for replacements in vascular surgery.

The range of potential applications for this new area of science seems very wide. At Imperial College London, my colleagues are researching synthetic biological devices as detectors for bladder infections and for use in biologically constructed computers. Various researchers are exploring the use of the technology for making new tissues, for detecting poisons in drinking water, for food production, and for synthesizing useful compounds such as starch cheaply. In the long term, synthetic biology may help with carbon capture and the manufacture of new energy sources, and should be an important way of producing many new drugs with few side-effects.

EUGENICS

'These officers will take the children of the better Guardians to a nursery . . . the children of the inferior Guardians and any defective offspring of others will be quietly got rid of.'[39] So affirms Plato in his fourth-century BCE view of the ideal society. And 400 years later, at the time of Nero, Seneca asserts the eugenic justification for infanticide once more: 'We drown the weakling and the monstrosity. It is not passion, but reason, to separate the useless from the fit.'[40]

Almost every human society – except the ancient Hebrews – has practised some form of eugenics at a point during its history. The word 'eugenics' itself was coined in 1883 by an Englishman, Francis Galton, a cousin of Charles Darwin. Galton, who was born in the same year as Gregor Mendel, came from a well-heeled Birmingham family of gun-makers and bankers. His word 'eugenics' (from the Greek, meaning 'good in birth') meant a way of improving society by strengthening its stock.

Francis Galton is particularly relevant to this book. He, above all, was utterly confident about science – seeing it as 'progress'. He first published some ideas about better breeding in 1865 in *MacMillan's Magazine*, one of the most significant of the British intellectual periodicals flourishing in the second half of the nineteenth century. Galton studied the pedigrees of the 'great and the good' – for example, distinguished poets, soldiers, statesmen and scientists – in various biographical encyclopaedias, and came to the conclusion that 'families of reputation' were much more likely than others to produce offspring with talent and ability. It does not seem to have occurred to him that privilege or wealth might be relevant to the achievement he was assessing; he was convinced that we could produce 'a highly gifted race of men by judicious marriages during consecutive generations'. Eventually, he proposed that the state rank people by ability and authorize those of the higher classes to have more children. His social aspirations plainly influenced his views on genetics. 'Certainly we led a life that many in our social rank might envy,' Galton mused after his marriage. 'Among our friends were not a few notable persons.' It has often been suggested that his own infertile marriage may also have played its part in helping to formulate his opinions about heredity.

Galton was a polymath of high intellectual gifts. He could read fluently at two years old, and enjoyed Latin and Greek literature at five; by the age of seven he was reading Shakespeare for pleasure. He was also a good mathematician, being profoundly interested in statistical method and the measurement of probability. He considered that the complex conflict about the importance of nature against nurture could be best unravelled by comparisons of twins, and hoped to evaluate whether twins who were similar at birth diverged when reared in dissimilar environments, and whether twins dissimilar at birth converged when reared in similar environments.

In this he anticipated the modern field of behavioural genetics and the work of Tom Bouchard in Minnesota, which relies strongly on twin studies.[41]

In the United States the biologist Charles Davenport, a driven man from a repressed background and oppressive upbringing, was hugely impressed by Galton and his disciples. In 1898 he persuaded the Carnegie Institution of Washington to set up a laboratory for evolutionary study at Cold Spring Harbor on Long Island, an institution which has since become one of the world's leading research centres for the study of molecular genetics. At that time Davenport considered that the protoplasm of the United States was being threatened by unrestricted immigration and advocated screening all immigrants to assess them for 'imbecile, epileptic, insane, criminalistic, alcoholic and sexually immoral tendencies', all of which he believed were genetically determined.[42] He had strong views on sexuality; prostitutes (or wayward girls, as he called them) were motivated not, he thought, by economic necessity but by an innate eroticism determined by a Mendelian dominant. He was not totally fascist as he did not advocate sterilization of the unfit – unless they were to be castrated as well. He felt that if the state could take a person's life then surely it could deny an individual the lesser right of reproduction; he thus supported the notion of state-enforced sterilization.

One should not think of Davenport, or for that matter Galton, as odd figures on the fringe of either science or society. They represented a growing movement of considerable importance. By 1911, six American states had enacted sterilization laws. By 1924, marriage between blacks and whites (miscegenation) was a criminal offence in Virginia – and this law, incidentally, was not overturned until 1967, when the United States Supreme Court ruled that anti-miscegenation laws were unconstitutional, compelling sixteen states to abandon similar laws. Nor was the pursuit of 'racial hygiene' a solely North American phenomenon. It became pretty well international after the end of the nineteenth century. The movement took off in Scandinavia, Switzerland, Germany, Poland, Russia, France and Italy, and by 1920 it was important in Latin America and Japan. Its English supporters included Rufus Isaacs, 1st Marquess of Reading and Lord Chief Justice of England; Winston Churchill; the

Bishop of Ripon, Thomas Strong; George Bernard Shaw; Beatrice and Sidney Webb; William Beveridge; H. G. Wells; Ottoline Morrell; J. B. S. Haldane; Julian Huxley; and Havelock Ellis. In the United States, Alexander Graham Bell, Charles Eliot (the President of Harvard University), John D. Rockefeller, Theodore Roosevelt, Margaret Sanger and George Eastman all gave strong support to the eugenic movement.

In Canada, as in several other countries, there was a strong accent on intelligence in the eugenics movement. The Canadian Sexual Sterilization Act, which was passed in 1928, applied to mentally deficient people, identified as such by means of IQ testing using the Stanford–Binet test. But this required considerable linguistic ability, which was a difficulty for those immigrants with limited English. As a result – leaving aside the much larger question of whether the 'mentally deficient' should be allowed to reproduce or not – some of those sterilized under the Act were unfairly classified. As the economy grew more depressed in the 1930s, 'defective breeding' became a key political issue. Many individuals blamed people with lesser intelligence for their predicament, calling them subhuman. The Act was not repealed until 1972.

In Australia at the end of the nineteenth century, aboriginal children were separated from their natural parents. Aboriginals were thought to be physically and mentally weak and inferior to citizens coming from Europe. Legislation passed early in the twentieth century in all Australian states gave white 'protectors' guardianship over many aboriginal children up to the age of sixteen, or twenty-one if they were of mixed parentage. Policemen and social workers had the power to locate children of mixed descent, take them from their parents and transfer them into institutions. The film *Rabbit-Proof Fence*, made in 2002, depicts the horrifying consequences of this inhumanity.

The Swedish Sterilization Act of 1934 provided for the 'voluntary' sterilization of 'the deviant' and the mentally ill. It was primarily targeted at women, and, in practice, 'voluntary' meant no such thing: there was massive coercion in most cases. This law was supported by all political parties as well as the Lutheran Church and my Swedish medical colleagues. Mental incapacity remained an indication for sterilization in Sweden without proper consent until

1975, by which time over 62,000 people there had been under the knife – more than in any other European state except Nazi Germany. Remarkably, forced sterilization was actually more common after the Second World War than it was before it. It was only in 1996 that the Swedish government agreed, reluctantly, to pay modest compensation to the victims.

Successive Japanese governments implemented eugenic policies limiting the birth of children to parents with 'inferior' traits. The Race Eugenic Law, which continued in force after the Second World War, meant that sterilization could be forced on criminals 'with genetic predisposition to commit crime', people with colour blindness, albinism or ichthyosis, schizophrenics and depressed patients. The Leprosy Prevention Law (which was not repealed until 1996) allowed for abortion without consent, and sterilizations of people with leprosy were common. Marriage between Japanese women and Korean men was actively discouraged. Koreans had been recruited following the wars of 1910, and some stayed to live in Japan. But as late as 1942, a report stated: 'Korean labourers brought to Japan, where they have established permanent residency, are of the lower classes and therefore of inferior constitution ... By fathering children with Japanese women, these men could lower the calibre of our women.' The notorious Japanese prostitution measures were also influenced by eugenic thinking: Japanese soldiers could have state-supported sex to protect the purity of the Japanese race. Laws to maintain this position were actually enacted in 1945, and even now the Japanese government is reluctant to offer apologies or compensation to the (mostly Chinese) women involved.

In America, a particularly influential and repellent character was Dr Harry Laughlin. Born in Iowa to a religious family, he worked as a high-school teacher before his interest turned to breeding. He contacted Charles Davenport at Cold Spring Harbor, and in 1910 Davenport appointed him superintendent of his new Eugenics Record Office there. In 1917 he gained a doctorate at Princeton in cytology. Subsequently, Laughlin testified to Congress in support of the conservative Johnson–Reed Immigration Act of 1924, alleging that immigrants from eastern and southern Europe were very likely to be insane. He impressed the Secretary of State sufficiently to be appointed eugenics agent to the Committee on Immigration and Naturalization.

Laughlin wanted compulsory sterilization legislation to be enacted throughout the United States: in his view, most states (with the exception of California) did not appear to be employing sterilization with enough enthusiasm, and did not have well-drafted legislation: so he drew up his own 'model law',[43] which advocated the establishment of a State Eugenicist to judge who should be sterilized. His criteria for compulsory sterilization were:

(1) Feeble-minded; (2) Insane (including the psychopathic); (3) Criminalistic (including the delinquent and wayward); (4) Epileptic; (5) Inebriate (including drug habitues); (6) Diseased (including the tuberculous, the syphilitic, the leprous, and others with chronic infections and legally segregable diseases); (7) Blind (including those with seriously impaired vision); (8) Deaf (including those with seriously impaired hearing); (9) Deformed (including the crippled); and (10) Dependent (including orphans, ne'er-do-wells, the homeless, tramps and paupers.

In this very brief description of the eugenic movement, I have not discussed Nazi Germany; its crimes against humanity are too well known to need repeating. But in 1933 the Reichstag passed its Law for the Prevention of Hereditarily Diseased Offspring, based partly on (and excused by) Laughlin's work; and that, over time, the German government sterilized some 350,000 persons. One document, which sends a shiver down the spine, is the archived photo of the honorary doctorate that Laughlin was awarded by the University of Heidelberg in 1936 for his 'scientific' work in promoting racial cleansing. There is a strange irony about the story of Harry Laughlin: towards the end of his life he developed epilepsy, a condition that would have made him a candidate for sterilization under his own law. After forty years of marriage he died in Missouri in 1943, without leaving any genetic heritage as he was childless.

In the introduction to this book I point out that, with the invention of the hand-axe, humans changed their own evolution, surely the only species to have done so. So there are other ironies about Harry Laughlin's work and the outstanding research that continues at Cold Spring Harbor that contributes to our knowledge of how genes work.

With such knowledge, humanity now stands on the brink of another opportunity to manipulate its own evolution. Our understanding of genetics means it will undoubtedly become possible, not merely to undertake gene therapy, but to alter – or 'enhance' – the human genome so that genetic characteristics can be passed on down the generations. Scientists acquired the power to make transgenic animals about thirty years ago. It has been a most valuable advance in genetic science, one that has greatly increased our understanding of genes and their function. The ability to undertake genetic manipulation, though currently very imperfect, will almost certainly be refined in future with the potential to make it 'safe' to be used in our own species.

Eventually, people may deem it reasonable to increase intelligence, to intensify brute human strength, to enhance longevity or physical appearance, or to give our descendants other characteristics that we regard, in our generation and in our environment, as 'desirable'. But the history of Harry Laughlin, Frances Galton, Trofim Lysenko and so many others like them is not reassuring.

It is not merely that mistakes will inevitably happen, nor that the changes we force on our descendants will be indelible, irreversible. Neither is this just a question of increasing the inequalities which are present in various human societies, serious though that is. A species is defined essentially by its genome. If we decide to modify some members of our species, *Homo sapiens*, what price our humanity then? And what happens to our central moral values, based so largely on the sanctity and quality of *human* life?

Chapter 12

Scientists and Citizens: Twelve Aphorisms and a Manifesto

SOME OF THE promise of scientific knowledge and many of the threats it poses have been laid out in this book. The benefits of human inventiveness have never been greater, and very few of us would rather live in the eighteenth, nineteenth or twentieth centuries – or earlier – than the twenty-first. We have many reasons to be optimistic about the future; but at the same time we must accept that the threats now facing humanity as a result of our cleverness have probably never been more serious. If our ingenuity as a species is to continue to bring great benefit, and if the dangers latent in nearly every technology are to be minimized, we all need to take greater care over how we handle our knowledge.

Most scientists do their work because they believe firmly that the information they will acquire and publish is likely to bring benefits to other humans. While being a scientist may be intellectually stimulating – even exciting from time to time – and may on rare occasions even bring huge reward and considerable acclaim, the great majority of my colleagues hope that, by pursuing their science, they may do some good. The benefit to society is high on their agenda.

The pursuit of science is not always easy; it requires mental discipline and long training, and is not usually particularly well rewarded financially. It frequently requires long periods spent doing tedious or mundane stuff during antisocial hours. Taking employment in research frequently means accepting considerable insecurity for oneself and one's family. The great majority of research scientists live on their wits, having to raise grant funding for every project they undertake. And applying for research grants is increasingly competitive – there is now no country in the world where more than a small proportion of carefully thought-through research grant applications achieve support. Scientific research is always demanding and sometimes lonely: even though most science is nowadays done by groups of scientists working as teams, every scientist knows that his or her intellectual worth is continuously under scrutiny.

The tensions that develop from the difficulties outlined above can lead to quite intellectually aggressive attitudes, as we have seen from some of the historical examples I have related. Moreover, with the possible exception of some areas of medicine, the ethical implications of the science the undergraduate learns are little discussed and not much taught. And because most scientists often talk a strange and complex language replete with arcane jargon, the research they are doing can be unintelligible to people around them. This sometimes inhibits good communication and prevents constructive debate with the very people who may be most affected by the adverse consequences of what they do. These are important issues we need to address in our educational system.

But scientists are not generally responsible for the misuse of the knowledge they generate. To some extent, all of the public are answerable for how science is used.[1] If we benefit from scientific progress, we need also to understand that we have a duty to learn more about science so that we can exercise a voice in how it is used, and be aware of the implications if it is misused. The obligation to use scientific knowledge wisely cannot be met without the widest involvement of society.

Wisdom requires knowledge. So in the first place there is a need, more pressing than ever before, for as many people as possible to be scientifically literate. Without that literacy, it will be very difficult for citizens to take wise decisions and for political representatives and policy-makers to make good choices. Science is in many ways a most important part of human culture. In a civilized, cultured society, there is an obligation for everybody to understand more about science and to recognize that ignorance of science is unacceptable.

So every society needs to ensure that the scientific education it provides is of the highest quality. And governments need to reflect on a current trend and question whether investing in science education simply because it is valuable to the economy is sound policy. Rather, they should consider investment in science education vital because it is the best way of ensuring that we, our children and our grandchildren live in a safer and healthier society.

A recurring theme throughout this book has been how we think about science. Most of us learn about new scientific developments from the media and, of course, the media often has its own agenda. So I suggest twelve aphorisms that might usefully be taken to heart: twelve points that are not always obvious when science is taught and discussed, or when it is presented in the written or broadcast media.

Twelve aphorisms about science

- The announcement of any new discovery is almost always heralded by exaggerated claims for its immediate or imminent value.
- Nearly all technological advances have threatening or negative aspects which usually are not fully recognized or predicted at the time of their invention.
- Many, if not most, human discoveries have beneficial applications that are not envisaged when the discovery is first made.
- Many human technological advances are made more or less simultaneously and independently in unconnected places around the globe.
- We constantly reinvent the same technological advances and rediscover the same discoveries.
- Most scientific advances are made by gentle, incremental progress. There is rarely such a thing as a 'breakthrough'.

- Scientific knowledge may be growing at an exponential rate, but the exploitation of scientific knowledge is usually much slower than is generally anticipated.
- Many really important discoveries and some inventions are arrived at by serendipity.
- Even 'good' democratic governments frequently misuse scientific knowledge.
- Scientists are human and therefore may not always be entirely objective.
- Scientists are no better than anybody else at forecasting the future. In fact, their predictions are usually widely inaccurate.
- The great majority of scientists do their work because they believe that the advance of knowledge brings benefits which should be shared by everybody.

What about the role of the scientist? If scientists are to deserve the trust of society, they need to see their work in a broad cultural context. They need to hear and respond to people's concerns, hopes and ideas, and to reflect on the ways their science may possibly be used. And because society should be more closely involved in the important decisions about how scientific knowledge is used, there is a need for scientists to evaluate how they pursue that knowledge and what its implications might be. This is surely good citizenship. Such an approach can best be fostered by greater engagement between scientists and the public. If that relationship is strong, then it is more likely that science will be used increasingly wisely and for the greatest benefit. It is good to reflect that in the United Kingdom we have already started this process of engagement, which is, I believe, one of the best ways of securing our future.

With these thoughts in mind, I offer the following manifesto. It may seem presumptuous, but it is offered as a starting point for a helpful change in thinking. These principles could, I believe, go far towards ensuring a better and safer relationship between scientist and society.

A SCIENTIST'S MANIFESTO

1.   We should try to communicate our work as effectively as possible, because ultimately it is done on behalf of society and because its adverse consequences may affect members of the society in which we all live.[2] We need to strive for clarity not only when we make statements or publish work for scientific colleagues, but also in making our work intelligible to the average layperson. We may also reflect that learning to communicate more effectively may improve the quality of the science we do and make it more relevant to the problems we are attempting to solve.

2.   Communication is a two-way process. Good engagement with the public is not merely a case of imparting scientific information clearly. It involves listening to and responding to the ideas, questions, hopes and concerns the public may have. We should accept that this kind of engagement with the public is a matter of good citizenship. We should reflect that sometimes proper dialogue with various sections of the public may inform some aspects of our work. Moreover, it can make any technology that is developed from our work more relevant to the needs of the public and less likely to be dangerous.

3.   The media, whether written, broadcast or web-based, play a key role in how the public learn about science. We need to share our work more effectively by being as clear, honest and intelligible as possible in our dealings with journalists. We also need to recognize that misusing the media by exaggerating the potential of what we are studying, or belittling the work of other scientists working in the field, can be detrimental to science.

4.   We need to recognize that the science we do is not entirely our property. Whether the taxpayer helps fund our scientific education or not, most of our training and research is paid for by the public – in grants from the research councils or charities. The public has a major stake in the ownership of what we do.

5.   Wherever possible, we should always consider the ethical problems that may be raised by the applications of our work. Some scientists have claimed that science does not have a moral value; but while pure knowledge may be ethically neutral, the way this knowledge is gained and the use to which it is put can involve many difficult ethical issues.

6.   We should reflect that science is not simply 'the truth' but merely a version of it. A scientific experiment may well 'prove' something, but a 'proof' may change with the passage of time as we gain better understanding. Mere assertion that something is fact will not persuade many people of the rightness of what we say. It is worth bearing in mind that sometimes two well-conducted experiments can give conflicting results that are equally valid. Science is not absolute; it is often about uncertainty.

7.   It is understandable and proper that we scientists are immensely proud of what we discover, but it is easy to forget that this special knowledge can sometimes breed a culture of assumed omnipotence and arrogant assertion. We need to avoid arrogance because it can lead to misinterpretation of data and to conflict instead of collaboration with colleagues. Moreover, arrogance is likely to damage the reputation of science by increasing public mistrust.

8.   Scientists are regularly called upon to assess the work of other scientists or review their reports before publication. While such peer review is usually the best process for assessing the quality of scientific work, it can be abused. When conducting peer review, we should try to ensure that we are fair and scrupulous and not acting out of a vested interest.

9.   We should try to see our science in a broad context, but also be aware of the limitations of our personal expertise. We should consider that, when talking outside our own subject, we may be more likely to mistake the facts of a case. We should be particularly cautious about making predictions about the future of science, not least because creating unrealistic expectations can be damaging.

10.   Governments, whether totalitarian, oligarchic or demo-
cratically elected, usually have vested interests. Such interests are not
necessarily conducive to good research or to good use of the fruits
of knowledge. Government control of science can have malign
influence. This is certainly true of totalitarian governments, but misuse of science is very common in virtually all liberal democracies,
including our own.

  It is difficult for scientists to retain independence from politicians,
because politicians ultimately make many key funding decisions. But
we need to keep some distance from politicians, and should not
avoid criticizing their decisions where we feel they are wrong or
dangerous.

11.   Commercial interests, so often promoted by governments and
universities, cannot be disregarded if technology is to be exploited
for public good. But scientists need to be aware of the dangers of
conflicts of interest and to retain a sense of balance, because
commercial interests can be a bad influence on scientific endeavour.
The history of science shows that the over-eager or narrow-minded
pursuit of commercial interests can lead to the loss of public trust.

12.   In the Western world, most of our best basic science is done in
universities. But historically universities have been élite and
mysterious institutions, and even today they are sometimes perceived as rather threatening places where the complex and
unintelligible takes place. Those of us working in universities should
try to help foster a new culture of open access to our institutions
and, where we can, help strengthen activities which involve
community service and outreach. Where possible we should do our
best to support whatever aspect of public engagement is undertaken
by the university.

13.   Schools have the most vital role to play in encouraging young
people to see the magnificence of the natural world. But sadly, at
present, many schools actively discourage children from appreciation of the wonders of science. We should try to support initiatives
that may promote more practical and experimental work for
children, and show our appreciation of inspirational teachers and

their teaching. If we are in a position to do so, we should promote stronger connections and collaborations between schools, school-children and universities, because this is likely to help produce a healthier, safer society.

14. Just a generation ago, the mark of a civilized person was an appreciation of Shakespeare, Milton, Goethe, Thucydides, Rembrandt and Beethoven. But the pursuit of science has become so intense and demanding that today's scientists are more likely to neglect our cultural inheritance. We may wish to reflect that by broadening our own interests; thus we may help non-scientists to see science as part of our culture. Shakespeare, Thucydides, Goethe or even Milton may not be directly relevant to our scientific research, but the cultural values such authors represent are universal and deeply important. The words of the Roman poet Terence are of particular relevance: *Homo sum: humani nil a me alienum puto* – 'I am a man: nothing human is foreign to me.'

Notes

Introduction: Too clever by half?

1 Hartmut Thieme, 'Lower Palaeolithic hunting spears from Germany', *Nature*, 385: 807–10 (1997).

2 Radiocarbon dating is possible only for the last 60,000 years or so (roughly since modern humans evolved) because of the relatively short half-life of carbon-14. If an object requires dating before that, other elements such as potassium or thorium may be used.

3 Robert Winston, *The Human Mind*, Bantam Press, London (2003).

4 Jean Clotte and Jean Courtin, *La Grotte Cosquer*, Éditions du Seuil, Paris (1994).

5 L. Bachechi, P.-F. Fabri and F. Mallegni, 'An arrow-caused lesion in a late Upper Paleolithic human pelvis', *Current Anthropology*, 38: 135–40 (1997).

6 Thor Gjerdrum, Philip Walker and Valerie Andrushko, 'Humeral retroversion: an activity pattern index in prehistoric southern California', *American Journal of Physical Anthropology*, 36 (suppl.): 100–1 (2003).

7 Sir Martin Rees, *Our Final Century? Will the Human Race Survive the Twenty-first Century?* Heinemann, London (2005).

8   Alec Broers (Lord Broers), *The Triumph of Technology: The Reith Lectures*, BBC Radio 4 (2005).

1: *The brighest heaven of invention*

1   Michael Frank, Daniel Everett, Evalina Fedorenko and Edward Gibson, 'Number as a cognitive technology: evidence from Pirahã language and cognition', *Cognition*, 108: 819–24 (2008).

2   Michael Sockol, David Raichlen and Herman Pontzer, 'Chimpanzee locomotor energetics and the origin of human bipedalism', *Proceedings of the National Academy of Sciences of the USA*, 134: 12265–9 (2007).

3   Robert Seyfarth and Dorothy Cheney, 'The acoustic features of vervet monkey grunts', *Journal of the Acoustical Society of America*, 75: 1623–8 (1984).

4   Robin Dunbar, 'Psychology: evolution of the social brain', *Science*, 302: 1160–1 (2003).

5   Patricia M. Greenfield, 'Language, tools and brain: the ontogeny and phylogeny of hierarchically organized sequential behavior', *Behavioral and Brain Sciences*, 14: 531–95 (1991).

6   Cecilia S. L. Lai, Simon E. Fisher, Jane A. Hurst, Faraneh Vargha-Khadem and Anthony P. Monaco, 'A forkhead-domain gene is mutated in a severe speech and language disorder', *Nature*, 412: 519–23 (2001).

7   Richard G. Klein, 'Whither the Neanderthals?', *Science*, 299: 1525–9 (2003).

8   Wolfgang Enard, Molly Przeworski, Simon E. Fisher, Cecilia S. L. Lai, Victor Wiebe, Takashi Kitano, Anthony P. Monaco and Svante Pääbo, 'Molecular evolution of *FOXP2*, a gene involved in speech and language', *Nature*, 418: 869–72 (2002).

9   B. Arensburg, A. M. Tillier, B. Vandermeersch, H. Duday, L. A. Schepartz and Y. Rak, 'A Middle Palaeolithic human hyoid bone', *Nature*, 338: 758–60 (1989).

10   M. Hauser, N. Chomsky and W. T. Fitch, 'The language faculty: what is it, who has it, and how did it evolve?', *Science*, 298: 1569–79 (2002).

11   S. Leitner and C. K. Catchpole, 'Female canaries that respond and discriminate more between male songs of different quality have a

larger song control nucleus (HVC) in the brain', *Journal of Neurobiology*, 52: 294–301 (2002).

12   Nancy Mitford and Osbert Lancaster, *Noblesse Oblige: An Enquiry into the Identifiable Characteristics of the English Aristocracy*, Penguin Books, London (1959).

13   Judg. 12: 5.

14   Seyfarth and Cheney, 'The acoustic features of vervet monkey grunts'.

15   K. Arnold and K. Zuberbühler, 'Language evolution: semantic combinations in primate calls', *Nature*, 441: 303, publ. online (2006).

16   Juliane Kaminski, Julia Fischer and Josep Call, 'Prospective object search in dogs: mixed evidence for knowledge of *What and Where*', *Animal Cognition*, 11: 1435 (2008).

17   Hauser et al., 'The language faculty'.

2: Appetite for destruction

1   William Bateson, *Gregor Mendel. Mendel's Principles of Heredity: A Defence*, Cambridge University Press, Cambridge (1902).

2   Jan Witkowski, 'Stalin's war on genetic science', *Nature*, 454: 577–9 (2008).

3   David Joravsky, *The Lysenko Affair*, University of Chicago Press, Chicago (1970).

4   Peter Pringle, *The Murder of Nikolai Vavilov*, Simon & Schuster, New York (2008).

3: A grinding existence

1   Gen. 3: 17–20.

2   Theya Molleson, 'The eloquent bones of Abu Hureyra', *Scientific American*, 271: 70–5 (1994).

3   K. N. Schneider, 'Dental caries, enamel composition, and subsistence among prehistoric Amerindians of Ohio', *American Journal of Physical Anthropology*, 71: 95–102 (1986).

4   Kim Hill, A. M. Hurtado and R. S. Walker, 'High adult mortality among Hiwi hunter-gatherers: implications for human evolution', *Journal of Human Evolution*, 52: 443–54 (2007).

5   Richard B. Lee and Irven DeVore, *Man the Hunter*, Aldine, New York (1968).

6   Mark Nathan Cohen, *Health and the Rise of Civilization*, Yale
     University Press, New Haven (1989).
7   Ofer Bar-Yosef, Avi Gopher, Eitan Tchernov and Mordechai E.
     Kislev, 'Netiv Hagdud: an early Neolithic village site in the Jordan
     Valley', *Journal of Field Archaeology*, 18: 405–4 (1991).
8   D. W. Gaylor, J. A. Axelrad, R. P. Brown, J. A. Cavagnaro,
     W. H. Cyr, K. L. Hulebak, R. J. Lorentzen, M. A. Miller,
     L. T. Mulligan and B. A. Schwetz, 'Health risk assessment practices
     in the U.S. Food and Drug Administration', *Regulatory Toxicology
     and Pharmacology*, 26: 307–21 (1997).

4: Animal farm

1   Margaret Jolly, *Women of the Place: Kastom, Colonialism and
     Gender in Vanuatu* (Studies in Anthropology and History),
     Harwood Academic, Newark, NJ (1994).
2   Matthew Kluger, 'Fever revisited', *Pediatrics*, 90: 846–50
     (1992).
3   http://www.goveg.com/factoryFarming_pigs_farms.asp.
4   A. Voss, F. Loeffen, J. Bakker, C. Klaassen and M. Wulf,
     'Methicillin-resistant *Staphylococcus aureus* in pig farming',
     (2005); N. Nitatpattana, A. Dubot-Pérès, M. Ar Gouilh, M. Souris,
     P. Barbazan, S. Yoksan et al., 'Incidence of Japanese encephalitis
     virus genotype, Thailand' (2008); S. AbuBakar, L. Y. Chang, A. R.
     Mohd Ali, S. H. Sharifah, K. Yusoff and Z. Zamrod, 'Isolation and
     molecular identification of Nipah virus from pigs' (Dec. 2004);
     W. Chen, M. Yan, L. Yang, B. Ding, B. He, Y. Wang et al., 'SARS-
     associated coronavirus transmitted from human to pig' (2005). All
     in Center for Disease Control and Prevention, *Emerging Infectious
     Diseases* (online journal).
5   Stephen Morse, 'Factors in the emergence of infectious diseases',
     Emerging Infectious Diseases, 1: 7–15 (1995).
6   http://www.smithfield.com/about/advertising.php.
7   M. Shun-Shin, M. Thompson, C. Heneghan, R. Perera, A.
     Harnden and D. Mant, 'Neuraminidase inhibitors for treatment
     and prophylaxis of influenza in children: systematic review and
     meta-analysis of randomised controlled trials', *British Medical
     Journal* (online edition), 10 Aug. 2009, 339:b3172.
8   *The Times*, 11 Aug. 2009.

9   Larry Pope is almost certainly right when he argues that the virus
     does not transmit through meat. And, on 14 May 2009
     (www.smithfieldfood.com/media/news.aspx), he reported the com-
     pany's cooperation with the Mexican government: 'The results of
     the testing process conducted by the Mexican government have
     confirmed that no virus, including the human strain of A(H1N1)
     influenza is present in the pig herd at Granjas Carroll de Mexico
     (GCM).' Consequently, he argues that 'the recent subtype of H1N1
     influenza virus affecting humans did not originate from GCM'.
     Whatever the source of this outbreak, given that it is well docu-
     mented that pigs are very frequently infected with a number of
     viruses, it is good that this company has seen fit to improve its
     record by cleansing the environment around many of its farms.

10   J. A. Hutchings, 'Collapse and recovery of marine fishes', *Nature*,
     406: 882–5 (2000).

11   Isa. 19: 6ff.

12   R. L. Naylor, R. J. Goldburg, J. H. Primavera, N. Kautsky,
     M. C. Beveridge, J. Clay, C. Folke, J. Lubchenco, H. Mooney and
     M. Troell, 'Effect of aquaculture on world fish supplies', *Nature*,
     405/6790: 1017–24 (2000).

13   Philip Lymbery, 'In too deep: the welfare of intensively farmed fish'
     (2002), paper commissioned for Compassion in World Farming
     (www.ciwf.org.uk).

14   pH is the measure of acidity. A solution which has a pH of 7.0 is
     ten times more acid than a solution with a pH of 8.0.

15   James C. Orr, Victoria J. Fabry, Olivier Aumont, Laurent Bopp,
     Scott C. Doney, Richard A. Feely et al., 'Anthropogenic ocean
     acidification over the twenty-first century and its impact on
     calcifying organisms', *Nature*, 437: 681–6 (2005).

16   Interacademy Panel on International Issues, *IAP Statement on
     Ocean Acidification* (2009).

17   S. B. Prusiner, 'Novel proteinaceous infectious particles cause
     scrapie', *Science*, 216: 136–44 (1982).

5: Wild and whirling words

1   Denise Schmandt-Besserat, *How Writing Came About*, University
     of Texas Press, Austin, Tex. (1996).

2   Ronald S. Stroud, 'The art of writing in ancient Greece', in Wayne

M. Senner, ed., *The Origins of Writing*, University of Nebraska Press, Lincoln, NE (1989).

3   G. Alun Evans, 'Evidence for Indo-European language in the Minoan documents', *Journal of Hellenic Studies*, 73: 84–103 (1953).

4   Rex E. Wallace, *An Introduction to Wall Inscriptions from Pompeii and Herculaneum*, Bolchazy-Carducci Inc., Wauconda, Ill. (2005).

5   David Keightley, *Sources of Shang History: The Oracle-Bone Inscriptions of Bronze Age China*, University of California Press, Berkeley (1978).

6   Jack Goody, *The Domestication of the Savage Mind* (Themes in the Social Sciences), Cambridge University Press, Cambridge (1977).

7   Robert Winston, *Human Instinct: How our Primeval Impulses Shape our Modern Lives*, Bantam Press, London (2002), pp. 313ff.

8   Alberto Manguel, *The Library at Night*, Yale University Press, New Haven (2008).

9   Fernando Baez, *A Universal History of the Destruction of Books*, trans. Alfred MacAdam, Atlas & Co., New York (2008).

10   P. D. G. Thomas, 'The beginning of parliamentary reporting in newspapers, 1768–1774', *English Historical Review*, 74: 623 (1959).

11   *Journals of the House of Lords*, proceedings against Lord Lovat, 20 George II, 529ff. (1747).

12   *Journals of the House of Commons*, 38: 745 (1760).

13   P. L. Simmonds, 'Statistics of newspapers in various countries', *Journal of the Statistical Society of London*, 4: 111–36 (1841).

14   I. Kershaw, *Hitler, the Germans, and the Final Solution*, Yale University Press, New Haven (2008).

15   D. Welch, 'Nazi propaganda and the *Volksgemeinschaft*: constructing a people's community', *Journal of Contemporary History*, 39: 2 ('Understanding Nazi Germany'), 213–38 (2004).

16   J. Curtice, 'Was it *The Sun* that won it again? The influence of newspapers in the 1997 election campaign', CREST working papers no. 75 (1999), http://www.crest.ox.ac.uk.

6: Digital communication

1   Stephen van Dulken, *Inventing the 19th Century*, British Library, London (2001).

2   The letter 'C' was indicated by repeated movement of the second
      needle. The letters 'J' and 'Q' were denoted by substituting 'G' and
      'K'. (University of Salford:
      http://www.cntr.salford.ac.uk/comms/johntawell.php.)

3   O. C. Howes, 'Compulsions in depression: stalking by text mes-
      sage', *American Journal of Psychiatry*, 163: 1642 (2006).

4   D. James and J. Drennan, 'Exploring addictive consumption of
      mobile phone technology', *Proceedings of Australian and New
      Zealand Marketing Academy Conference* (2005).

5   *Orange County Register*, 7 Jan. 2009.

6   H. Luntiala, *The Last Message*, Tammi, Helsinki (2007).

7   J. Sutherland, 'Cn u txt?', *Guardian*, 11 Nov. 2002.

8   John Humphrys, 'I h8 txt msgs: how texting is wrecking our
      language', *Mail OnLine*, 24 Sept. 2007.

9   Will Self and Lynne Truss, 'The joy of text', *Guardian*, 5 July 2008.

10   *A Language for Life*, report of the Committee of Inquiry appointed
      by the Secretary of State for Education and Science under the
      chairmanship of Sir Alan Bullock FBA, HMSO, London (1975).

11   George Orwell, *Nineteen Eighty-four*, Penguin Modern Classics,
      London (1970; first publ. 1949).

12   E. Fromm, *The Anatomy of Human Destructiveness*, Fawcett,
      Greenwich, CT (1975).

13   K. Taylor, *Cruelty: Human Evil and the Brain*, Oxford University
      Press, Oxford (2009).

14   P. K. Smith, J. Mahdavi, M. Carvalho, S. Fisher, S. Russell and N.
      Tippett, 'Cyberbullying: its nature and impact in secondary school
      pupils', *Child Psychology and Psychiatry*, 49: 376–85 (2008).

15   J. F. Chisholm, 'Cyberspace violence against girls and adolescent
      females', *Annals of the New York Academy of Sciences*, 1087:
      74–89 (2006).

16   The National Campaign to Prevent Teen and Unplanned
      Pregnancy, survey 2008: http://www.thenationalcampaign.org.

17   Internet World Stats (http://www.internetworldstats.com/stats.htm)
      2008.

18   Francis Maude MP, Hansard (Commons), col. 756, 12 Nov. 2008.

7: The right Promethean fire

1   Stella Brewer, *The Forest Dwellers*, Fontana, London (1979).

2   N. Goren-Inbar, N. Alperson, M. E. Kislev, O. Simchoni, Y. Melamed, A. Ben-Nun and E. Werker, 'Evidence of hominin control of fire at Gesher Benot Ya'aqov, Israel', *Science*, 304: 725–7 (2004).

3   This refers to a method of making stone tools during the Lower Paleolithic Period, dating from around 1.5 million to 150,000 years ago, characterized especially by flaked bifacial hand-axes, cleavers and other core tools. The earliest Acheulean tools have been found at African sites associated with *Homo erectus* fossils.

4   C. K. Brain and A. Sillent, 'Evidence from the Swartkrans cave for the earliest use of fire', *Nature*, 336: 464–6 (1988).

5   Lewis Binford, 'Were there elephant hunters at Torralba?', in Matthew H. Nitecki and Doris V. Nitecki, eds, *The Evolution of Human Hunting*, Springer, Chicago (1987).

6   Alfred Reginald Radcliffe-Brown, *The Andaman Islanders*, Free Press, New York (repr. 1964).

7   *Thucydides*, vol. 2, trans. Benjamin Jowett, Clarendon Press, Oxford (1900).

8   'Sir Edward Kelle's Worke', in *The Alchemical Works of Sir Edward Kelley Including The Philosopher's Stone*. From *Theatrum Chemicum Britannicum*, ed. Elias Ashmole. Kessinger Publishing, Whitefish, Mont. (2006).

9   Charlotte Fell Smith, *John Dee 1527–1608*, Constable & Co., London (1909).

10   Richard Feynman, 'There's plenty of room at the bottom', paper presented to annual meeting of American Physical Society, 29 Dec. 1959, available online at www.zyvex.com/nanotech/feynman.html.

11   Jessica Ponti, Enrico Sabbioni, Barbara Munaro, Francesca Broggi, Patrick Marmorato, Fabio Franchini, Renato Colognato and François Rossi, 'Genotoxicity and morphological transformation induced by cobalt nanoparticles and cobalt chloride: an in vitro study in Balb/3T3 mouse fibroblasts', *Mutagenesis*, 24: 439–45 (2009).

12   Paresh Chandra Ray, Hongtao Yu and Peter P. Fu, 'Toxicity and environmental risks of nanomaterials: challenges and future needs', *Journal of Environmental Science and Health*, Part C (2009), available online at www.informaworld.com/smpp/title~content=t713597270.

13 R. J. Griffitt, J. Luo, J. Gao, J. C. Bonzongo and D. S. Barber, 'Effects of particle composition and species on toxicity of metallic nanomaterials in aquatic organisms', *Environmental Toxicology and Chemistry*, 27: 1972–8 (2008).

14 D. H. Lin and B. S. Xing, 'Phytotoxicity of nanoparticles: inhibition of seed germination and root elongation', *Environmental Pollution*, 150: 243–50 (2007).

8: Sulphurous and thought-executing fires

1 Michael Hunter, 'Robert Boyle for the twenty-first century', *Notes and Records of the Royal Society of London*, 59: 87–90 (2005). See also Michael Hunter's excellent essay on Boyle, 'The life and thought of Robert Boyle', http://www.bbk.ac.uk/Boyle/biog.html.

2 Donald Read, *The English Provinces*, c.1760–90, Hodder & Stoughton, London (1964).

3 Joseph Priestley, *Familiar Letters Addressed to the Inhabitants of the Town of Birmingham, by the Revd Mr Madan (1790–92)*, letter 4.6. Printed by J. Thompson, sold by J. Johnson.

4 Jenny Uglow, *The Lunar Men*, Faber & Faber, London (2002). A gripping and highly recommended read.

5 His book has recently been reprinted in paperback: John Robison, *Proofs of a Conspiracy: Against all the Religions and Governments of Europe, Carried on in the Secret Meetings of Freemasons, Illuminati and Reading Societies*, Forgotten Books (2008), available online from Amazon via www.forgottenbooks.org.

6 'The method to soften bones and to cook all kinds of meat in a short time and still fresh, with a description of the cooking vessel, its qualities and uses.'

7 Durham Mining Museum, 1812 Felling Colliery Disaster Memorial at St Mary's Churchyard, Heworth. The museum holds photographs and related material.

8 R. G. Carpenter and A. L. Cochrane, 'Death rates of miners and ex-miners with and without coalworker's pneumoconiosis in South Wales', *British Journal of Industrial Medicine*, 13: 102 (1956).

9 David Egan, *Coal Society: A History of the South Wales Mining Valleys 1840–1980*, Gomer Press, Dyfed (1987).

10 Thomas Crump, *A Brief History of the Age of Steam*, Robinson, London (2007).

11   The painting appears as an aquatint by Rowlandson in the contemporary book *Microcosm of London*, published by Rudolph Ackermann (1808).

12   Anthony Burton, *Richard Trevithick: The Man and his Machines*, Aurum Press, London (2000).

13   Alexis de Tocqueville, 'Memoir on pauperism: does public charity produce an idle and dependent class of society?', first publ. 1835; repr. by Cosimo Press, Alief, Tex. (2006).

14   Stockholm International Peace Research Institute, *SIPRI Year Book 2008*, Oxford, 2008, quoting US Center for Defense Information.

15   L. Calderon-Garciduenas, M. Franco-Lira, R. Torres-Jardón, C. Henriquez-Roldán, G. Barragán-Mejía, G. Valencia-Salazar, A. González-Maciel, R. Reynoso-Robles, R. Villarreal-Calderón and W. Reed, 'Pediatric respiratory and systemic effects of chronic air pollution exposure: nose, lung, heart, and brain pathology', *Toxicologic Pathology*, 35: 154–62 (2007).

9: Oil that maketh man's heart glad

1   Andrew Cook, *Reilly: Ace of Spies. The True Story of Sidney Reilly*, Tempus, Stroud (2004).

2   Andrew Lycett, *The Man behind James Bond*, Turner, Nashville (1996).

3   Thomas Jock Murray, 'Dr Abraham Gesner: the father of the petroleum industry', *Journal of the Royal Society of Medicine*, 86/1: 43–4 (1993).

4   David Leon Chandler, *Henry Flagler: The Astonishing Life and Times of the Visionary Robber Baron Who Founded Florida*, Macmillan, New York (2009).

5   Geoffrey Crowther, *Traffic in Towns: A Study of the Long Term Problems of Traffic Urban Areas*, report of the Steering Group and Working Group Appointed by the Minister of Transport, HMSO, London (1963).

6   Talmud Bavli, Tractate Avodah Zarah, 70*a*.

7   Elizabeth Monroe, *Philby of Arabia: St John Philby*, Faber & Faber, London (1973).

8   'The Ikhwan', http://www.eb.com:180/cgi-bin/g?DocF=micro/287/66.html.

9   James Howard Kunstler, *The Long Emergency: Surviving the End*

of Oil, Climate Change, and Other Converging Catastrophes of the Twenty-first Century, Grove/Atlantic, New York (2006).

10   M. King Hubbert, 'Nuclear energy and fossil fuels', Meeting Southern Division of Production, American Petroleum Institute, 1956, http://www.hubbertpeak.com/Hubbert/1956/1956.pdf.

11   Colin J. Campbell, *The Coming Oil Crisis*, Multi-Science Publishing Co. & Petroconsultants (1997).

12   Erik Davis, *TechGnosis: Myth, Magic and Mysticism in the Age of Information*, Serpent's Tail, London (1999).

13   *Renewable Energy*, report of House of Lords Select Committee on Science and Technology, HL paper 69 (2006).

14   *Non-food Crops*, report of House of Lords Select Committee on Science and Technology, HL paper 5 (2000).

15   Arnulf Jaeger-Waldau, in J. M. Kroon, G. Dennler, A. Jaeger-Waldau and A. Slaoui, eds, *Advanced materials and concepts for photovoltaics (AMPS)*, European Materials Research Society (E-MRS) Symposium Proceedings, vol. 215, Elsevier (2007).

16   J. Bostaph, 'Thin film fuel-cells for low-power portable applications', *Proceedings of the 39th Power-sources Conference*, Cherry Hill, NJ: 152–5 (2000).

17   Rattan Lal, 'Carbon sequestration', *Philosophical Transactions of the Royal Society*, 363: 815–30 (2008).

18   See n. 9 above.

19   Don Hopey, 'State sues utility for US pollution violations', *Pittsburgh Post-Gazette*, 29 June 2005.

20   J. P. McBride, R. E. Moore, J. P. Witherspoon and R. E. Blanco, 'Radiological impact of airborne effluents of coal and nuclear plants', *Science*, 202: 1045 (1978).

21   Alex Gabbard, 'Coal combustion: nuclear resource or danger?', Oak Ridge National Laboratory, available online at www.Mindfully.org/energy/coal-combustion-waste-ccw1jul93.htm.

22   Gordon J. Aubrecht, *Energy: Physical, Environmental, and Social Implications*, 3rd edn, Macmillan USA, New York (2005).

23   http://www.epa.gov/hiri/.

24   M. R. Raupach, G. Marland, P. Ciais, C. Le Quéré, J. G. Canadell, G. Klepper and C. B. Field, 'Global and regional drivers of accelerating global CO_2 emissions', *Proceedings of the National Academy of Sciences*, 104: 10288–93 (2007).

25   S. Rahmstorf, A. Cazenave, J. A. Church, J. E. Hansen, R. F.
      Keeling, D. E. Parker and R. C. J. Somerville, 'Recent climate
      observations compared to projections', *Science*, 316: 709 (2007).

10: Physician, heal thyself

1   Even today there is a widespread notion among some devoutly
      religious people that healing is 'playing God' and therefore
      ethically questionable. This view seems ridiculous, for it argues
      that any human intelligence – presumably God-given – should not
      be used.
2   Calcination is the process of heating a compound or element
      without actually melting it, to cause it to decompose or react with
      other substances.
3   Monica H. Green, trans., *The Trotula: A Medieval Compendium
      of Women's Medicine*, University of Pennsylvania Press,
      Philadelphia (2002).
4   Roy Porter, *Blood and Guts: A Short History of Medicine*,
      Penguin, London (2003).
5   1 Sam. 5: 9. The Hebrew word for bubo is *ophal*. 'Emerods' in the
      King James version of the Bible is 'haemorrhoids', but can mean
      any swelling or ulcer in or around the genitalia.
6   Such quarantine measures produced some wonderful cultural
      results, of course. Giovanni Boccaccio's *Decameron* depicts ten
      noble Florentines who occupied themselves with story-telling
      during their quarantine in the plague of 1348, which killed some
      three-quarters of the population.
7   Erasistratus (304–250 BCE) was a leading Greek anatomist.
      Together with Herophilus, his fellow doctor, he established the
      Alexandrian School of Anatomy and described the valves in
      the heart and the difference in structure between arteries and veins.
8   Benjamin Lee Gordon, *Medieval and Renaissance Medicine*,
      Philosophical Library, New York (1959).
9   Geoffrey Keynes, *The Life of William Harvey*, Clarendon Press,
      London (1966).
10   Schaudinn is another physician whose demise was a result of his
      own expertise: infected by one of the specimens he was studying,
      he died at the age of thirty-five with an intestinal amoebic abscess.
11   Robert D. Simoni, Robert L. Hill and Martha Vaughan, 'The

discovery of insulin: the work of Frederick Banting and Charles Best', *Journal of Biological Chemistry*, 26: 15 (2002) (JBC Centennial 1905–2005).

12   In this respect, a visit to the wonderful Hunterian Museum in the Royal College of Surgeons in Lincoln's Inn Fields, London, is very revealing. It demonstrates the remarkable prowess of John Hunter, its founder, as a truly great surgeon (some of his eighteenth-century dissections are preserved here in spirit), and I find the section on war wounds depicting the plastic surgical work of Gillies and McIndoe truly moving.

13   James Le Fanu, *The Rise and Fall of Modern Medicine*, Little, Brown, London (1999).

14   Maurice H. Pappworth, *Diagnostic Pitfalls: The Sin of Greed. A primer of Medicine*, Butterworths, London (1978).

15   Maurice Pappworth, *Human Guinea Pigs: Experimentation on Man*, Beacon Press, London (1968).

16   Admittedly, there were a number of political and financial reasons for this closure, but it is fair to say that the translational research (that is, research of real benefit to patients) that it was set up to do did not really make a huge impact.

17   Christopher C. Booth, 'Clinical research and the MRC', *Quarterly Journal of Medicine*, n.s. 59: 435–47 (1986).

18   Ivan Illich, *Limits to Medicine. Medical Nemesis: The Expropriation of Health* (Penguin Social Sciences), Penguin, London (1991).

19   This may sound cynical, but it is worth pointing out that it is usually much easier to give treatment than to refuse it: when a doctor refuses IVF treatment, most couples will expect long conversations and explanations of how a refusal can be justified.

20   Raymond Tallis, *Hippocratic Oaths: Medicine and its Discontents*, Atlantic Books, London (2004).

21   At Bristol, between 1984 and 1995, the mortality rate for babies undergoing complex cardiac surgery was substantially higher than the national average. An inquiry was held in 1999 and it was clear that, while there were many problems in the NHS which may have contributed to this, there was certainly no attempt at a cover-up by hospital medical staff, who published their results openly.

22   Thomas McKeown, *The Role of Medicine: Dream, Mirage, or*

Nemesis?, Nuffield Provincial Hospitals Trust, London (1976).

23   Richard Doll and F. Avery Jones, *Occupational Factors in the Aetiology of Gastric and Duodenal Ulcers, with an Estimate of their Incidence in the General Population*, HMSO, London (1951).

24   A. Keys, J. Brozek, A. Henschel, O. Mickelsen and H. L. Taylor, 'Role of dietary fat in human nutrition, III: Diet and the epidemiology of coronary heart disease', *American Journal of Public Health* (Nation's Health series), 47: 1520–30 (1957).

25   Multiple Risk Factor Intervention Trial Research Group. 'Multiple risk factor intervention trial: risk factor changes and mortality results', *Journal of the American Medical Association*, 24: 1465–77 (1982); WHO European Collaborative Group, 'Multi-factorial trial in prevention of heart disease incidence and mortality results', *European Heart Journal*, 4: 141–7 (1983).

11: Genetics: potent art or rough magic?

1   Miriam T. Jacobs, Yuan-Wei Zhang, Scott D. Campbell and Gary Rudnick, 'Ibogaine, a noncompetitive inhibitor of serotonin transport, acts by stabilizing the cytoplasm-facing state of the transporter', *Journal of Biological Chemistry*, 282: 29441–7 (2007).

2   Gen. 30: 40.

3   How curious that the science of epigenetics suggests there may be more in these old myths about experiences early in intra-uterine life than we thought.

4   Charles A. Shull and J. Fisher Stanfield, 'Thomas Andrew Knight: in memoriam', *Plant Physiology*, 14: 1–8 (American Society of Plant Biologists, 1939).

5   Arthur Darbishire, *Breeding and the Mendelian Discovery*, Cassell, London (1911).

6   A plant which is true breeding produces offspring with the same traits only when it is self-fertilized.

7   Ulf Lagerkvist, *DNA Pioneers and their Legacy*, Yale University Press, New Haven (1998).

8   Archibald Edward Garrod, 'The Croonian lectures on inborn errors of metabolism. Lecture II: alkaptonuria', *Lancet* , 2: 73–9 (1908).

9   Alexander G. Bearn, 'Archibald Edward Garrod, the reluctant geneticist', *Genetics*, 137: 1 (1995).

10   James Watson and Francis Crick, 'Molecular structure of nucleic

acids: a structure for deoxyribose nucleic acid', *Nature*, 171: 737–8 (1953).

11 Hans Eiberg, Jesper Troelsen, Mette Nielsen, Annemette Mikkelsen, Jonas Mengel-From, Klaus W. Kjaer and Lars Hansen, 'Blue eye color in humans may be caused by a perfectly associated founder mutation in a regulatory element located within the HERC2 gene inhibiting OCA2 expression', *Human Genetics*, 123: 177–87 (2008).

12 Robert Winston, *A Child Against All Odds*, Bantam Press, London (2006).

13 Alok Jha, 'From arthritis to diabetes: scientists unlock genetic secrets of diseases afflicting millions', *Guardian*, 7 June 2007.

14 D. M. Potts and W. T. W. Potts, *Queen Victoria's Gene: Haemophilia and the Royal Family*, Alan Sutton, Stroud (1995).

15 K. E. Davies, A. Speer, F. Herrmann, A. W. J. Spiegler, S. McGlade, M. H. Hofker, P. Briand, R. Hanke, M. Schwartz, V. Steinbicker, R. Szibor, H. Korner, D. Sommer, P. L. Pearson and Ch. Coutelle, 'Human X chromosome markers and Duchenne muscular dystrophy', *Nucleic Acids Research*, 13: 3419–26 (1985).

16 Hiromu Ito, Mette Koefoed, Prarop Tiyapatanaputi, Kirill Gromov, J. Jeffrey Goater, Jonathan Carmouche, Xinping Zhang, Paul T. Rubery, Joseph Rabinowitz, R. Jude Samulski, Takashi Nakamura, Kjeld Soballe, Regis J. O'Keefe, Brendan F. Boyce and Edward M. Schwarz, 'Remodeling of cortical bone allografts mediated by rAAV-RANKL and VEGF gene therapy', *Nature Medicine*, 11: 291–7 (2005).

17 E. C. Svensson, H. B. Black, D. L. Dugger, S. K. Tripathy, E. Goldwasser, Z. Hao, L. Chu and J. M. Leiden, 'Long-term erythropoietin expression in rodents and non-human primates following intramuscular injection of a replication-defective adenoviral vector', *Human Gene Therapy*, 8: 1797–1806 (1997).

18 Aubrey de Grey, Bruce Ames, Julie Andersen, Andrzej Bartke, Judith Campisi, Christopher Heward, Roger McCarter and Gregory Stock, 'Time to talk SENS: critiquing the immutability of human aging', *Annals of the New York Academy of Science*, 959: 452–62 (2002).

19 A. de Grey, L. Gavrilov, S. J. Olshansky, L. S. Coles, R. G. Cutler, M. Fossel and S. M. Harman, 'Antiaging technology and pseudoscience', *Science*, 296/5568: 656 (2002).

20   S. J. Hauck, W. S. Hunter, N. Danilovich, J. J. Kopchick and A. Bartke, 'Reduced levels of thyroid hormones, insulin, and glucose, and lower body core temperature in the growth hormone receptor/binding protein knockout mouse', *Experimental Biology and Medicine*, 226: 552–8 (2001).

21   A. J. French, C. A. Adams, L. S. Anderson, J. R. Kitchen, M. R. Hughes and S. H. Wood, 'Development of human cloned blastocysts following somatic cell nuclear transfer with adult fibroblasts', *Stem Cells*, 26: 485–93 (2008).

22   Editorial, *British Medical Journal*, 1 July 2000.

23   http://www.sanger.ac.uk/HGP/draft2000/future.shtml.

24   Peter Kraft and David Hunter, 'Genetic risk prediction – are we there yet?', *New England Journal of Medicine*, 360: 1701–3 (2009).

25   Jennifer Couzin, 'DNA test for breast cancer risk draws criticism', *Science*, 322: 357 (2008).

26   Marcus Pembrey, 'Human inheritance, differences and diseases: putting genes in their place, part I', *Paediatric and Perinatal Epidemiology*, 22: 497–504 (2008).

27   Gunnar Kaati, Lars Bygren and S. Edvinsson, 'Cardiovascular and diabetes mortality determined by nutrition during parents' and grandparents' slow growth period', *European Journal of Human Genetics*, 10: 682–8 (2002).

28   Francis Bacon, *Instauratio Magna* (1620), in *Translations of the Philosophical Works*, ed. James Spedding, Robert Ellis and Douglas Heath, vol. 5, Longman and Co., London (1858).

29   Gunnar Kaati, Lars Olov Bygren, Marcus Pembrey and Michael Sjostrom, 'Transgenerational response to nutrition, early life circumstances and longevity', *European Journal of Human Genetics*, 15: 784–90 (2007).

30   J. E. Cropley, C. M. Suter, K. B. Beckman and D. I. Martin, 'Germ-line epigenetic modification of the murine A vy allele by nutritional supplementation', *Proceedings of the National Academy of Sciences of the USA*, 103: 17308–12 (2006).

31   Frances A. Champagne, 'Epigenetic mechanisms and the transgenerational effects of maternal care', *Frontiers in Neuroendocrinology*, 29: 386–97 (2008).

32   Suzanne King, Adham Mancini-Mari, Alain Brunet, Elaine Walker,

Michael Meaney and David Laplante, 'Prenatal maternal stress from a natural disaster predicts dermatoglyphic asymmetry in humans', *Development and Psychopathology*, 21: 343–53 (2009).

33   David Laplante, Ronald Barr, Alain Brunet, Guillaume Galbaud du Fort, Michael Meaney and Suzanne King, 'Stress during pregnancy affects general intellectual and language functioning in human toddlers', *Pediatric Research*, 56: 400–10 (2004).

34   Talmud Babli: Sanhedrin 65*b*.

35   The DVD of this film, which is well worth watching just for the extraordinary expressionist set if not for some of the acting, is on www.eurekavideo.co.uk.

36   Daniel Gibson, Gwynedd Benders, Kevin Axelrod, Jayshree Zaveri, Mikkel Algire, Monzia Moodie, Michael Montague, Craig Venter, Hamilton Smith and Clyde Hutchison III, 'One-step assembly in yeast of 25 overlapping DNA fragments to form a complete synthetic *Mycoplasma genitalium* genome', *Proceedings of the National Academy of Sciences of the USA*, 105: 20404–9 (2008).

37   Shota Atsumil, Taizo Hanail and James C. Liao, 'Non-fermentative pathways for synthesis of branched-chain higher alcohols as biofuels', *Nature*, 451: 86–90 (2008).

38   D. K. Ro, E. M. Paradise, M. Ouellet, K. J. Fisher, K. L. Newman, J. M. Ndungu, K. A. Ho, R. A. Eachus, T. S. Ham, J. Kirby, M. C. Chang, S. T. Withers, Y. Shiba, R. Sarpong and J. D. Keasling, 'Production of the antimalarial drug precursor artemisinic acid in engineered yeast', *Nature*, 440: 940–3 (2006).

39   Plato, *The Republic*, book V, trans. H. D. P. Lee, Penguin Classics, London (1955).

40   Lucius Annaeus Seneca, *De Ira*, i. 18.

41   Winston, *Human Instinct*.

42   Charles Benedict Davenport, *Heredity in Relation to Eugenics*, H. Holt & Co., New York (1911).

43   Harry Hamilton Laughlin, 'Eugenical sterilization in the United States', Psychopathic Laboratory of the Municipal Court, Chicago (1922).

12: Scientists and citizens: twelve aphorisms and a manifesto

1   I deliberately use the word 'public' recognizing that there is no single 'public', which is why social scientists sometimes refer to the

'publics'. But this usage seems to me to be a clumsy expression so I have avoided it. By 'the public' I mean to include every echelon of human society.

2   My colleagues may argue that not every scientist has the capability of good communication with the public and that poor communicators are more likely to mislead, or do more damage than good, by making the attempt. So I must emphasize that I am not suggesting here that every scientist should be presenting his or her work to the public. Nevertheless, clarity in thinking about how research and results are presented inside the scientific community are a vital prerequisite for good science, and this important skill is sometimes lacking or not well taught.

Index

Robert Winston is one of Britain's best-known scientists. He holds the chair of Science and Society at Imperial College London, where he is Emeritus Professor of Fertility Studies and has been a leading voice in the debate on embryo research and genetic engineering. His television series include *Child of Our Time*, *Human Instinct* and *The Story of God*, which have made him a household name. He became a life peer in 1995.